Nanotechnology in Enhanced Oil Recovery

This book presents an in-depth analysis of the various nanotechnologies that have been developed and their potential application in enhanced oil recovery (EOR). It begins with an introduction to EOR, and the current state of the oil and gas industry followed by discussion of various nanoparticles used in EOR, including metal and metal oxide nanoparticles, carbon-based nanoparticles, and composite nanoparticles.

Features

- Explains the various mechanisms by which nanoparticles can enhance oil recovery, and the challenges and limitations of using nanotechnology in EOR.
- Provides a comprehensive analysis of how nanoparticles affect EOR, practical application, and simulation.
- Explores stepwise information about the utility of nanotechnology in EOR.
- Includes dedicated case study chapters to get an in-depth idea of EOR by using nanotechnology.
- Illustrates various essential parameters affecting the EOR.
- Discusses the challenges of nanotechnology-based EOR.

This book is aimed at graduate students, researchers, and professionals in Petroleum and Chemical Engineering, and Nanotechnology.

Krunal M. Gangawane is an Assistant Professor of Chemical Engineering at the Indian Institute of Technology Jodhpur, India. He holds a Ph.D. Degree in Chemical Engineering from the Indian Institute of Technology Roorkee in 2015. His current area of research is magneto-convection, nanofluidics, enhanced oil recovery, aerogels, etc. He has more than 20 publications in journals of repute. He had a research collaboration with the Firat University (Turkey), Ghent University (Belgium), and King Saud University (Saudi Arabia) on the topic of 'Convection heat transfer in enclosed bodies for different fluids.' He is also the recipient of the ISRO-RESPOND sponsored project in March 2020.

Abhishek Kumar is a Ph.D. student of Chemical Engineering at the National Institute of Technology (NIT) Rourkela, India. He graduated in Chemical Engineering (B.Tech.) from Monad University in 2014. Later, he received his M.Tech. Degree in Chemical Engineering from C.V. Raman College of Engineering & Technology, Bhubaneshwar, in 2018. His Ph.D. research is based on hybrid ferrofluids for energy and environmental applications; their focus area is enhanced oil recovery using hybrid ferrofluids.

Nanotechnology in Enhanced Oil Recovery

Edited by
Krunal M. Gangawane and Abhishek Kumar

CRC Press
Taylor & Francis Group
Boca Raton London New York

CRC Press is an imprint of the
Taylor & Francis Group, an **informa** business

Designed cover image: Shutterstock

First edition published 2025
by CRC Press
2385 NW Executive Center Drive, Suite 320, Boca Raton FL 33431

and by CRC Press
4 Park Square, Milton Park, Abingdon, Oxon, OX14 4RN

CRC Press is an imprint of Taylor & Francis Group, LLC

ISBN: 978-1-032-59989-2 (hbk)
ISBN: 978-1-032-65066-1 (pbk)
ISBN: 978-1-032-65068-5 (ebk)

DOI: 10.1201/9781032650685

Typeset in Times
By codeMantra

Contents

Preface....vii
Contributors List....ix

Chapter 1 Overview of Enhanced Oil Recovery (EOR).... 1
Nupur Srivastava, Veena Sumesh Nair, Santosh Kumar Senapati and Anupama Kumar

Chapter 2 Fundamentals of Nanotechnology for EOR.... 27
Krishna Raghav Chaturvedi, Abhishek Kumar, Anand Raghav Chaubey, Himanshu Kesharwani, Ravi Shankar Kumar and Tushar Sharma

Chapter 3 Wettability Alteration and Interfacial Tension Reduction in EOR....39
Ali Khalilnezhad, Lady J. Giraldo, Daniel Lopez, Camilo A. Franco, Farid B. Cortés and Masoud Riazi

Chapter 4 Factors Affecting Nanofluids by Nanotechnology for Enhanced Oil Recovery....58
Uma Sankar Behera and Ramanand Singh

Chapter 5 Nanoparticle-Assisted Foam for Enhanced Oil Recovery....74
N.S.V. Dinesh and P. Sivasankar

Chapter 6 Functional Groups and Functionalized Nanoparticles in EOR Techniques.... 92
Krishna Raghav Chaturvedi, Abhishek Kumar and Anand Raghav Chaubey

Chapter 7 Effects of Silica Nanoparticles Coupled with Low-Salinity Water on Oil Recovery in a Part of Upper Assam Basin, India....104
Nayan Medhi, Uttaron Goswami and Pushpa Sharma

Chapter 8 Case Studies of EOR by Using Nanoparticles 122

Krishna Raghav Chaturvedi, Abhishek Kumar and Anand Raghav Chaubey

Chapter 9 Synthesis of Iron Oxide Nanoparticles for Enhanced Oil Recovery 127

Aditi Pradeep Dahake and Nirmalya Bachhar

Chapter 10 CFD of Fluid Flow for EOR Using Nanoparticles 162

Krishna Raghav Chaturvedi, Abhishek Kumar and Anand Raghav Chaubey

Chapter 11 Simulation of Nanoparticle-Assisted Enhanced Oil Recovery: A Coupled Multiphase Fluid Flow and Nanoparticle Transport Model 174

P. Sivasankar

Chapter 12 Optimisation of Parameters for Enhanced Oil Recovery and the Role of Nanotechnology 190

Uma Sankar Behera

Chapter 13 Nanotechnology in Enhanced Oil Recovery: Challenges and Future Directions 215

Krishna Raghav Chaturvedi, Abhishek Kumar and Anand Raghav Chaubey

Index 227

Preface

In the dynamic landscape of oil recovery, the infusion of nanotechnology marks a pivotal turning point, reshaping traditional methodologies, and unlocking new horizons of efficiency and productivity. With rapid advancements in nanoscale science and engineering, enhanced oil recovery (EOR) has undergone a profound transformation, offering unprecedented opportunities for optimizing extraction processes. Positioned at the forefront of this transformative journey, the book *Nanotechnology in Enhanced Oil Recovery* serves as a beacon illuminating the path forward.

The primary goal of this book is to provide readers with an in-depth analysis of the ways that nanotechnology could enhance oil recovery methods. It explores the link between nanotechnology and EOR via a carefully chosen collection of chapters that concentrate on the underlying ideas, novel approaches, and practical applications.

The contributors to this volume represent a diverse array of expertise from esteemed institutions across the globe. Among them are renowned universities and research centres such as the Indian Institutes of Technology (IIT) in Jodhpur and Madras in India; Chonnam National University in Yeosu, Jeonnam, South Korea; and Sahand University of Technology in Tabriz, Iran. Additionally, institutions like Universidad Nacional de Colombia, Sede Medellín in Colombia, Nazarbayev University in Kazakhstan, and Dibrugarh University in India have also contributed their expertise. Notable Indian institutions, including the Indian Institute of Petroleum & Energy (IIPE) in Visakhapatnam, Rajiv Institute of Petroleum Technology, National Institutes of Technology (NIT) Rourkela and Nagpur, and UPES Dehradun are also part of this collaborative effort. This diverse representation underscores the interdisciplinary nature of the subject matter, enriching the discourse with a mosaic of perspectives and insights from experts across various cultural, educational, and professional backgrounds.

As you go through each chapter, you'll have a better grasp of the complex relationship between oil recovery and nanotechnology. The chapters cover all aspects of EOR, from theoretical foundations to real-world applications, revealing the revolutionary power of nanomaterials, computational modelling, and innovative techniques. Through its integration of theory and practice, this book seeks to provide scholars, practitioners, and students with the understanding and resources needed to traverse the intricacies of EOR skilfully.

As readers embark on this intellectual voyage through *Nanotechnology in Enhanced Oil Recovery,* they will embark on a journey of discovery, exploring the boundless possibilities that nanotechnology offers in reshaping the future of oil extraction. This book stands as a testament to the collective endeavour of the scientific community, striving towards progress and sustainability in the energy sector.

Contributor List

Nirmalya Bachhar
Department of Chemical Engineering
Anantharaman Building
Indian Institute of Technology Jodhpur
Jodhpur, India

Uma Sankar Behera
Department of Chemical and Biomolecular Engineering
Chonnam National University
Yeosu, South Korea
Krishna Raghav Chaturvedi
Rajiv Institute of Petroleum Technology
Jais, Amethi, India

Anand Raghav Chaubey
Udai Pratap Autonomous College
Varanasi, India

Farid B. Cortés
Grupo de Investigación en Fenómenos de Superficie – Michael Polanyi, Facultad de Minas
Universidad Nacional de Colombia Sede Medellín
Medellín, Colombia

Aditi Pradeep Dahake
Department of Chemical Engineering
Anantharaman Building
Indian Institute of Technology Jodhpur
Jodhpur, India

N.S.V. Dinesh
Indian Institute of Petroleum & Energy
Visakhapatnam, India

Camilo A. Franco
Grupo de Investigación en Fenómenos de Superficie – Michael Polanyi, Facultad de Minas
Universidad Nacional de Colombia Sede Medellín
Medellín, Colombia

Lady J. Giraldo
Grupo de Investigación en Fenómenos de Superficie – Michael Polanyi, Facultad de Minas
Universidad Nacional de Colombia Sede Medellín
Medellín, Colombia

Uttaron Goswami
Department of Petroleum Technology
Dibrugarh University
Dibrugarh, India

Himanshu Kesharwani
Presidency University
Bengaluru, India

Ali Khalilnezhad
Faculty of Petroleum and Natural Gas Engineering
Sahand University of Technology
Tabriz, Iran

Abhishek Kumar
National Institute of Technology
Rourkela, India

Anupama Kumar
Visvesvaraya National Institute of Technology
Nagpur, India

Ravi Shankar Kumar
Laboratory Division
Al-Arfaj Engineering Co.
WLL, Kuwait

Daniel Lopez
Grupo de Investigación en Fenómenos de Superficie – Michael Polanyi, Facultad de Minas
Universidad Nacional de Colombia Sede Medellín
Medellín, Colombia

Nayan Medhi
Department of Petroleum Engineering
Dibrugarh University
Dibrugarh, India

Veena Sumesh Nair
Anna University
Chennai, Tamil Nadu, India

Masoud Riazi
School of Mining and Geosciences
Nazarbayev University
Astana, Kazakhstan

Santosh Kumar Senapati
Visvesvaraya National Institute of Technology
Nagpur, India

Pushpa Sharma
Department of Petroleum Engineering
University of Petroleum and Energy Studies
Dehradun, India

Tushar Sharma
Rajiv Institute of Petroleum Technology
Jais, Amethi, India

Ramanand Singh
Design and Simulation Laboratory
Petroleum Engineering Section
Department of Ocean Engineering
Indian Institute of Technology Madras
Chennai, India

P. Sivasankar
Indian Institute of Petroleum & Energy
Visakhapatnam, India

Nupur Srivastava
Visvesvaraya National Institute of Technology
Nagpur, India

1 Overview of Enhanced Oil Recovery (EOR)

Nupur Srivastava, Veena Sumesh Nair, Santosh Kumar Senapati and Anupama Kumar

1.1 INTRODUCTION

Enhanced Oil Recovery (EOR) represents the method of recovering crude oil from oil fields that resist extraction through conventional means. Also referred to as tertiary recovery, EOR distinguishes itself from primary and secondary recovery processes by introducing a change in the chemical composition of the oil, thereby facilitating a more efficient extraction process. Unlike primary and secondary methods, which rely on the differential pressure between the field surface and the well, EOR focuses on altering the properties of the oil to enhance recoverability. While primary and secondary recovery typically yields 20%–40% extraction rates, EOR boasts a higher range of 30%–60%. Key points to highlight about EOR include:

- EOR entails the extraction of oil from a well that has undergone both primary and secondary stages of oil recovery.
- The economic feasibility of EOR techniques is contingent upon the fluctuating price of oil.
- Despite the potential adverse environmental impacts associated with EOR techniques, ongoing innovations in the sector aim to mitigate these effects in the future.

The application of nanotechnology in EOR aims to address and eliminate the negative aspects mentioned in the second and third points. However, the initiation of oil extraction from a specific reservoir requires in-depth knowledge of the reservoir properties, which significantly influences the selection of extraction methods. Further insights into the rock/reservoir properties are expounded upon by Ganat (2020).

I. *Rock Permeability*: In reservoir engineering, the flow of hydrocarbons is contingent on adequate permeability in the reservoir rocks, a crucial parameter described by Henry Darcy.
II. *Porosity*: Defined as the measure of void spaces within rocks, porosity encompasses void fractions between particles or inside cavities and cracks in soil or rock. It is expressed as a unit fraction between 0 and 1 or as a

DOI: 10.1201/9781032650685-1

percentage ranging from 0% to 100%. Typically, porosity in rocks varies from less than 1% to 40%.

III. *Wettability*: When two immiscible fluids come into contact with a solid surface, one phase tends to adhere more to the solid due to the balance of intermolecular forces and surface energies between the fluids and the solid.

IV. *Saturation*: Described as the ratio of fluid volume in the porous medium to the pore volume of the rock, saturation represents the proportion of interrelated apertures filled by a specified phase.

V. *Relative Permeability*: Darcy's law for fluid flow in permeable media is based on the assumption of complete saturation with the flowing fluid, a condition rarely found in hydrocarbon reservoirs. Reservoirs often contain a mix of gas, oil, and water occupying the pores.

More advanced EOR techniques, sometimes termed quaternary recovery, go beyond conventional methods. The three primary EOR techniques include gas injection (using gases like nitrogen and carbon dioxide), thermal injection (requiring high heat levels), and chemical injection (utilizing polymers to enhance waterflood effectiveness). A recent development, plasma pulse technology (PPT), has demonstrated a 50% improvement in the production of existing wells. Further exploration of these methods follows. Top of Form

1.1.1 Gas Injection

As described by Li et al. (2022), less permeability, as well as tightness of reservoirs, plays a significant function in the development of gas and oil reservoirs because of the size of these reserves. Sometimes, instead of water, gases are preferred because gas is more straightforward to inject in tight spaces that are present in the reservoirs. Currently, gas injection, commonly called the miscible flooding method, is the most common technique used in EOR (Hoffman, 2023; Pal & Sen, 2024). This process involves the introduction of miscible gases into the reservoir. The miscible displacement process improves oil displacement as there is a reduction in the tensions between the oil and the gas in the interface while maintaining the reservoir pressure. For total displacement efficiency, complete removal of the interface between the two interacting fluids is necessary. Gases, such as carbon dioxide, nitrogen, or natural gas, are generally used. Carbon dioxide is widely used as the fluid that induces miscible displacement as it is cheaper than liquified petroleum gas and can potentially reduce the viscosity of oil. However, N_2 is not only chemically stable but also insoluble in water compared to other gases. In addition to this, it has a slight compressibility factor as well, resulting in N_2 gas that can replenish the formation energy and thus maintain the formation pressure. The diagram presented below explains the gas injection method by using carbon dioxide gas (Figure 1.1).

1.1.2 Thermal Injection

In this method, several techniques can be employed to increase the temperature of the crude oil to decrease its viscosity or vaporize a portion of the oil, which causes

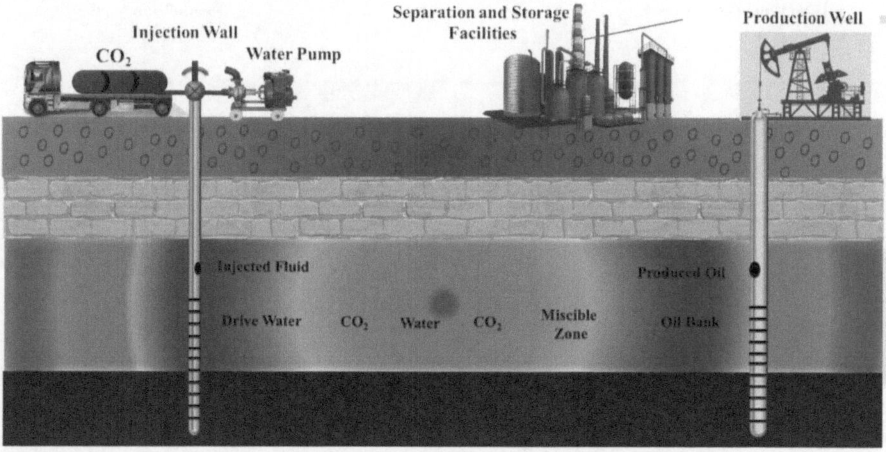

FIGURE 1.1 Illustration showing the process of gas injection (Hoffman, 2023).

the decline of the mobility ratio. As elucidated by Mokheimer et al. (2019), the thermal power used for thermal recoveries can lower the viscosity of oil, interfacial tension (IFT), and specific gravity, which enables the flow of oil in the production well at high rates. In addition to this, the chemical properties of oil also change by reactions such as dehydrogenation and cracking.

The high quantity of heat elevates the permeability of oil while bringing down the surface tension. The vaporization process leads to condensation, which forms improved oil production. The techniques include steam flooding as well as fire flooding, which are proven to increase displacement and sweep efficiency.

1.1.2.1 Steam Flooding

The process of steam flooding involves forcing steam into the well to heat the reservoir. Warm water is formed as the steam slowly condenses. This results in the evaporation of oil, which will cause the oil to expand, its viscosity to lower, and its permeability to increase. These processes must be conducted cyclically in order for them to be successful (Figure 1.2).

1.1.2.2 Fire Flooding

This technique involves the ignition of some portion of the oil within the reservoir, leading to the creation of a combustion front that moves through the reservoir, heating the surrounding oil and lowering its viscosity. Fire flooding can be used in reservoirs with heavy or high-viscosity crude oil as these oils are less responsive to conventional EOR methods and may benefit from the high-temperature combustion process, which reduces oil viscosity and causes an improvement in its mobility (He et al., 2022).

There are three ways for the conduction process to occur: wet combustion, reverse, and dry forward. In wet combustion, the water that is injected right behind the front is transformed into steam by the hot rocks. The heat spreads evenly using this method

FIGURE 1.2 Process of steam flooding (Likhacheva & Kashin, 2020; Mokheimer et al., 2019).

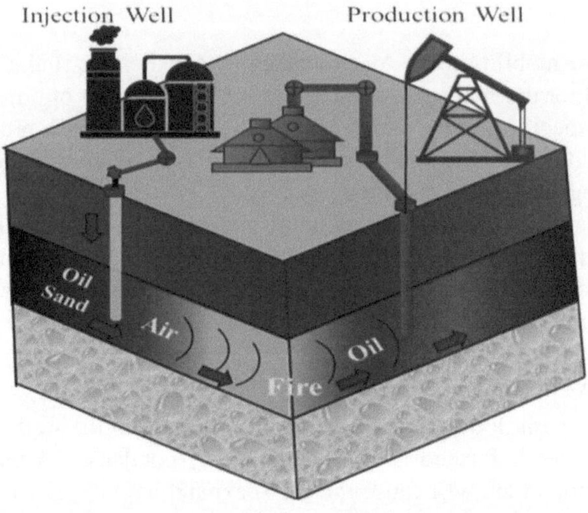

FIGURE 1.3 The technique of fire flooding (Likhacheva & Kashin, 2020).

as the fire gets quenched (Likhacheva & Kashin, 2020). In the reverse method, the ignition and the injection of air occur in opposite directions. A dry forward igniter was used to light the oil (Figure 1.3).

1.1.3 Chemical Injection

The introduction of several chemicals (usually in diluted solution form) by injection is known to reduce the surface tension and help with mobility. Lowering the IFT is a significant factor as it increases the production. This can be achieved by injecting caustic/alkaline solutions into oil reservoirs, preferably those reservoirs whose oils naturally produce organic acids. It leads to the production of soap, which might lead to the lowering of the surface tension and increase overall production. In some formations, the amount of recovered oil could be improved by increasing the viscosity of the injected water, which can be achieved by injecting water-soluble polymers into a dilute solution. Capillary pressure or IFT (which is analyzed in terms of bond number; the bond number is a dimensionless number measure and compares the gravitational forces and surface tension forces of the liquid movement in the front) restricts the movement of oil droplets in the reservoir, in order to reduce them, dilute surfactants such as rhamnolipids or petroleum sulphonates can be injected. Particular formulations of oil, water surfactant, and microemulsions are also capable of lowering IFT. It is to be noted that the application of these methods is low as the absorption of these chemicals onto the rock of the oil may be low, and losses may occur while the cost of these chemicals is high (Alvarado & Manrique, 2010; Pal & Sen, 2023b).

1.1.3.1 Polymer Flooding

According to Cheraghian and Hendraningrat (2016), polymer flooding focuses on increasing the water viscosity by mixing long-chain polymer molecules with the injected water (Setiati et al., 2021) (Figure 1.4). This technique helps to improve the areal and vertical sweep efficiency as the water/oil mobility ratio is getting improved. By Nwidee et al. (2016), the mobility ratio (M) should have values less than one ($M\,26 < 1$) to prevent unswept regions, where oil is left behind the displacement front. The overall mechanisms of oil recovery by polymer flooding are as follows:

- To increase the water viscosity.
- To decrease the effective permeability of water due to polymer retention.
- To decrease the water/oil mobility ratio that improves the macroscopic sweep efficiency.

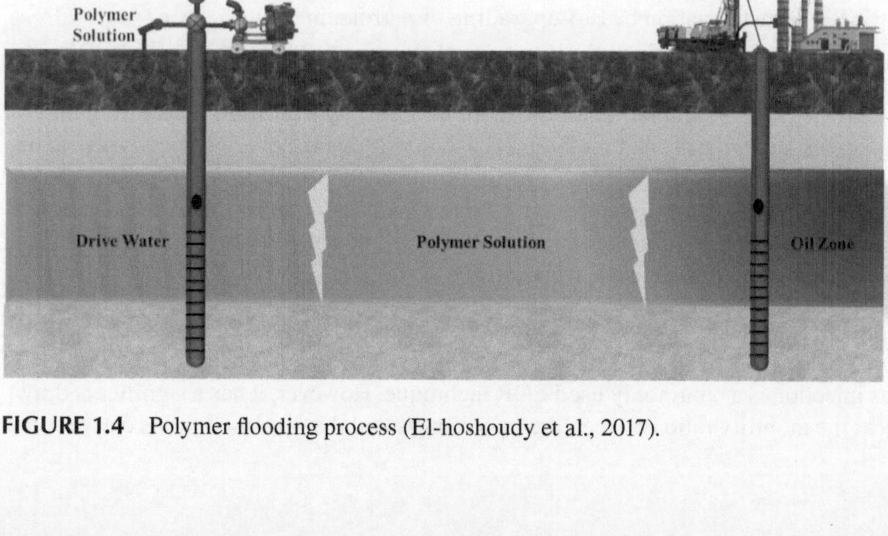

FIGURE 1.4 Polymer flooding process (El-hoshoudy et al., 2017).

Surfactants can be used in cooperation with hyperbranched polyglycerols and other polymers to decrease the IFT between the oil and water. This results in the reduction of the residual oil saturation and enhances the efficiency of the process from a macroscopic point of view. Co-surfactants, co-solvents, and activity boosters should be added to primary surfactants to increase the stability of the formulation.

Caustic flooding, as the name suggests, is the addition of sodium hydroxide (prevalently known as caustic soda) to the injection water. It causes the lowering of surface tension and reverses the rock's wettability, and it also is capable of emulsifying oil and its mobilization so that the oil can be drawn out of the rock.

1.1.3.2 Surfactant Flooding

Surfactant flooding is an important method used in EOR to lower the amount of oil in the pore space of matrix rock. Surfactants are injected to lower the IFT between oil and water or alter the wettability to water-wet from oil-wet and, thus, give motility to the residual oil and, therefore, changes in the mobility ratio. Important information by Nwidee et al. (2016) states that the amphiphilic structure of surfactant molecules improves the:

- Inclination of surfactant molecules to adsorb at interfaces.
- Changes in the parallel interfacial energies.
- Direction of hydrophobic groups away from the bulk solvent phase due to the agreeable orientation of adsorbed molecules.

1.1.4 Microbial Injection (MEOR)

According to Saravanan et al. (2020), MEOR processes are mainly applied with two categories in the reservoirs for EOR, such as in-situ and ex-situ mechanisms (Figure 1.5). In petroleum-refining industries, petroleum will be purified from their impurities by the following biological processes:

- Bio-desulfurization
- Bio-demetallation
- Bio-denitrogenation
- Bio-transformation or Bio-upgrading of petroleum

The microbial products are surfactants, solvents, acids, emulsifiers, polymers, and carbon dioxide. Two major factors account for EOR: hydrocarbon-consuming microorganisms and a nutritional medium. These microorganisms are of two types: exogenous or indigenous. Also, industrial by-products such as corn-steep liquor, cheese whey, and molasses are injected into the reservoir as a natural nutritional medium to reduce the cost; nitrogen and phosphorous are added to improve microorganism growth and microbial activity, respectively.

1.1.5 Water-Alternating Gas

Gas injection is a commonly used EOR technique. However, it has a significant drawback; the mobility ratio between the injected gas and the displaced fluids can become

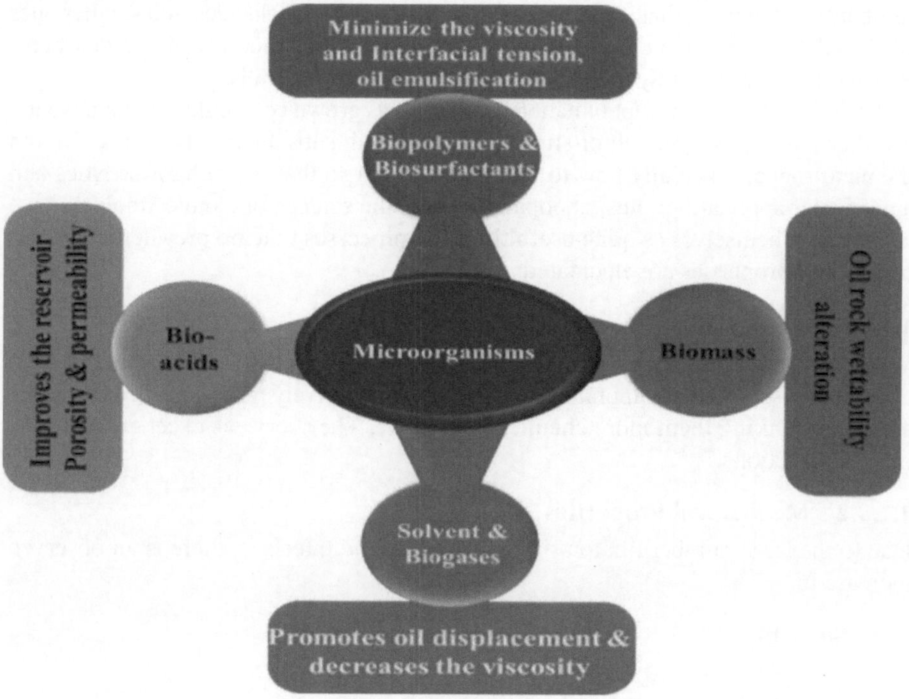

FIGURE 1.5 Microbial injection process (Saravanan et al., 2020).

undesirable. It might result in instability that causes viscous fingering, gas channelling in high-permeability streaks, and gravity segregation, all leading to an early release of gas. According to Janssen et al. (2020), to overcome this drawback, one might inject the gas phase in slugs alternated by slugs of water, i.e., water-alternating-gas (WAG) flooding. But, the early breakthrough of gas could be an issue here, too.

An alternative method is the foaming of the gas. Foam can reduce gas mobility to a large extent by trapping the gas in separate gas bubbles within a continuous aqueous phase. Foam stability is primarily a function of the thickness of thin liquid films or lamellae. The lamellae get stabilized by selected surfactants that are present in the liquid phase (Wojnicki et al., 2022).

1.1.6 Plasma Pulse

According to Patel et al. (2018), PPT-aided EOR treatment is carried out with an electric wireline conveyed plasma pulse generator tool that is run in the well and positioned alongside the perforations. By the use of energy stored in the generator's capacitors, a plasma arc is generated that releases a tremendous amount of heat and pressure for a very short duration of time. This results in the creation of a broad band of hydraulic impulse acoustic waves that are powerful enough to clean perforations and near wellbore damage. These waves continue to pass into the reservoir depths, exciting the fluid molecules across the expanse of the reservoir and increasing the reservoir's natural resonance to the point that it can break larger hydrocarbon molecules into smaller ones while reducing adhesion tension, which results in increased mobility of hydrocarbons. PPT has been successfully used in production and injection wells.

Let's see why the concept of nanotechnology has grown so popular in recent years. Nanotechnology is a branch of study that deals with substances whose size falls in the nano-range, especially how to manipulate them so that their characteristics suit a particular application; thus, nanoparticles and the science of nanotechnology have established themselves as quite useful in many processes that are prevalent in industries. Their properties are elucidated as follows.

1.1.6.1 Surface Area

When macromolecules and nanomaterials of the same masses are compared, it is generally observed that nanomaterials have a comparatively more significant surface area, thus making them more chemically reactive. They serve as excellent catalysts in many reactions.

1.1.6.2 Mechanical Properties

Due to the large number of atoms on the surface and interface, there is an observed increase in:

- Strength
- Hardness
- Toughness
- Elasticity
- Ductility

1.1.6.3 Thermal Properties

Nanoparticles have a higher density of defects compared to macromolecules or bulk materials, resulting in a higher thermal expansion coefficient, where the coefficient is a measure of how the size of an object changes with the alteration of temperature.

1.1.6.4 Melting Point

There is a higher number of atoms on the surface of a nanomaterial than in its interior. Melting occurs when the vibrational amplitude exceeds the bond length between two atoms, and when heat is applied, the outer atoms melt first and spread to the interior. It causes the melting point of nanomaterials to be much lower than that of bulk materials.

1.1.6.5 Electrical Properties

As nanomaterials have a high density of grain boundaries, electric-phonon and phonon-phonon scattering are more successful. It causes the nanomaterials to have lower thermal and electrical activities than bulk materials.

1.1.6.6 Magnetic Properties

Nanomaterials have higher magnetic coercivity values and higher saturation magnetization as they have small particle sizes (each particle will behave as a magnetic domain), which enables an external magnet to align them easily.

1.1.6.7 Diffusivity

When compared to bulk materials, nanomaterials have a lot more interfaces, which are perfect for diffusion. More pathways allow diffusion in nanomaterials than their counterparts, and thus, it is said they have high diffusivity.

We have seen the general properties that make nanomaterials very useful for industrial applications. Let's bring our attention to the characteristics of nanomaterials that make them favourable for the process of EOR.

a. They can be easily modified to enhance their properties in order to help with oil recovery.
b. They are economically cheap.
c. They can move with ease in porous media without clogging the pore throats.
d. They are capable of modifying oil/water IFT.
e. Metallic nanoparticles are separated easily by magnetic separation.
f. They help in creating a wedge film or structural disjoined pressure to sweep the droplets of oil from the rock surface.
g. It has a low cost of installation.
h. It has a high surface-to-volume ratio for enhanced thermal properties.

1.2 EFFECT OF NANOTECHNOLOGY ON EOR

The demand for and investment in renewable energy is increasing day by day as the threat of global warming and its consequences are looming towards us at an alarming speed. However, the development of energy from these sources fails to match these increasing demands; thus, we still stick to oil as our primary resource. One might know that the oil barrel prices in UAE play a massive role in the world's economy. Thus, it is imperative that we recover the oil to its fullest before our existing reserves run out and everyone gets a chance to explore new fields. It is common knowledge that tertiary methods make up around two-thirds of the oil produced of the total oil that is extracted initially. Recently, a new issue arose: the drastic increase in the cost of chemicals while oil prices saw a reduction due to the recession. This is where the brilliance of the implementation of nanotechnology in the oil industry comes into play; the introduction of nanotechnology can make many recovery processes more economical and cost-effective, as proven by many extensive experiments conducted on the lab scale (Negin et al., 2016).

Nanomaterials can be used in recovery processes in three significant forms, which are discussed below.

1.2.1 Nanoemulsions

Nanoemulsions are liquid-in-liquid dispersions that have high kinetic stability, which is provided by nanoparticles. This stability solves many problems posed by surfactant-stabilized emulsions (they are conventional in their composition) and colloidal solids. The size of nanoemulsion droplets is comparable with that of a typical macroemulsion droplet, both falling in the range of 100 nm or less, and the main difference does not lie in their size or shape. A nanoemulsion can retain its morphology when a change occurs in the volume fraction of oil because it's a kinetically controlled system; this ability

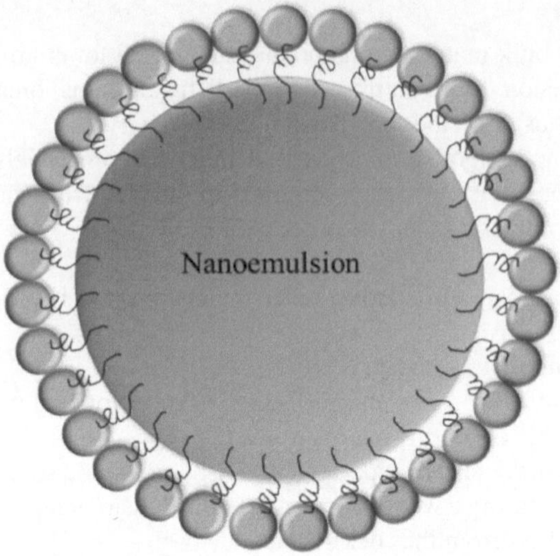

FIGURE 1.6 Illustration of nanoemulsion formed by mechanical shearing (Mariyate & Bera, 2022).

grants it the capacity to withstand harsh conditions such as high pressure, temperature, salinity, and sheer. Besides, nanoemulsions can easily control the mobility ratio during the flooding process. Let's briefly see what the mobility ratio is. The mobility ratio is the ratio that measures the rate and displacement of other immiscible fluids.

It is given by $M_r = M_{rw}/M_{ro} = (K_{rw} \times \mu_o)/(K_{ro} \times \mu_w)$. Here, K_{rw} is the relative permeability of water, K_{ro} is the relative permeability of oil, μ_o is the viscosity of oil, and μ_w is the viscosity of water.

High values of mobility ratios are not preferred because they result in poor displacement and sweep efficiencies, which causes the injected water to be released quickly (an early breakthrough of injected water). However, if the mobility ratio is reduced, this breakthrough of water can be delayed, causing effective displacement and sweep efficiencies (aerially and vertically); this results in the recovery of more oil at any site where a water cut has been introduced (Figure 1.6).

Lastly, the size of nanoemulsions makes them ideal for penetration into pore throats whilst not staying inside for a long time. Thus, they can be used instead of polymers, which have very high retention on reservoir rocks. It makes them useful in field-scale applications in a large scale. These factors are hugely advantageous for oil recovery and are of superior interest to researchers, academicians, and as well as oil companies.

1.2.2 Nanocatalysts

Nanocatalysts, as described by Sun et al. (2017), are metal particles that fall in the nano-size range and are used as catalysts during the steam injection process in heavy oil reservoirs. The question that many people may ask is: How are nanocatalysts better than regular catalysts? Two chief points come to the forefront. Firstly, nanocatalysts can efficiently catalyze the breaking of carbon-sulphur bonds in asphaltenes, thus increasing the level of aromatics and saturation in heavy oils. Secondly, they are helpful in the conduction of a set of reactions called aquathermolysis. Aquathermolysis reaction involves the conversion of bitumen to other lightweight products in heavy oil reservoirs. The reaction is described below:

R-S-R (organic sulphur compound) + H_2O → CO + H_2S + lighter organic compound

Aqua thermolysis results in the irreversible reduction of the viscosity of heavy oil.

1.2.3 Nanofluids

In the most basic sense, a nanofluid can be defined as a colloidal suspension of nanoparticles having an average size range of less than 100 nm present in a base fluid. The base fluid can be any liquid, such as oil, water, or gas. Generally, nanofluids formed by adding different nanoparticles in water or brine are used to enhance water-flooding recovery.

The EOR mechanisms of nanofluids have already been investigated in the literature, which have been briefly discussed below.

1.2.3.1 Reduction of IFT

IFT is the force between two different phases. According to Sun et al. (2017), it can be measured using a force or an optical tensiometer. The reduction of IFT reduces the magnitude of capillary pressure, thus releasing the trapped oil.

Nanoparticles mix with surfactants to form a mixed layer at the interface between the oil and the fluid that is injected. Formation of an emulsion of oil/fluid takes place due to shear, and once this dispersion stabilizes, the nanoparticles are adsorbed in the interface. Both the nanoparticles and the surfactants help each other to stabilize the newly formed emulsion and themselves; the nanoparticle improves the stability of sedimentation while the surfactant helps stabilize the nanoparticles. These activities reduce the capillary forces, and the capillary number will experience a hike. However, it must be known that the concentration of the nanoparticles plays a massive role in this process: at higher concentrations, the nanoparticles can completely wipe off the surfactant from the layer, while, at lower concentrations, IFT is reduced. The diagram presented below describes the process of using zinc oxide (ZnO) nanoparticles (Figure 1.7).

1.2.3.2 Wettability Alteration

As mentioned before, wettability is the tendency of the liquid to spread on a solid surface. The interactions of rocks and liquids in the reservoir, such as brine or crude oil, are determined by the wettability in EOR. Contact angles are used as a measure of wettability.

Reservoirs (rocks) can be classified as oil-wet, water-wet, and mixed-wet. These different wetting states have different contact angle ranges.

a. Oil-wet: Oil is in contact with a particular mineral. Contact angles range between 110 and 180.

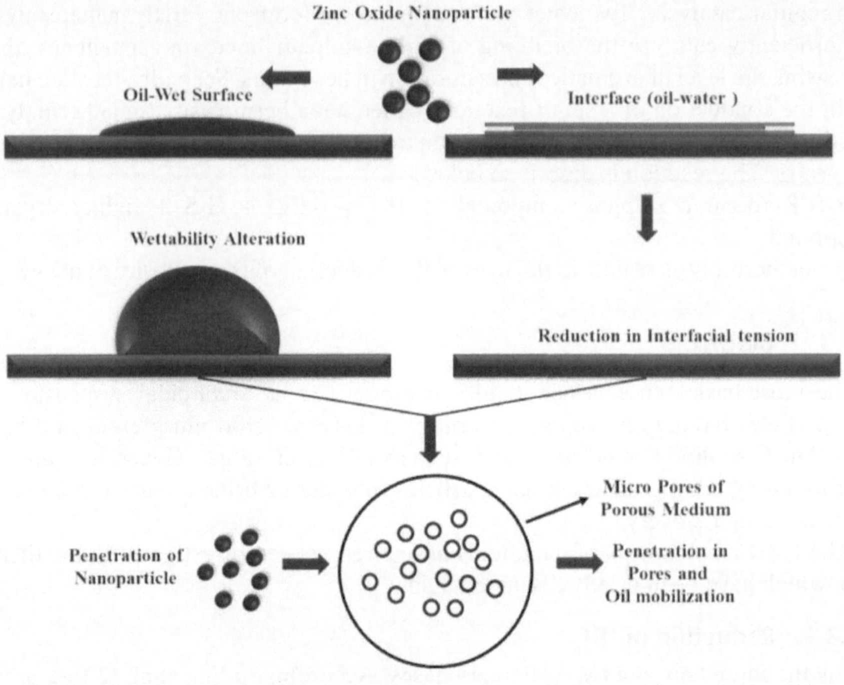

FIGURE 1.7 The process of EOR with the usage of zinc oxide nanoparticles (Soleimani et al., 2018).

b. Water-wet: Water is in contact with the mineral, while oil is the surrounding phase. Contact angles range between 0 and 70 (Francisca et al., 2003).
c. Mixed-wet/Intermediate-wet: The state in between the two states previously mentioned. Contact angles range between 70 and 110.

In the simplest of terms, wettability alteration refers to making a reservoir water-wet. It is done because a reservoir responds differently to water flooding based on its wettability. The recovery rate reduces as the rock behaves more oil-wet. Many of the succeeding EOR applications target improving oil recovery by altering the wettability to more water-wet (Figure 1.8).

1.2.3.3 Disjoining Pressure

As described by Sun et al. (2017), when in contact with the oil phase, the nanoparticles present in the nanofluids can form a film that is shaped like a wedge; this film is self-assembled. This film is formed due to the pressure that is present, namely, structural disjoining pressure or, simply, disjoining pressure. This film causes the oil droplets to separate from the surface of the rock, which increases oil recovery (Figure 1.9).

The generation of the structural disjoining pressure is attributed to osmotic reasons. Pressure is used to inject nanofluids. These pressurized nanofluids, in turn, exert pressure on nanoparticles, forcing them to move forward. Then, they proceed

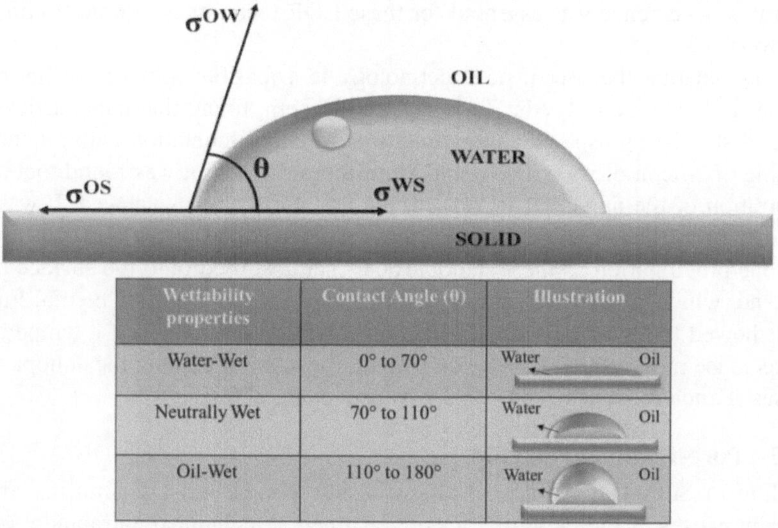

FIGURE 1.8 Wettability properties of liquids in enhanced oil recovery (Francisca et al., 2003).

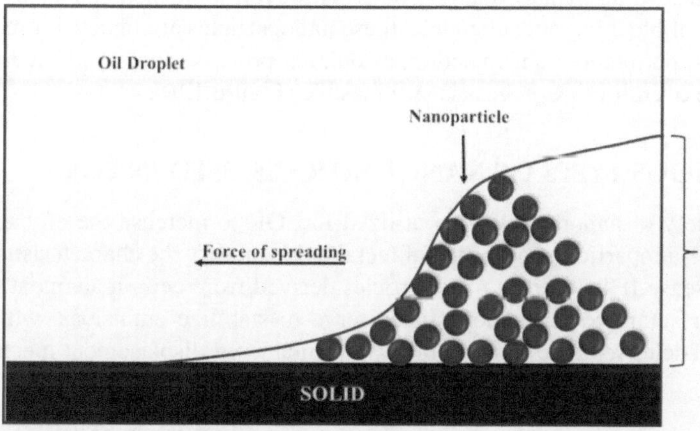

FIGURE 1.9 Disjoining-pressure oil recovery (Rath et al., 2023).

to arrange themselves in layers, causing an increase in the entropy; this leads to the exertion of more disjoining pressure at that particular interface than in the bulk liquid (Rath et al., 2023).

1.2.3.4 Preventing Asphaltene Precipitation

Sun et al. (2017) mentioned that techniques in EOR, such as carbon dioxide flooding, are known to cause significant asphaltene precipitation due to special reservoir conditions, which leads to unfavourable wettability alteration, reduction of permeability in formations, and blockage of transportation pipeline. Thus, finding a solution to prevent this occurrence was essential for these EOR techniques in order to improve oil recovery.

It turns out that the use of nanotechnology is a feasible solution to this pressing issue. Two researchers, Abu Tarboush and Husein, found that nanoparticles are capable of stabilizing asphaltene precipitation. In experimentation with a nanofluid consisting of silicon dioxide (SiO_2) and aluminium dioxide, it was found that as the concentration of the nanofluid increased, the precipitation was delayed. It was also found that SiO_2, nickel oxide (Ni_2O_3), and ferric oxide nanoparticles can prevent asphaltene precipitation as these nanoparticles get absorbed onto the surface of the asphaltene, which reduces the aggregation of asphaltene in porous media. Further studies showed that with an increase in contact time, a decrease in the temperature, and a decrease in the amount of existing water, the adsorption onto the nanoparticles increases (Panda & Gangawane, 2023b; Vargas et al., 2009).

1.2.3.5 Pore-Channel Plugging

According to Sun et al. (2017), mechanical entrapment and log-jamming are the significant causes of the plugging of pore channels. Mechanical entrapment occurs because the diameter of injected components is larger than the pore channels through which they are made to flow. However, the size range of nanoparticles is much smaller than the pore channels, which are predominantly on the macroscale and are at least a thousand times larger than the nanoparticles. However, some metal nanoparticles are still capable of blocking pore channels; these nanoparticles are injected directly into the reservoir, but before their injection, evaluation processes are conducted to make sure that the overall EOR process is still feasible (Figure 1.10).

1.3 VARIOUS TYPES OF NANOPARTICLES USED IN EOR

A wide variety of nanoparticles are utilized in EOR to increase the efficacy of oil production. Nanoparticles composed of metal oxide modify the characteristics of fluids and decrease IFT, whereas nanoparticles derived from carbon augment the flow of fluid. Surfactant-coated nanoparticles serve to stabilize emulsions, while polymer nanoparticles regulate fluid mobility. By enhancing displacement mechanisms,

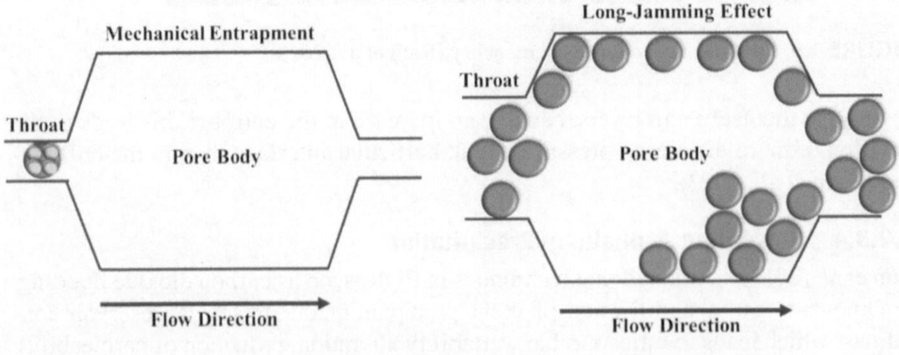

FIGURE 1.10 Pore-channel mechanical entrapment for EOR (Sun et al., 2017).

modifying fluid behaviour, and improving rock–fluid interactions, these varied nanoparticles optimize hydrocarbon recovery.

1.3.1 Metal Oxide Nanoparticles

The major constituents of these nanoparticles are metal oxides, and several types of metal oxide nanoparticles need to be studied.

a. Aluminium oxide (Al_2O_3) nanoparticles: It was found that Al_2O_3 nanofluids can reduce IFT and viscosity of the oil; there is a more substantial effect displayed in the latter than in the former. The best oil recovery is brought about when brine or distilled water is employed as the dispersing agent: 0.05% (by weight) of Al_2O_3 is dispersed in 3% sodium chloride (NaCl; by weight) brine. Stringent experimentation on the emulsion stabilities of Al_2O_3, titanium oxide, and silica was done at temperatures of 80°C, and it was found that Al_2O_3 showed maximum recovery from the intermediate-wet rocks (as compared to oil-wet and water-wet rocks). It was concluded that Al_2O_3 can be of good use to EOR from a sandstone reservoir (Amrouche et al., 2021).
b. Copper oxide (CuO) nanoparticles: The knowledge of the usage of these nanoparticles in EOR is limited. The little research that was conducted involved the gas injection of CuO nanoparticles to see whether it improved EOR or not; this was done in the presence of API gravity. The results showed that CuO nanoparticles in the size range of 23–40 nm increase the density and viscosity of the gas injected and reduce the viscosity of the heavy oil simultaneously. These results were even more improved when 1% of CuO nanoparticles were mixed with a concoction of surfactants [Poly-Di-Methyl-Siloxane (PDMS)] and carbon dioxide. After thorough research and optimization procedures, it was found that the best results were obtained when 30 mol% carbon dioxide, PDMS, and 1% (by weight) CuO nanoparticles were used at 122°F (Cao et al., 2019; Dabhade et al., 2021; Sahoo & Mishra, 2023).
c. Iron oxide (Fe_2O_3/Fe_3O_4) nanoparticles: These nanoparticles are widely used as data storage, imaging, and sensors as they possess unique magnetic characteristics. Research showed that iron oxide mixed with brine displayed promising results in oil recovery. However, when combined with diesel, these nanoparticles showed 82.5% oil recovery. Yet, nanoparticles of oxides of aluminium and nickel gave better results (Popescu et al., 2019).
d. Ni_2O_3 nanoparticles: These nanoparticles can increase the viscosity of the displacing fluid while decreasing the viscosity of the displaced oil. These nanoparticles showed a recovery of 85% of heavy oil, even in the presence of high asphaltene precipitation. Ni_2O_3 nanoparticles are great for oil recovery when dispersed in diesel, but the recovery was found to be more significant when brine was used as a dispersing agent instead (Hashemi et al., 2016).
e. Magnesium oxide (MgO) nanoparticles: The experiments that were performed using MgO nanoparticles showed that their use in EOR would bring

about positive effects. However, the research on these nanoparticles barely cut the tip of the iceberg. However, it was found that MgO nanoparticles dispersed in ethanol can cause a significant reduction in oil viscosity, yet the recovery fell short. It goes without saying that more research and experimentation need to be conducted for more conclusive results.

f. Tin oxide nanoparticles: These nanoparticles are used extensively in applications like gas sensors, gas discharge and display, transparent conductive electrodes for solar cells, n-type semiconductors, lithium-ion batteries, and so on. Thus, it spiked the curiosity about whether their field of application could be expanded to EOR as well. It was found that these nanoparticles perform reasonably well in sandstone cores when dispersed in distilled water as compared to ethanol and brine at room temperature.

g. Titanium dioxide (TiO_2) nanoparticles: In the absence of these nanoparticles, the oil recovery was 49%, and after the injection of TiO_2 nanoparticles, the numbers went up to 80%. The strangest fact discovered was that these nanoparticles don't affect the IFT and the viscosity; their mechanism lies in their capability to alter the wettability drastically. It was also found that 1% (by weight) of Povidone acts as an excellent stabilizer for the emulsion that is formed compared to 3% (by weight) of NaCl brine solution.

h. ZnO nanoparticles: These nanoparticles are used as stabilizers for ceramics, food, semiconductors, photocatalysts, and ointments. Their applications in EOR have been limited mainly because they posed a few disadvantages that reduced oil recovery; they agglomerate at the injection point and cause the blockage of pores, and it had a degradative effect on the permeability of the samples (Minz & Mishra, 2023; Santhosh & Sarkar, 2022; Soleimani et al., 2018).

i. Zirconium oxide (ZrO_2) nanoparticles: Research showed that these nanoparticles are not suited for oil recovery; injection of these nanoparticles didn't make much difference compared to the injection of distilled water. In addition to this, the recovery saw a reduction when ethanol and brine were used as dispersed agents. Their uses are much better in applications such as solid oxide fuel components, pressure and temperature transmitters, drug delivery, and ceramics.

1.3.2 Magnetic Nanoparticles

Magnetic nanoparticles are particles derived from either a natural or synthetic source whose size is in the nano-range. They are known to be able to demulsify crude oil and cause a significant reduction of the total organic carbon levels in the water layer (Kumar et al., 2023b; Kumar & Gangawane, 2022a, 2022b).

a. Ferro nanofluids: Initially, research was focused on the rheological properties of ferro-fluids rather than their applications in EOR processes. However, it was found later on that when these ferro nanofluids are combined with appropriate surfactants, a reduction of IFT takes place. This mixture can be used for both oil-wet and water-wet reservoirs. In addition to this, there is a recurring problem in the water-wet reservoirs that these nanofluids help overcome: oil bubbles get trapped in the centre of the pores, and ferro nanofluids make these bubbles collapse. Ferro nanofluids have a significant dipole moment, which causes the alignment of the reservoir fluid, which, in turn, makes recovery very easy. The advantages of these nanofluids have indeed been noted, but not enough research has been done to understand whether or not the application of ferro nanofluids will be applicable. However, this has a lot of scope. These particles have the formula of XY_2O_4, where X and Y are metal atoms, and X, Y, or both are Fe in this case (Chand et al., 2017; Kumar et al., 2022, 2023a).

b. Cobalt ferrite nanoparticles: The molecular formula is $CoFe_2O_4$. These particles are generated via the sol-gel method. Two experiments were conducted separately for adequate comparison. In both the experiments, it is applied to magnetic feeders along with an antenna to enhance the magnetic field; however, in the first one, it is used as particles, whereas in the second, it is applied as a nanofluid. The experiments also involved the addition of cobalt ferrite nanoparticles to nanofluid, and this mixture was injected into an apparatus consisting of a PVC column packed with glass beads. This sample is then exposed to electromagnetic waves for 24 hours at 55°C temperature. The results showed that residual oil decreased to 8.7% from 31.58%. It should be noted that a stabilizer is used: 1% sodium dodecyl sulphate (SDS); this concentration creates the highest suspension layer in an emulsion. These nanoparticles are not being used widely as there is a lack of research done in this particular area (Pourgolmohammad et al., 2017).

1.3.3 Organic Nanoparticles

Organic nanoparticles are collections of organic molecules whose overall size falls in the nano-range.

a. Carbon nanoparticles: They are usually found in the form of black powder consisting of many spherically shaped nanoparticles; these nanoparticles are normally synthesized by hydrothermal methods. These particles can be made to undergo surface modifications, for example: organic molecules or even polymers can be chemically bonded to the surface of these particles. After thorough research, it was found that improved retention can be achieved by modifying the surface and the surface properties of nanoparticles. In order to exactly determine how much carbon nanoparticles are suitable for EOR, an extensive experiment was conducted using the difficult atmosphere of a carbonate reservoir having a temperature higher than 100°C and with 12,000 ppm salinity formation (this value is very high). The results were the most favourable; these nanoparticles increased the oil recovery by 96% (Panda et al., 2023).
b. Carbon nanotubes: The use of carbon nanotubes in EOR has only recently come to attention. They have shown some characteristics that are desirable for EOR. These characteristics include:

- They are lightweight.
- They are strong.
- They are resistant to corrosion.

These nanotubes could be single-walled or multi-walled, where each wall is made up of graphene. Their surprising electrical, mechanical, and thermal conductivity is attributed to the carbon atom of the graphene. Every carbon atom forms three sp^2 bonds (brought about by hybridization) and an extra electron, which is free to move throughout the network. It also must be known that these nanotubes are highly hydrophobic.

It was found that after the injection of nanofluids containing multi-walled nanotubes (MWNTs), the oil recovery was 36%. This percentage almost doubled when the same was done in the presence of electromagnetic waves. In addition to this, nanotubes are known to increase the efficiency of drilling fluids as they cause a reduction in viscosity.

1.3.4 Inorganic Nanoparticles

Unlike organic nanoparticles, these nanoparticles do not consist of carbon molecules in their composition.

1.3.4.1 Silica-Containing Nanoparticles

a. SiO_2 nanoparticles: SiO_2 or silica is obtained very quickly, and thus, it is one of the most cost-effective nanoparticles that exist. They can be obtained naturally (from quartz) or artificially synthesized. Experimental studies showed that the wettability changed immediately from water-wet to intermediate-wet, and there was a reduction in IFT when these nanoparticles dispersed in ethanol were applied. The dominant mechanism of these nanoparticles is the wettability alteration. It was also found that the specific area of silica barely changed when subjected to temperatures as high as 650°C, which proves that these nanoparticles have good thermal stability. When SiO_2 nanoparticles were mixed with 3% (by weight) NaCl brine, they formed better emulsions (stabilizer is not required) than a mixture of SiO_2 nanoparticles and metal oxides. The surface properties of SiO_2 nanoparticles can be switched from hydrophobic and hydrophilic states and vice-versa by the silanization method, which makes use of hydrophilic hydroxyl groups, hydrophobic sulphonic acid, and hydrophilic polyethene glycol. The disadvantage of these nanoparticles is that they tend to form aggregates and cause clogging of the pore channels (El-hoshoudy et al., 2017; Panda et al., 2024; Panda & Gangawane, 2023a, 2023c).
b. Alumina-coated silica nanoparticles: The alumina coat causes a drastic change in the SiO_2 nanoparticles; this coating generates a positive charge on the surface of the nanoparticles. These coated nanoparticles seem to have a greater surface area and are more environment friendly than bare SiO_2 nanoparticles. Research on these nanoparticles was conducted, and the modified surface made more stable foams and increased oil recovery in sandstone cores. Under this category, there is a specific type of nanoparticles,

which is hydrophobic SiO_2 nanoparticles; they are known to increase the oil recovery when dispersed in ethanol.

c. Spherical, fumed silica nanoparticles: The size range of these particles is several to tens of nanometres; therefore, clogging of the pore channels is not an issue. However, the direct application of these nanoparticles in EOR is limited. It is known that these nanoparticles can act as stabilizing agents for oil/water emulsions.
d. Inorganic silica-core/polymer-shell nanocomposite: These nanoparticles consist of silica nanoparticles in the core and a synthetic polyacrylamide polymer shell. Their suitability in applications that involve environments (offshore reservoirs) having high temperatures, high salinity, and high amounts of complex ions is attributed to their composite nature. These particles can reduce the IFT and increase the viscosity at critical concentrations. These nanoparticles increased the oil recovery by 6.2% in granite reservoirs at 92°C.
e. Silicon oxide treated with silane nanoparticles: These nanoparticles can serve excellently in EOR in sandstone reservoirs. The best results are obtained when these nanoparticles are dispersed in ethanol.
f. Polysilicon nanoparticles (PSNP): SiO_2 nanoparticles that constitute a central component in polysilicon are known as polysilicon. There are three categories of these nanoparticles: Neutrally Wet Polysilicon Nanoparticles (NWPN), Hydrophobic and Lipophilic Polysilicon Nanoparticles (HLPN), and Lipophobic and Hydrophilic Polysilicon Nanoparticles (LHPN).

- *NWPN*: The dominant mechanism for these nanoparticles is wettability alteration and IFT reduction. The best results are displayed when these nanoparticles are dispersed in ethanol.
- *HLPN*: Research shows that wettability alteration and IFT reduction are the dominant mechanisms with an edge towards the latter. For EOR, 4 g/L of these nanoparticles is the optimum concentration, but concentrations around 3 g/L are also acceptable. Oil (in-situ) recovery increased up to 80% at room temperature with 2,000 psi in water-wet sandstone. Ethanol acts as a dispersing agent.
- *LHPN*: When γ-rays are added to SiO_2, it leads to the formation of ultra-fine powder. This powder is the desired nanoparticle. It acts as a good EOR agent for wet-oil formations as the dominant mechanism. The major disadvantages are that there is a chance that these nanoparticles alter the wettability in such a way that the conditions become unfavourable and that they can cause clogging of the pore channels due to their minimal size.

1.3.4.2 Non-silica Nanoparticles

These nanoparticles do not possess silica as their main component.

a. Nano-structured zeolite: The porous structure of zeolite and its ion-exchanging property have made them a point of interest in EOR. Recent studies have shown a 40% increase in the extraction of gasoline than the other catalysts, but there has been no direct research on their application in EOR.

b. Nano-sensors: The concept of development and usage of nanosensors are relatively new; they seem to have massive benefits as they have a thousand times more resolution than EM or RF. Due to the high resolution it provides, imaging of the reservoir interiors can be done quickly, the mechanisms that take place there can be understood more quickly, and more precise data can be procured.
c. Nano-sized colloidal dispersion gels (CDG): Research has shown that the usage of nano-sized colloidal dispersion gels can increase oil recovery by up to 10.5%. However, they are usually an agent to improve the sweep in order to displace oil when the mobility ratio is not favourable.
d. Polymer nanoparticles: It is interesting to know that many methods, such as interfacial polymerization, solvent evaporation, micro- and mini-emulsion, dialysis, and supercritical fluid technology, can produce these nanoparticles. Properties of the nanoparticles, such as particle size, particle size distribution, and the area of application, will differ with the method of generation. However, there have been no studies regarding the use of polymer nanoparticles in EOR.
e. Polymer-coated nanoparticles: Metal oxide cores covered/coated with polymers are used as they can resist electrostatic interaction. Polymer-coated nanoparticles are known to alter the wettability of the surface, reduce IFT, increase the viscosity of the dispersing agent, and stabilize emulsions and foams. Still, they pose challenges such as aggregation and adsorption.
f. Polyacrylamide micro-gel nanospheres: They can and are used for EOR, but they can't be applied directly. These nanoparticles need to be dispersed in a mixture containing two emulsifiers: polyethene glycol sorbitan monostearate with sodium hydroxide and sorbitan monooleate. The sodium hydroxide plays a significant role in decreasing the IFT, which can cause an increase in oil recovery.

1.4 CHALLENGES

- As nanoparticles tend to form aggregations because of the Van der Waals interactions in the harsh conditions of the reservoir, it is difficult to create a homogeneous suspension of nanoparticles. It has posed a technical challenge to create a nanofluid with accurate proportions.
- The research on nanofluids and their effects on EOR remains a new field to delve into; thus, due to a lack of actual experimental research, we are unable to see the wide variety of applications that these nanofluids can be used for EOR.
- Extrapolating on the previous point, to have a basic comprehension of EOR with the assistance of nanotechnology requires the performance of precise calculations and modelling based on these calculations. These continue to be a challenge as they involve following complex procedures on mechanisms that have not been fully understood yet. In addition to this, there are many conflicts among the various types of nanofluids; people are researching whether SiO_2 nanoparticles would be better than metal nanoparticles. Thus, more thorough research is required.

- To this day, there is not much clarity on the interactions between nanoparticles and how they will affect rock properties.

1.5 OPPORTUNITIES FOR FUTURE RESEARCH

- To overcome the challenges mentioned above, further experimental research needs to be conducted. Ideally, this research should focus on improving the stability of nanofluids, controlling their dispersion state, and understanding the mechanisms better.
- Theoretical studies and mathematical models need to be conducted for the various nano-assisted EOR processes to comprehend the mechanisms better. This reduces the risk when the developed methods are actually applied in the fields.
- Development of environment-friendly and cost-effective nanoparticles needs to be done. In addition to this, nano-assisted EOR applications need to be explored in pilot scales.

1.6 POTENTIAL HEALTH RISKS

According to Davoodi et al. (2022), various characteristics pose nanoparticles as a threat to health. The specific morphologies of nanoparticles, their high surface area to volume ratio, and their contents (that include metallic substances) cause nanoparticles to be very harmful when inhaled; they can give rise to serious respiratory disorders. Complex reactions occurring between these nanoparticles and pollutants can give rise to stable-derivative air pollutants that are detrimental to the current ecosystems, animal health, and the environment overall. Stringent laboratory tests need to be conducted to understand how adverse the outcomes can be if nanoparticles (of different compositions) are used in a full-fledged manner in oil-based industries. Despite all these concerns, many industries have started to use nanoparticles in multiple-scale applications. The used nanoparticles get collected in the soil, atmosphere, rivers, and oceans as they are being employed in waste-water treatment and manufacturing plants (Kumar & Gangawane, 2023a, 2023b; Pal & Sen, 2023a).

It seems that these nanoparticles are more of a potential health risk than micro- or macro-particles; in addition to this, the accumulation and movement of nanoparticles in the environment remains an enigma as there are no standard tests that are sensitive enough to quantify their presence. Nanoparticles are being used as adsorbents and catalysts for heavy oil recovery, even though tests regarding detection and recovery of nanoparticles from soils and groundwater using zerovalent metals have just begun. These tests have concluded that there is a high possibility that some of those nanoparticles will remain in rock formations for years on end. No studies regarding the monitoring of the movement of these nanoparticles in the sub-surface and their integration into the formations have been conducted. Thus, the detrimental effects of nanoparticles cannot be ruled out. Recyclable, reusable, recoverable, cost-effective, durable, sustainable, and selective nanocatalysts need to be developed.

1.7 CONCLUSION

Traditionally used EOR methods pose specific challenges that the use of nanoparticles can overcome. Thus, the effect of nanotechnology in EOR has been the focus of extensive research lately. Nanoparticles possess unique characteristics such as low costs, very small size, and high surface-area-to-volume ratio, which prove to be very useful in EOR.

Nanoparticles can be used for EOR in three forms: nanofluids, nanoemulsions, and nanocatalysts. Their EOR mechanisms depend on their respective application processes. Nanoemulsions can be used for water, gas, and chemical flooding. Nanocatalysts are always used together with thermal EOR methods to conduct in-situ upgrading inside reservoirs by aquathermolysis. Nanofluids can be used as a technique to increase oil recovery of water or gas flooding. Application of nanofluids in EOR is known to positively affect the IFT, wettability, disjoined pressure, asphaltene precipitation, and clogging of the pore channels, which significantly increase the oil recovery. However, many factors, such as types and sizes of nanoparticles, temperature, and injection time, have enormous effects on the performance of nanofluid flooding. The ageing of nanofluid contributes to increased oil recovery.

Various types of nanoparticles are used in EOR, which can be divided into five categories: metal oxide nanoparticles, magnetic nanoparticles, organic nanoparticles, inorganic nanoparticles, and non-silica nanoparticles. Thorough studies and experimental research gave us information on nanoparticles and their dominant mechanisms in increasing the oil recovered; this information has been tabulated below (Table 1.1).

There are challenges that the usage of nanoparticles in EOR poses: their tendency to aggregate and the deficiency of experimental research on this new topic, and they can be a potential risk to health: their ultra-small size and high surface area to volume ratio are attributed to various respiratory-related disorders. Areas such as the development of accurate mathematical models that predict the EOR mechanisms of

TABLE 1.1
Various types of nanoparticles and their dominant mechanism

Dominant mechanism	Type of nanoparticles
Reduction of viscosity	Aluminium oxide, copper oxide, iron oxide, nickel oxide, magnesium oxide, and polymer-coated nano particles
Reduction of interfacial tension	SiO_2 nanoparticles, polyacrylamide micro-gel nanospheres, polymer-coated nano particles, ferro-fluids
Wettability alteration	SiO_2 nanoparticles, alumina-coated silica nanoparticles, hydrophobic silicon oxide, spherical fumed silica nanoparticles, NWP, LHPN, polymer-coated nano particles
Sweep and displacement efficiency	Nano-sized colloidal dispersion gels (CDG), polymer nano particles, polymer-coated nano particles
Requires more research and investigation	Zinc oxide, zirconium oxide, ferro-fluids, carbon nanoparticles, carbon nanotubes

Abbreviation: SiO_2, silicon dioxide.

nanoparticles and the generation of nanoparticles that are environment friendly must be focused on so that the usage of nanotechnology in EOR can be viable in every aspect.

ACKNOWLEDGEMENTS

The author acknowledges financial support from the Indian Council for Medical Research (ICMR), New Delhi (grant no. 3/1/2/228/2021-Nut.) and Visvesvaraya National Institute of Technology, Nagpur for providing basic infrastructure facility.

REFERENCES

Alvarado, V., & Manrique, E. (2010). Enhanced oil recovery: An update review. *Energies*, *3*(9), 1529–1575. https://doi.org/10.3390/en3091529.

Amrouche, F., Gomari, S. R., Islam, M., & Xu, D. (2021). Improving thermal conductivity phase change materials—A study of paraffin nanomagnetite composites. *Journal of Cleaner Production*, *281*, 124891.

Cao, K., Liu, H., Li, W., Han, Q., Zhang, Z., Huang, K., Jing, Q., & Jiao, L. (2019). CuO nanoplates for high-performance potassium-ion batteries. *Small*, *15*(36), 1901775. https://doi.org/10.1002/smll.201901775.

Chand, P., Vaish, S., & Kumar, P. (2017). Structural, optical and dielectric properties of transition metal (MFe2O4; M = Co, Ni and Zn) nanoferrites. *Physica B: Condensed Matter*, *524*(August), 53–63. https://doi.org/10.1016/j.physb.2017.08.060.

Cheraghian, G., & Hendraningrat, L. (2016). A review on applications of nanotechnology in the enhanced oil recovery part A: Effects of nanoparticles on interfacial tension. *International Nano Letters*, *6*(2), 129–138. https://doi.org/10.1007/s40089-015-0173-4.

Dabhade, A., Jayaraman, S., & Paramasivan, B. (2021). Development of glucose oxidase-chitosan immobilized paper biosensor using screen-printed electrode for amperometric detection of Cr(VI) in water. *3 Biotech*, *11*(4), 183. https://doi.org/10.1007/s13205-021-02736-5.

Davoodi, S., Al-Shargabi, M., Wood, D. A., Rukavishnikov, V. S., & Minaev, K. M. (2022). Experimental and field applications of nanotechnology for enhanced oil recovery purposes: A review. *Fuel*, *324*, 124669. https://doi.org/10.1016/j.fuel.2022.124669.

El-hoshoudy, A. N., Desouky, S. E. M., Elkady, M. Y., Al-Sabagh, A. M., Betiha, M. A., & Mahmoud, S. (2017). Hydrophobically associated polymers for wettability alteration and enhanced oil recovery – Article review. *Egyptian Journal of Petroleum*, *26*(3), 757–762. https://doi.org/10.1016/j.ejpe.2016.10.008.

Francisca, F. M., Rinaldi, V. A., & Santamarina, J. C. (2003). Instability of hydrocarbon films over mineral surfaces: Microscale experimental studies. *Journal of Environmental Engineering*, *129*(12), 1120–1128. https://doi.org/10.1061/(ASCE)0733-9372(2003)129:12(1120).

Ganat, T. A.-A. O. (2020). *Fundamentals of Reservoir Rock Properties*. Springer International Publishing. https://doi.org/10.1007/978-3-030-28140-3.

Hashemi, M., Mohandes, F., & Salavati-Niasari, M. (2016). Application of [M(en)3]3[Fe(ox)3]2 (M = Zn, Cd, Ni) complexes as new precursor for the synthesis of ferrite micro/nanostructures. *Advanced Powder Technology*, *27*(2), 388–394. https://doi.org/10.1016/j.apt.2016.01.013.

He, H., Liu, P., Li, Q., Tang, J., Guan, W., & Chen, Y. (2022). Experiments and simulations on factors affecting the stereoscopic fire flooding in heavy oil reservoirs. *Fuel*, *314*, 123146. https://doi.org/10.1016/j.fuel.2022.123146.

Hoffman, B. T. (2023). Enhanced oil recovery in unconventional reservoirs. In *Recovery Improvement* (pp. 365–426). Elsevier. https://doi.org/10.1016/B978-0-12-823363-4.00004-2.

Janssen, M. T. G., Torres Mendez, F. A., & Zitha, P. L. J. (2020). Mechanistic modeling of water-alternating-gas injection and foam-assisted chemical flooding for enhanced oil recovery. *Industrial & Engineering Chemistry Research*, *59*(8), 3606–3616. https://doi.org/10.1021/acs.iecr.9b06356.

Kumar, A., & Gangawane, K. M. (2022a). Effect of precipitating agents on the magnetic and structural properties of the synthesized ferrimagnetic nanoparticles by co-precipitation method. *Powder Technology*, *401*, 117298. https://doi.org/10.1016/j.powtec.2022.117298.

Kumar, A., & Gangawane, K. M. (2022b). Synthesis and effect on the surface morphology & magnetic properties of ferrimagnetic nanoparticles by different wet chemical synthesis methods. *Powder Technology*, *410*(June), 117867. https://doi.org/10.1016/j.powtec.2022.117867.

Kumar, A., & Gangawane, K. M. (2023a). Advanced computational approaches for water treatment. In *Advanced Computational Approaches for Water Treatment* (pp. 1–12). CRC Press. https://doi.org/10.1201/9781003325147-1.

Kumar, A., & Gangawane, K. M. (2023b). Molecular dynamics for the membrane process. In *Advanced Computational Approaches for Water Treatment* (pp. 13–38). CRC Press. https://doi.org/10.1201/9781003325147-2.

Kumar, A., Gangawane, K. M., & Ramanjaneyulu, B. (2022). Ferrofluids for waste-water treatment. *International Conference on Chemical, Bio and Environmental Engineering*, 723–744. https://link.springer.com/chapter/10.1007/978-3-030-96554-9_48.

Kumar, A., Jena, H. M., & Gangawane, K. M. (2023a). Effect of ferrimagnetic nanoparticles as an oxygen absorber on the shelf life of tomatoes. *9th International Food Convention (9th IFCoN)*, CSIR-CFTRI, Mysuru, India, 9 December 2023. https://doi.org/10.13140/RG.2.2.15144.14088.

Kumar, A., Jena, H. M., & Gangawane, K. M. (2023b). Variations in Magnetic Behavior of Ferrimagnetic Nanoparticles with Temperature. *75th CHEMCON 2023*. Dr. H L Roy Building, Jadavpur University Campus, Kolkata, India. https://doi.org/10.13140/RG.2.2.18714.62403.

Li, X., Xiao, K., Wang, R., & Li, X. (2022). Experimental research on enhanced oil recovery methods for gas injection of fractured reservoirs based on microfluidic chips. *ACS Omega*, *7*(31), 27382–27389. https://doi.org/10.1021/acsomega.2c02390.

Likhacheva, O., & Kashin, G. (2020). Use of fire flooding in the Elnikovsky field of the Udmurt Republic, Russia. *XIV International Scientific Conference "Monitoring of Geological Processes and Ecological Condition of the Environment"*, 1–5. https://doi.org/10.3997/2214-4609.202056012.

Mariyate, J., & Bera, A. (2022). A critical review on selection of microemulsions or nanoemulsions for enhanced oil recovery. *Journal of Molecular Liquids*, *353*, 118791. https://doi.org/10.1016/j.molliq.2022.118791.

Minz, P., & Mishra, S. (2023). Jackfruit peel derived ZnCl2-impregnated activated carbon: Optimization, characterization, and application in dye removal. *Biomass Conversion and Biorefinery*, 0123456789. https://doi.org/10.1007/s13399-023-04592-0.

Mokheimer, E. M. A., Hamdy, M., Abubakar, Z., Shakeel, M. R., Habib, M. A., & Mahmoud, M. (2019). A comprehensive review of thermal enhanced oil recovery: Techniques evaluation. *Journal of Energy Resources Technology*, *141*(3), 030801. https://doi.org/10.1115/1.4041096.

Negin, C., Ali, S., & Xie, Q. (2016). Application of nanotechnology for enhancing oil recovery – A review. *Petroleum*, *2*(4), 324–333. https://doi.org/10.1016/j.petlm.2016.10.002.

Nwidee, L. N., Theophilus, S., Barifcani, A., Sarmadivaleh, M., & Iglauer, S. (2016). EOR processes, opportunities and technological advancements. In *Chemical Enhanced Oil Recovery (cEOR) - A Practical Overview*. InTech. https://doi.org/10.5772/64828.

Pal, D., & Sen, S. (2023a). Emerging petroleum pollutants and their adverse effects on the environment. In I. D. Behera, & A. P. Das (Eds.), *Impact of Petroleum Waste on Environmental Pollution and its Sustainable Management Through Circular Economy. Environmental Science and Engineering* (pp. 103–137). Springer Nature Switzerland. https://doi.org/10.1007/978-3-031-48220-5_5.

Pal, D., & Sen, S. (2023b). In-depth coverage of petroleum waste sources, characteristics, environmental impact, and sustainable remediation process. In I. D. Behera, & A. P. Das (Eds.), *Impact of Petroleum Waste on Environmental Pollution and Its Sustainable Management through Circular Economy* (pp. 1–38). Springer Nature Switzerland. https://doi.org/10.1007/978-3-031-48220-5_1.

Pal, D., & Sen, S. (2024). Optimal synthesis of dolochar derived faujasite zeolite X for highly effective Cd(II) removal. *Environmental Research*, *240*, 117494. https://doi.org/10.1016/j.envres.2023.117494.

Panda, D., & Gangawane, K. M. (2023a). Development of superhydrophobic hybrid silica-cellulose aerogel as promising thermal insulation and sound absorption. *International Journal of Materials Research*, *114*(4–5), 405–417. https://doi.org/10.1515/ijmr-2022-0066.

Panda, D., & Gangawane, K. M. (2023b). Hybrid NiFe2O4-NiCO4O4 nanoparticles-based eutectic phase change materials for enhancement of thermal efficiency of pin-fin heat sink arrangement. *Journal of Energy Storage*, *60*(November 2022), 106644. https://doi.org/10.1016/j.est.2023.106644.

Panda, D., & Gangawane, K. M. (2023c). Recycled cellulose–silica hybrid aerogel for effective oil adsorption: Optimization and kinetics study. *Sadhana - Academy Proceedings in Engineering Sciences*, 48(3). https://doi.org/10.1007/s12046-023-02161-9.

Panda, D., Sahu, A. K., & Gangawane, K. M. (2024). Effect of molar concentration and drying methodologies on monodispersed silica sol for synthesis of silica aerogels with temperature-resistant characteristics. *Materials Science and Engineering: B*, *299*, 117062. https://doi.org/10.1016/j.mseb.2023.117062.

Panda, D., Sharma, S., & Gangawane, K. M. (2023). Synthesis of novel biocatalyst from organic waste protonated with acid treatment for hydrogen production. *International Journal of Hydrogen Energy*, *48*(63), 24242–24254. https://doi.org/10.1016/j.ijhydene.2023.03.117.

Patel, K., Shah, M., & Sircar, A. (2018). Plasma pulse technology: An uprising EOR technique. *Petroleum Research*, *3*(2), 180–188. https://doi.org/10.1016/j.ptlrs.2018.05.001.

Popescu, R. C., Andronescu, E., & Vasile, B. S. (2019). Recent advances in magnetite nanoparticle functionalization for nanomedicine. *Nanomaterials*, *9*(12), 1–31. https://doi.org/10.3390/nano9121791.

Pourgolmohammad, B., Masoudpanah, S. M., & Aboutalebi, M. R. (2017). Effect of starting solution acidity on the characteristics of CoFe2O4 powders prepared by solution combustion synthesis. *Journal of Magnetism and Magnetic Materials*, *424*(September 2016), 352–358. https://doi.org/10.1016/j.jmmm.2016.10.073.

Rath, G. K., Singh, S., Pandey, G., Poothia, T., & Kumar, A. (2023). Role of nanofluids in chemical enhanced oil recovery. In *Challenges and Recent Advances in Sustainable Oil and Gas Recovery and Transportation* (pp. 81–119). Elsevier. https://doi.org/10.1016/B978-0-323-99304-3.00016-9.

Sahoo, A., & Mishra, S. (2023). Synthesis of microporous-activated carbon via direct H3PO4 and KOH activation of Pterospermum acerifolium fruits: process optimization and characterization. *Biomass Conversion and Biorefinery*, 0123456789. https://doi.org/10.1007/s13399-023-05072-1.

Santhosh, R., & Sarkar, P. (2022). Jackfruit seed starch/tamarind kernel xyloglucan/zinc oxide nanoparticles-based composite films: Preparation, characterization, and application on tomato (Solanum lycopersicum) fruits. *Food Hydrocolloids*, *133*(April), 107917. https://doi.org/10.1016/j.foodhyd.2022.107917.

Saravanan, A., Kumar, P. S., Vardhan, K. H., Jeevanantham, S., Karishma, S. B., Yaashikaa, P. R., & Vellaichamy, P. (2020). A review on systematic approach for microbial enhanced oil recovery technologies: Opportunities and challenges. *Journal of Cleaner Production*, *258*, 120777. https://doi.org/10.1016/j.jclepro.2020.120777.

Setiati, R., Malinda, M. T., & Sabrina, J. (2021). The potential of polymer for enhanced oil recovery process on oil refinery: A literature research. *IOP Conference Series: Earth and Environmental Science*, *737*(1), 012046. https://doi.org/10.1088/1755-1315/737/1/012046.

Soleimani, H., Baig, M. K., Yahya, N., Khodapanah, L., Sabet, M., Demiral, B. M. R., & Burda, M. (2018). Synthesis of ZnO nanoparticles for oil–water interfacial tension reduction in enhanced oil recovery. *Applied Physics A*, *124*(2), 128. https://doi.org/10.1007/s00339-017-1510-4.

Sun, X., Zhang, Y., Chen, G., & Gai, Z. (2017). Application of nanoparticles in enhanced oil recovery: A critical review of recent progress. *Energies*, *10*(3). https://doi.org/10.3390/en10030345.

Vargas, F. M., Gonzalez, D. L., Hirasaki, G. J., & Chapman, W. G. (2009). Modeling asphaltene phase behavior in crude oil systems using the perturbed chain form of the statistical associating fluid theory (PC-SAFT) equation of state. *Energy & Fuels*, *23*(3), 1140–1146. https://doi.org/10.1021/ef8006678.

Wojnicki, M., Lubaś, J., Gawroński, M., Szuflita, S., Kuśnierczyk, J., & Warnecki, M. (2022). An experimental investigation of WAG injection in a carbonate reservoir and prediction of the recovery factor using genetic programming. *Energies*, *15*(6), 2127. https://doi.org/10.3390/en15062127.

2 Fundamentals of Nanotechnology for EOR

Krishna Raghav Chaturvedi, Abhishek Kumar, Anand Raghav Chaubey, Himanshu Kesharwani, Ravi Shankar Kumar and Tushar Sharma

2.1 INTRODUCTION

Nanomaterials have been explored for subsurface applications in enhanced oil recovery (EOR). Researchers are investigating their potential to improve oil displacement efficiency and enhance recovery rates in reservoirs (Al-Anssari et al., 2017; Yuan & Wang, 2018). Nanoparticles can alter the properties of fluids and interact with reservoir rock surfaces, offering new possibilities for optimizing EOR processes (Bera et al., 2020; Dehaghani & Daneshfar, 2019; Kumar & Gangawane, 2022a, 2022b). Nanomaterials offer a range of unique properties that can be harnessed to enhance oil recovery processes. Their application in EOR is a multidisciplinary approach that involves modifying the physical and chemical properties of reservoir fluids, altering rock-fluid interactions, and activated carbon, and improving overall oil displacement efficiency (Kumar et al., 2023b, 2023a; Minz & Mishra, 2023; Sahoo & Mishra, 2023; Sun et al., 2017) (Figure 2.1).

Here are some key ways in which nanomaterials can be employed for EOR:

- Interfacial Tension (IFT) Reduction: Nanoparticles can be used to reduce the IFT between oil and water, making it easier for oil to displace from the reservoir rock (Dabhade et al., 2023; Kumar et al., 2021). By modifying the IFT, nanomaterials help overcome capillary forces that trap oil in the rock pores.
- Wettability Alteration: Nanomaterials can alter the wettability of the reservoir rock surface. Engineered nanoparticles can promote a shift from oil-wet to water-wet conditions, enhancing the ability of water or other displacing fluids to move through the porous rock and mobilize trapped oil.
- Capillary Force Modification: The capillary forces within the porous rock play a crucial role in oil retention (Bayat et al., 2014). Nanomaterials can modify these forces, reducing capillary trapping and allowing for more effective displacement of oil by injected fluids.
- Improved Mobility of Fluids: Nanoemulsions and nanofluids can be introduced into the reservoir to alter the viscosity and mobility of the fluids (Mady & Kelland, 2020). This modification can lead to better sweep

DOI: 10.1201/9781032650685-2

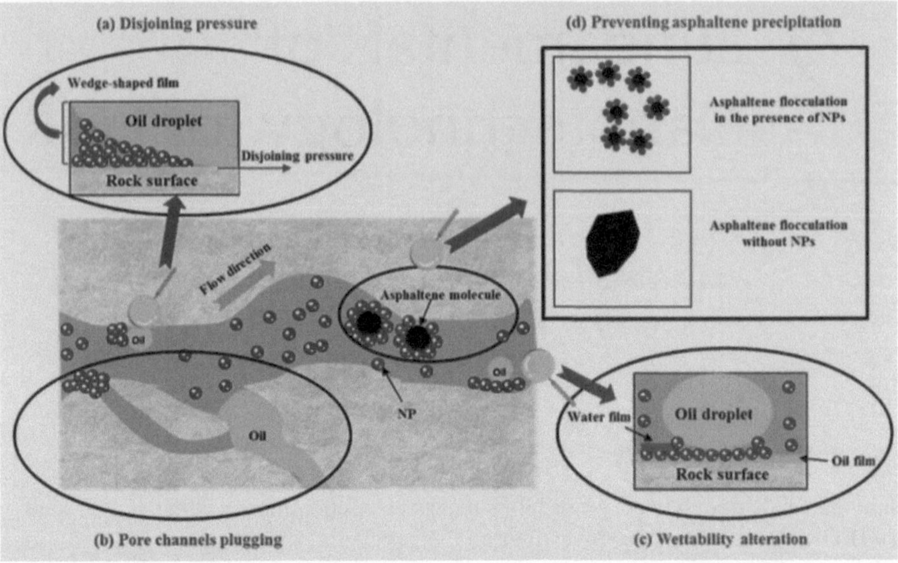

FIGURE 2.1 Mechanisms influencing oil recovery on nanomaterial application. Adapted from a study by Sun et al. (2017).

efficiency, ensuring that the injected fluid reaches and displaces oil from previously unrecovered regions (Kumar et al., 2022).

- Enhanced Reservoir Sweep Efficiency: Nanomaterials can improve the efficiency of fluid sweep within the reservoir. By mitigating fingering effects and improving the overall displacement of oil, nanomaterials contribute to a more uniform and thorough recovery of oil from the reservoir.
- Reservoir Conformance Control: Nanoparticles can be utilized for reservoir conformance control, helping to divert injected fluids into less-swept zones and ensuring a more even sweep of the reservoir (Bazazi et al., 2017). This is particularly important in heterogeneous reservoirs where certain areas may be bypassed during traditional recovery methods.
- Improved Oil Mobility in Heavy Oil Reservoirs: In heavy oil reservoirs, nanomaterials can reduce the viscosity of the oil, making it more mobile and easier to displace. This is achieved through the addition of nanoparticles that interact with heavy oil components, altering their rheological properties.
- In-Situ Gelling and Plugging: Nanostructured materials can be designed to form in-situ gels or plugs, controlling fluid flow and diverting injected fluids to specific zones within the reservoir (Reena et al., 2020; Wu et al., 2016). This helps manage reservoir heterogeneity and improve the effectiveness of the EOR process.
- Smart Nanoparticles and Monitoring: Nanomaterials with responsive properties can be designed to undergo changes in behavior in response to specific

reservoir conditions (Li & Meyyappan, 2006). Additionally, nanosensors can be employed to monitor the progress of EOR operations in real-time, providing valuable feedback for optimizing the process.

While the potential benefits of nanomaterials in EOR are substantial, challenges such as cost, scalability, and environmental considerations must be carefully addressed to ensure the practical and sustainable implementation of these technologies in the oil and gas industry (Pal & Sen, 2023a, 2023b).

2.2 DISJOINING PRESSURE APPLICATION BY NANOMATERIALS

Disjoining pressure refers to the pressure acting between two closely spaced surfaces in a confined fluid film. It arises due to the interactions between the fluid molecules and the surfaces, and it plays a crucial role in governing the stability and behavior of thin films or layers of fluids in confined spaces (Choi et al., 2017; Hendraningrat et al., 2013). Disjoining pressure is typically described in the context of thin films, such as those found between solid surfaces and liquid or gas phases. In the realm of EOR, disjoining pressure becomes particularly relevant when considering the interactions between nanomaterials, reservoir rocks, and fluids. Nanomaterials can influence disjoining pressure, thereby impacting the stability and behavior of fluid films within porous rock formations.

Nanomaterials, especially nanoparticles and nanoemulsions, can be introduced into the reservoir fluids. These nanomaterials interact with the fluids and can alter their rheological properties, affecting the disjoining pressure within the porous rock. Nanomaterials can modify the surface properties of reservoir rocks. By interacting with the rock surfaces, nanomaterials influence the wetting behavior and adhesion of fluids. This modification in surface properties can, in turn, affect the disjoining pressure in thin fluid films within the rock pores. Disjoining pressure is closely related to capillary forces, which play a significant role in trapping and retaining oil within the rock pores. Nanomaterials can influence capillary forces by modifying the wetting properties of the reservoir rocks, making it easier for fluids to displace and recover oil. The introduction of nanomaterials can lead to controlled changes in the stability of fluid films within the reservoir. By altering the disjoining pressure, nanomaterials can affect the thickness and stability of thin films, influencing the efficiency of oil displacement (Chengara et al., 2004; Wasan et al., 2011). Nanomaterials can improve the overall sweep efficiency in the reservoir by modifying the disjoining pressure. This is particularly crucial in heterogeneous reservoirs where certain regions may have different capillary forces affecting fluid flow and oil recovery.

2.3 WETTABILITY ALTERATION BY NANOMATERIALS

Wettability alteration refers to the change in the affinity of a solid surface, such as reservoir rock, for a particular fluid (e.g., oil or water). In the context of EOR applications, adjusting wettability is important because it influences the interactions between the reservoir rock, oil, and injected fluids, impacting the efficiency of oil recovery. Wettability alteration can enhance spontaneous imbibition, the process where the rock absorbs fluid without external force. If the rock becomes more water-wet, it promotes the imbibition of water into the rock matrix, helping displace and recover trapped oil (Pal & Sen, 2024; Rahimi & Adibifard, 2015). Wettability influences capillary forces within the rock pores. Changing the wettability can modify these forces, aiding in the detachment of oil from pore surfaces and facilitating its movement toward production wells. Wettability alteration can also reduce the residual-oil saturation in the reservoir. Residual oil is the portion of oil that remains trapped in the rock pores after primary recovery methods. Modifying the wettability helps release and recover this residual oil (Panda & Gangawane, 2023c, 2023a).

Nanomaterials with specific surface chemistries can be designed to interact with the rock surface. By attaching to the rock, they can influence its wettability (Jiang et al., 2003; Rahimi & Adibifard, 2015). Nanoparticles can be incorporated into nanoemulsions, which are stable dispersions of nanoscale droplets. These nanoemulsions can alter the wettability of the rock surfaces as they contact and spread over them. Nanoparticles can adsorb onto rock surfaces and change the rock's affinity for water or oil. This adsorption can be manipulated to shift wettability in a way that favors improved oil recovery. Research is ongoing to understand the specific mechanisms by which nanomaterials influence wettability in different reservoir conditions, and this area holds promise for enhancing the effectiveness of EOR techniques.

2.4 MOBILITY CONTROL BY NANOMATERIALS

Mobility control in EOR refers to the regulation of fluid flow through the reservoir to ensure that injected fluids, such as water or polymers, effectively displace and recover oil. It is a critical factor for optimizing the sweep efficiency of the injected fluids and preventing premature breakthrough, where the injected fluid bypasses significant portions of the reservoir without efficiently contacting and displacing oil (Barrabino et al., 2018). Efficient mobility control ensures that the injected fluid sweeps through a larger portion of the reservoir, contacting more oil and improving recovery rates. Without effective mobility control, injected fluids can create preferential pathways (channels) through the reservoir, leading to uneven displacement and leaving substantial oil volumes unrecovered. Mobility control helps achieve better conformance in the reservoir, meaning that the injected fluids are distributed more uniformly, reducing the risk of leaving oil pockets untouched.

Nanoparticles can be dispersed in injected fluids to alter their rheological properties. This modification influences the flow behavior, viscosity, and mobility of the fluids, allowing for better control of their movement through the reservoir. Nanoparticles can enhance the viscosity of injected fluids. This increased viscosity helps in controlling fluid mobility by reducing the tendency for rapid breakthrough and improving the fluid's ability to displace oil effectively. Nanoparticles can selectively plug high-permeability zones in the reservoir. By doing so, they divert the injected fluids to less-permeable regions, ensuring a more even sweep of the reservoir and preventing early breakthroughs in certain areas. The colloidal stability of nanoparticle dispersions is crucial for maintaining their effectiveness in controlling fluid mobility throughout the EOR process. Stable dispersions ensure that nanoparticles remain evenly distributed and do not aggregate, preserving their mobility control properties. By leveraging these mechanisms, nanomaterials offer a means to enhance mobility control in EOR, contributing to more effective and uniform oil recovery from reservoirs.

2.5 IFT REDUCTION BY NANOMATERIALS

IFT is a measure of the force per unit length at the interface between two immiscible fluids, such as oil and water. It quantifies the energy required to increase the surface area of the interface between the two fluids. In the context of oil recovery, the IFT between oil and water is a critical parameter that influences the ability to displace oil from reservoir rocks. IFT plays a crucial role in capillary-driven processes within porous rock formations. When oil is trapped in the pores of the reservoir rock, the capillary forces tend to retain the oil, making it difficult to displace. Higher IFT exacerbates these capillary forces, hindering the ability of injected fluids (e.g., water or surfactant solutions) to penetrate the rock pores and mobilize the trapped oil. Reducing IFT is thus an essential strategy for enhancing oil recovery (Le et al., 2011).

Nanomaterials contribute to the reduction of IFT through several mechanisms, including by acting as surfactants, adsorbing at the oil-water interface, and altering the surface tension (Figure 2.2) (Chaturvedi et al., 2021). This reduces the IFT and enhances the spreading of the displacing fluid, making it more effective in pushing and displacing oil from the reservoir. Engineered nanoparticles can form a coating on the surfaces of oil droplets, reducing the IFT between the oil and surrounding water. This coating can stabilize emulsions and improve the miscibility of oil with the displacing fluid. Amphiphilic nanoparticles have both hydrophobic and hydrophilic regions, allowing them to interact simultaneously with oil and water. These nanoparticles can be designed to adsorb at the oil-water interface, lowering the IFT and facilitating the displacement of oil. Nanoparticles can influence the wettability of the reservoir rock surface, promoting a shift from oil-wet to water-wet conditions. This alteration in wettability contributes to a decrease in IFT, making it easier for water to wet the rock surface and displace oil. Nanoparticles can modify the thermodynamic properties of the fluid interface, impacting the intermolecular forces between oil and water molecules (Kumar & Gangawane, 2023a, 2023b).

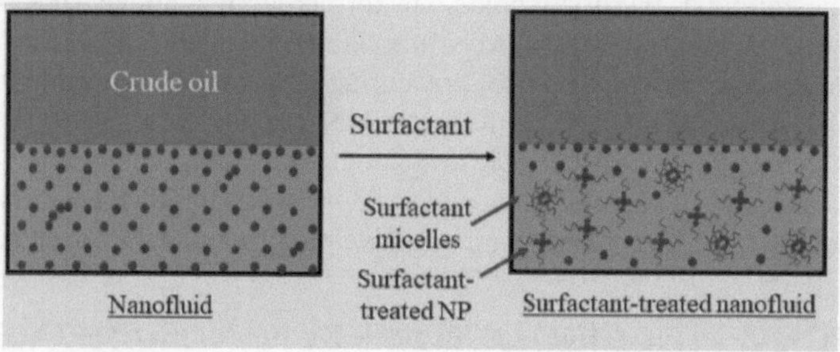

FIGURE 2.2 Competitive surfactant-nanoparticle adsorption at the crude oil-injected fluid interface lowers IFT values and enhances oil recovery. Adapted with permission from Chaturvedi et al. (2021).

Additionally, nanofluids can alter the rheological properties of fluids, affecting their flow behavior and IFT. By leveraging these mechanisms, nanomaterials effectively reduce the IFT between oil and water, improving the efficiency of the displacement process and contributing to EOR. The ability of nanomaterials to fine-tune their properties for specific reservoir conditions makes them valuable tools in addressing the challenges associated with IFT during oil-recovery processes.

2.6 EMULSION FORMATION BY NANOMATERIALS

Emulsions are colloidal systems consisting of two immiscible liquids, typically oil and water, stabilized by an emulsifying agent (Panda et al., 2024; Radnia et al., 2018; Sahoo & Mishra, 2023; Worthen et al., 2013). The emulsifying agent, often a surfactant, facilitates the dispersion of one liquid phase in the other, resulting in the formation of small droplets. Emulsions can be formed by mechanical agitation, i.e. agitating two immiscible liquids mechanically, either through stirring or other means, which can break the more extensive liquid phase into smaller droplets or disperse them in the continuous phase. Here, surfactants or emulsifying agents reduce the IFT between the immiscible liquids, stabilizing the emulsion by preventing the coalescence of droplets and promoting their dispersion.

Nanomaterials, including nanoparticles, can act as stabilizing agents for emulsions. Their surface properties and ability to interact with both oil and water phases contribute to the stabilization of emulsion droplets. Nanoparticles can be used to create nanoemulsions, which are emulsions with droplet sizes in the nanometer range. These nanoemulsions exhibit increased stability and may have enhanced properties for various applications, including EOR.

Emulsions formed with nanomaterials can alter the rheological properties of the injected fluid, influencing its flow behavior through the reservoir. This alteration can lead to improved displacement efficiency and EOR. Emulsions formed with nanomaterials may have reduced the IFT between the oil and water phases. Lower IFT facilitates the mobilization of oil droplets and aids in their displacement from reservoir rock pores. Nanomaterial-stabilized emulsions can be designed for selective plugging of high-permeability zones in the reservoir. It helps in diverting the injected fluids to areas with lower permeability, contributing to a more uniform sweep of the reservoir. The colloidal stability of emulsions formed with nanomaterials is essential for maintaining their effectiveness over time. Stable emulsions ensure that the dispersed phase remains well-distributed, preventing phase separation and maintaining the desired properties during the EOR process (Narukulla et al., 2019; Panda et al., 2023). Nanomaterials can play a significant role in forming stable emulsions, and these emulsions can be tailored to enhance various aspects of oil-recovery processes, including fluid flow, IFT reduction, and selective plugging, for improved efficiency.

2.7 SELECTIVE PLUGGING OF PORES BY NANOMATERIALS

Selective plugging in EOR involves the controlled obstruction or reduction of fluid flow in high-permeability zones within a reservoir. The objective is to divert the injected fluids away from these zones and channel them into less-permeable regions.

It promotes a more even sweep of the reservoir, preventing premature breakthroughs and increasing the efficiency of oil recovery. Nanoparticles dispersed in injected fluids can be designed to transport and accumulate selectively in high-permeability zones. This transport can be influenced by factors such as fluid velocity and the permeability contrast between high- and low-permeability zones. Once nanoparticles reach high-permeability zones, they can interact with the rock surfaces and clog pore spaces. This selective plugging reduces the permeability of these zones, limiting the flow of injected fluids and diverting them toward less-permeable areas. The properties of nanomaterials, such as size, surface charge, and surface chemistry, can be tailored to control their migration within the reservoir. It allows for a targeted approach in selectively plugging specific regions (Risal et al., 2019; Son et al., 2015).

By selectively plugging high-permeability fractures or channels, nanomaterials help ensure a more uniform sweep of the reservoir. It prevents the preferential flow of injected fluids through fractures, allowing for better contact with oil in less-permeable matrix rock. Selective plugging with nanomaterials improves the displacement efficiency of injected fluids. The controlled diversion of fluids facilitates a more thorough sweep of the reservoir, reaching and recovering oil from areas that might be bypassed in the absence of selective plugging. Nanomaterial-induced selective plugging minimizes the risk of early breakthroughs of injected fluids. It is crucial for achieving a more comprehensive and efficient recovery of oil from the reservoir. Nanomaterials contribute to selective plugging by being engineered to migrate and accumulate in high-permeability zones. This controlled obstruction redirects injected fluids to less-permeable areas, enhancing the uniformity of the sweep and ultimately increasing the recovery of oil from fractures and other high-permeability regions within the reservoir.

2.8 NANOPARTICLE AGGLOMERATION: MECHANISMS AND DRAWBACKS

Nanoparticle agglomeration refers to the process by which individual nanoparticles come together and form clusters or aggregates. Instead of existing as isolated particles, agglomeration results in the formation of larger assemblies due to various forces acting on the nanoparticles (Bouguerra et al., 2018; Chaturvedi & Sharma, 2021; Panda & Gangawane, 2023b). This phenomenon can significantly impact the properties and performance of nanomaterials, influencing their applications in EOR (Figure 2.3).

The various causes for nanomaterial agglomeration have been listed as under:

- *Van der Waals Forces:* Nanoparticles are subject to attractive Van der Waals forces, which can lead to their agglomeration. As particles approach each other, these forces become stronger, causing them to stick together (Gudarzi, 2016).
- *Electrostatic Interactions:* Nanoparticles may carry charges, and electrostatic interactions between particles can result in agglomeration. Repulsive or attractive forces between charged nanoparticles influence their spatial arrangement.

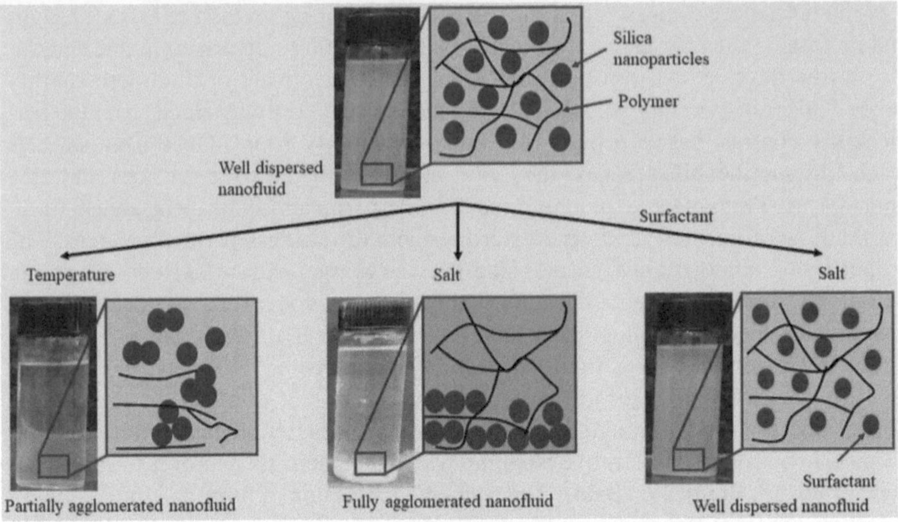

FIGURE 2.3 Visual appearance of nanofluid solutions in different stages of agglomeration. Adapted with permission from Chaturvedi and Sharma (2021).

- *Brownian Motion:* While Brownian motion tends to keep nanoparticles in constant motion, collisions between particles may occur, leading to agglomeration. This is more likely for nanoparticles with weak repulsion forces.
- *High Surface Energy:* Nanoparticles often have high surface energy, making them prone to sticking together. Factors like surface functionalization and the presence of impurities can influence this.
- *pH and Ionic Strength:* The pH and ionic strength of the solution can affect nanoparticle stability. Changes in these factors may alter the electrostatic interactions between particles, leading to agglomeration.

The decreased stability of nanomaterial solutions adversely impacts the EOR application potential. The reasons for the same have been listed as under:

- *Reduced Colloidal Stability:* Agglomeration compromises the colloidal stability of nanomaterial solutions. Well-dispersed particles characterize stable dispersions; agglomeration disrupts this stability, leading to phase separation.
- *Increased Viscosity:* Agglomerated nanoparticles contribute to increased solution viscosity. This can impede the flow of the solution and affect its performance in applications such as EOR or drug delivery.
- *Altered Rheological Properties:* Nanomaterial agglomeration can alter the rheological behavior of the solution. Changes in viscosity, shear-thinning, or shear-thickening properties may impact the solution's applicability in various industrial processes.

- *Diminished Surface Area:* Agglomeration reduces the effective surface area of nanoparticles. This is critical in applications where a high surface area is desired for interactions with other materials, such as in catalysis or sensors.
- *Settling and Sedimentation:* Agglomerated nanoparticles are more prone to settling and sedimentation. This can lead to uneven distribution of nanoparticles in a solution and compromise their effectiveness in certain applications.

Understanding the causes and consequences of nanomaterial agglomeration is crucial for designing strategies to mitigate it, ensuring the stability and efficacy of nanomaterial solutions in various practical applications.

2.9 CONCLUSION

The chapter delves into the cutting-edge realm of EOR through the utilization of nanomaterials, providing a comprehensive exploration of the underlying mechanisms that govern their efficacy. With the global demand for energy escalating and conventional oil-recovery methods reaching their limits, the integration of nanotechnology offers a promising avenue to revolutionize oil extraction processes. This chapter reviews the latest advancements in nanomaterial-enabled EOR and presents a detailed analysis of the fundamental mechanisms at play. A critical aspect of this chapter involves a discussion of the challenges and prospects associated with the practical implementation of nanomaterials in field-scale EOR projects.

REFERENCES

Al-Anssari, S., Nwidee, L. N., Ali, M., Sangwai, J. S., Wang, S., Barifcani, A., & Iglauer, S. (2017). Retention of silica nanoparticles in limestone porous media. In *Society of Petroleum Engineers - SPE/IATMI Asia Pacific Oil and Gas Conference and Exhibition 2017, 2017-Janua* (pp. 17–19). https://doi.org/10.2118/186253-ms.

Barrabino, A., Holt, T., & Lindeberg, E. (2018). An evaluation of graphene oxides as possible foam stabilizing agents for CO2 based enhanced oil recovery. *Nanomaterials*, *8*(8). https://doi.org/10.3390/nano8080603.

Bayat, A. E., Junin, R., Samsuri, A., Piroozian, A., & Hokmabadi, M. (2014). Impact of metal oxide nanoparticles on enhanced oil recovery from limestone media at several temperatures. *Energy and Fuels*, *28*(10), 6255–6266. https://doi.org/10.1021/ef5013616.

Bazazi, P., Gates, I. D., Nezhad, A. S., & Hossein Hejazi, S. (2017). Silica-based nanofluid heavy oil recovery a microfluidic approach. In *Society of Petroleum Engineers - SPE Canada Heavy Oil Technical Conference 2017* (pp. 438–452). https://doi.org/10.2118/185008-ms.

Bera, A., Shah, S., Shah, M., Agarwal, J., & Vij, R. K. (2020). Mechanistic study on silica nanoparticles-assisted guar gum polymer flooding for enhanced oil recovery in sandstone reservoirs. *Colloids and Surfaces A: Physicochemical and Engineering Aspects*, *598*, 124833. https://doi.org/10.1016/j.colsurfa.2020.124833.

Bouguerra, N., Poncet, S., & Elkoun, S. (2018). Dispersion regimes in alumina/water-based nanofluids: Simultaneous measurements of thermal conductivity and dynamic viscosity. *International Communications in Heat and Mass Transfer*, *92*, 51–55. https://doi.org/10.1016/j.icheatmasstransfer.2018.02.015.

Chaturvedi, K. R., Narukulla, R., Amani, M., & Sharma, T. (2021). Experimental investigations to evaluate surfactant role on absorption capacity of nanofluid for CO2 utilization in sustainable crude mobilization. *Energy*, *225*, 120321. https://doi.org/10.1016/j.energy.2021.120321.

Chaturvedi, K. R., & Sharma, T. (2021). Rheological analysis and EOR potential of surfactant treated single-step silica nanofluid at high temperature and salinity. *Journal of Petroleum Science and Engineering*, *196*, 107704. https://doi.org/10.1016/j.petrol.2020.107704.

Chengara, A., Nikolov, A. D., Wasan, D. T., Trokhymchuk, A., & Henderson, D. (2004). Spreading of nanofluids driven by the structural disjoining pressure gradient. *Journal of Colloid and Interface Science*, *280*(1), 192–201. https://doi.org/10.1016/j.jcis.2004.07.005.

Choi, S. K., Son, H. A., Kim, H. T., & Kim, J. W. (2017). Nanofluid enhanced oil recovery using hydrophobically associative zwitterionic polymer-coated silica nanoparticles. *Energy and Fuels*, *31*(8), 7777–7782. https://doi.org/10.1021/acs.energyfuels.7b00455.

Dabhade, A. H., Verma, R. P., Paramasivan, B., Kumawat, A., & Saha, B. (2023). Development of silver nanoparticles and aptamer conjugated biosensor for rapid detection of E. coli in a water sample. *3 Biotech*, *13*(7), 244. https://doi.org/10.1007/s13205-023-03663-3.

Dehaghani, A. H. S., & Daneshfar, R. (2019). How much would silica nanoparticles enhance the performance of low-salinity water flooding? *Petroleum Science*. https://doi.org/10.1007/s12182-019-0304-z.

Gudarzi, M. M. (2016). Colloidal stability of graphene oxide: Aggregation in two dimensions. *Langmuir*, *32*(20), 5058–5068. https://doi.org/10.1021/ACS.LANGMUIR.6B01012/SUPPL_FILE/LA6B01012_SI_001.PDF.

Hendraningrat, L., Li, S., & Torsæter, O. (2013). A coreflood investigation of nanofluid enhanced oil recovery. *Journal of Petroleum Science and Engineering*, *111*, 128–138. https://doi.org/10.1016/j.petrol.2013.07.003.

Jiang, L., Gao, L., & Sun, J. (2003). Production of aqueous colloidal dispersions of carbon nanotubes. *Journal of Colloid and Interface Science*, *260*(1), 89–94. https://doi.org/10.1016/S0021-9797(02)00176-5.

Kumar, R. S., Al-Arbi Ganat, T., & Sharma, T. (2021). Performance evaluation of silica nanofluid for sand column transport with simultaneous wettability alteration: An approach to environmental issue. *Journal of Cleaner Production*, *303*, 127047. https://doi.org/10.1016/j.jclepro.2021.127047.

Kumar, A., & Gangawane, K. M. (2022a). Effect of precipitating agents on the magnetic and structural properties of the synthesized ferrimagnetic nanoparticles by co-precipitation method. *Powder Technology*, *401*, 117298. https://doi.org/10.1016/j.powtec.2022.117298.

Kumar, A., & Gangawane, K. M. (2022b). Synthesis and effect on the surface morphology & magnetic properties of ferrimagnetic nanoparticles by different wet chemical synthesis methods. *Powder Technology*, *410*(June), 117867. https://doi.org/10.1016/j.powtec.2022.117867.

Kumar, A., & Gangawane, K. M. (2023a). Advanced computational approaches for water treatment. In *Advanced Computational Approaches for Water Treatment* (pp. 1–12). CRC Press. https://doi.org/10.1201/9781003325147-1.

Kumar, A., & Gangawane, K. M. (2023b). Molecular dynamics for the membrane process. In *Advanced Computational Approaches for Water Treatment* (pp. 13–38). CRC Press. https://doi.org/10.1201/9781003325147-2.

Kumar, A., Gangawane, K. M., & Ramanjaneyulu, B. (2022). Ferrofluids for waste-water treatment. In *International Conference on Chemical, Bio and Environmental Engineering* (pp. 723–744). https://link.springer.com/chapter/10.1007/978-3-030-96554-9_48.

Kumar, A., Jena, H. M., & Gangawane, K. M. (2023a). Effect of ferrimagnetic nanoparticles as an oxygen absorber on the shelf life of tomatoes. Poster presented at the *9th International Food Convention (IFCoN)*, CSIR-CFTRI, Mysuru, India, 9 December 2023.

Kumar, A., Jena, H. M., & Gangawane, K. M. (2023b). Variations in magnetic behavior of ferrimagnetic nanoparticles with temperature. Poster presented at the *75th Indian Chemical Engineering Congress (CHEMCON 2023)*, Dr. H L Roy Building, Jadavpur University Campus, Kolkata, India, 28 December 2023.

Le, N. Y. T., Pham, D. K., Le, K. H., & Nguyen, P. T. (2011). Design and screening of synergistic blends of SiO2 nanoparticles and surfactants for enhanced oil recovery in high-temperature reservoirs. *Advances in Natural Sciences: Nanoscience and Nanotechnology*, *2*(3). https://doi.org/10.1088/2043-6262/2/3/035013.

Li, J., & Meyyappan, M. (2006). *(12) United States Patent.*

Mady, M. F., & Kelland, M. A. (2020). Review of nanotechnology impacts on oilfield scale management. *ACS Applied Nano Materials*, *3*(8), 7343–7364. https://doi.org/10.1021/acsanm.0c01391.

Minz, P., & Mishra, S. (2023). Jackfruit peel derived ZnCl2-impregnated activated carbon: Optimization, characterization, and application in dye removal. *Biomass Conversion and Biorefinery*, 0123456789. https://doi.org/10.1007/s13399-023-04592-0.

Narukulla, R., Ojha, U., & Sharma, T. (2019). Enhancing the stability and redispersibility of o/w Pickering emulsion through polyacryloyl hydrazide-tannic acid synergy. *Colloids and Surfaces A: Physicochemical and Engineering Aspects*, *568*, 204–215. https://doi.org/10.1016/j.colsurfa.2019.02.012.

Pal, D., & Sen, S. (2023a). Emerging petroleum pollutants and their adverse effects on the environment. In I. D. Behera, & A. P. Das (Eds.), *Impact of Petroleum Waste on Environmental Pollution and Its Sustainable Management through Circular Economy. Environmental Science and Engineering* (pp. 103–137). Springer Nature Switzerland. https://doi.org/10.1007/978-3-031-48220-5_5.

Pal, D., & Sen, S. (2023b). In-depth coverage of petroleum waste sources, characteristics, environmental impact, and sustainable remediation process. In I. D. Behera, & A. P. Das (Eds.), *Impact of Petroleum Waste on Environmental Pollution and its Sustainable Management Through Circular Economy* (pp. 1–38). Springer Nature Switzerland. https://doi.org/10.1007/978-3-031-48220-5_1.

Pal, D., & Sen, S. (2024). Optimal synthesis of dolochar derived faujasite zeolite X for highly effective Cd(II) removal. *Environmental Research*, *240*, 117494. https://doi.org/10.1016/j.envres.2023.117494.

Panda, D., & Gangawane, K. M. (2023a). Development of superhydrophobic hybrid silica-cellulose aerogel as promising thermal insulation and sound absorption. *International Journal of Materials Research*, *114*(4–5), 405–417. https://doi.org/10.1515/ijmr-2022-0066.

Panda, D., & Gangawane, K. M. (2023b). Hybrid NiFe2O4-NiCO4O4 nanoparticles-based eutectic phase change materials for enhancement of thermal efficiency of pin-fin heat sink arrangement. *Journal of Energy Storage*, *60*(November 2022), 106644. https://doi.org/10.1016/j.est.2023.106644.

Panda, D., & Gangawane, K. M. (2023c). Recycled cellulose–silica hybrid aerogel for effective oil adsorption: optimization and kinetics study. *Sadhana - Academy Proceedings in Engineering Sciences*, *48*(3). https://doi.org/10.1007/s12046-023-02161-9.

Panda, D., Sahu, A. K., & Gangawane, K. M. (2024). Effect of molar concentration and drying methodologies on monodispersed silica sol for synthesis of silica aerogels with temperature-resistant characteristics. *Materials Science and Engineering: B*, *299*, 117062. https://doi.org/10.1016/j.mseb.2023.117062.

Panda, D., Sharma, S., & Gangawane, K. M. (2023). Synthesis of novel biocatalyst from organic waste protonated with acid treatment for hydrogen production. *International Journal of Hydrogen Energy*, *48*(63), 24242–24254. https://doi.org/10.1016/j.ijhydene.2023.03.117.

Radnia, H., Rashidi, A., Solaimany Nazar, A. R., Eskandari, M. M., & Jalilian, M. (2018). A novel nanofluid based on sulfonated graphene for enhanced oil recovery. *Journal of Molecular Liquids*, *271*, 795–806. https://doi.org/10.1016/j.molliq.2018.09.070.

Rahimi, K., & Adibifard, M. (2015). Experimental study of the nanoparticles effect on surfactant absorption and oil recovery in one of the Iranian oil reservoirs. *Petroleum Science and Technology*, *33*(1), 79–85. https://doi.org/10.1080/10916466.2014.950382.

Reena, Kumar, A., Mahto, V., & Choubey, A. K. (2020). Synthesis and characterization of cross-linked hydrogels using polyvinyl alcohol and polyvinyl pyrrolidone and their blend for water shut-off treatments. *Journal of Molecular Liquids*, *301*, 112472. https://doi.org/10.1016/j.molliq.2020.112472.

Risal, A. R., Manan, M. A., Yekeen, N., Azli, N. B., Samin, A. M., & Tan, X. K. (2019). Experimental investigation of enhancement of carbon dioxide foam stability, pore plugging, and oil recovery in the presence of silica nanoparticles. *Petroleum Science*, *16*(2), 344–356. https://doi.org/10.1007/s12182-018-0280-8.

Sahoo, A., & Mishra, S. (2023). Synthesis of microporous-activated carbon via direct H3PO4 and KOH activation of Pterospermum acerifolium fruits: Process optimization and characterization. *Biomass Conversion and Biorefinery*, 0123456789. https://doi.org/10.1007/s13399-023-05072-1.

Son, H. A., Yoon, K. Y., Lee, G. J., Cho, J. W., Choi, S. K., Kim, J. W., Im, K. C., Kim, H. T., Lee, K. S., & Sung, W. M. (2015). The potential applications in oil recovery with silica nanoparticle and polyvinyl alcohol stabilized emulsion. *Journal of Petroleum Science and Engineering*, *126*, 152–161. https://doi.org/10.1016/j.petrol.2014.11.001.

Sun, X., Zhang, Y., Chen, G., & Gai, Z. (2017). Application of nanoparticles in enhanced oil recovery: A critical review of recent progress. *Energies*, *10*(3). https://doi.org/10.3390/en10030345.

Wasan, D., Nikolov, A., & Kondiparty, K. (2011). The wetting and spreading of nanofluids on solids: Role of the structural disjoining pressure. *Current Opinion in Colloid & Interface Science*, *16*(4), 344–349. https://doi.org/10.1016/j.cocis.2011.02.001.

Worthen, A. J., Bagaria, H. G., Chen, Y., Bryant, S. L., Huh, C., & Johnston, K. P. (2013). Nanoparticle-stabilized carbon dioxide-in-water foams with fine texture. *Journal of Colloid and Interface Science*, *391*(1), 142–151. https://doi.org/10.1016/j.jcis.2012.09.043.

Wu, Z., Liu, H., Pang, Z., Wu, Y., Wang, X., & Gao, M. (2016). *Visualized Study on Blocking Characteristics and EOR Mechanisms of High-Temperature Gel During Steam Injection Process*. https://doi.org/10.2523/iptc-18622-ms.

Yuan, B., & Wang, W. (2018). Using nanofluids to control fines migration for oil recovery: Nanofluids co-injection or nanofluids pre-flush? - A comprehensive answer. *Fuel*, *215*, 474–483. https://doi.org/10.1016/j.fuel.2017.11.088.

3 Wettability Alteration and Interfacial Tension Reduction in EOR

Ali Khalilnezhad, Lady J. Giraldo, Daniel Lopez, Camilo A. Franco, Farid B. Cortés and Masoud Riazi

3.1 INTRODUCTION

Enhanced Oil Recovery (EOR) processes are implemented as alternatives to obtain more hydrocarbon from reservoirs after primary and secondary production procedures. The main purpose of EOR processes is to activate some mechanisms between reservoir fluids and porous media, like wettability alteration and reduction of interfacial tension (IFT) and fluid viscosity, among others, to improve the oil-sweeping conditions in the reservoir. Within EOR processes, the no-thermal or chemical-EOR (CEOR) classification is one of the most applied processes (Green & Willhite, 1998; Lake & Venuto, 1990). CEOR or no-thermal processes can be considered a continuation of secondary recovery like a water flooding process with the application of chemical agents such as alkali, polymers, surfactants, and mixtures between them to increase the sweep efficiency driven by macroscopic and microscopic displacement modifying the viscosity of the injected aqueous phase (Betancur et al., 2020), reducing the IFT, and leading to wettability alteration (Sheng, 2010, 2013).

In CEOR processes, the dimensionless expression of capillary number (N_c) is widely recognized as the most important parameter in EOR, which relates the changes of residual-oil saturation on porous media driven by viscous and capillary forces between the interaction of displaced and displacing fluids and porous media (Schramm et al., 2003; Sheng, 2013). Based on this, to modify the capillary and viscous forces, the addition of chemicals to injection fluids is proposed, deploying different techniques of CEOR processes such as polymer flooding, surfactant flooding, and alkali-polymer-surfactant flooding, among other combinations. Guo, Song, and Hilfer (2020) present a review of the different N_c ways used in EOR processes, remarking on their correlation with the residual-oil saturation, also known as capillary desaturation curve (CDC), being Eq. (3.1) the most representative expression used. Equation (3.1) describes the N_c expression where the IFT values and contact angle (as a measurement of wettability alteration) are highlighted as the main variables involved.

DOI: 10.1201/9781032650685-3

$$N_c = \frac{\mu v}{\sigma * \cos(\theta)} \tag{3.1}$$

Where N_c is the capillary number, μ is the brine viscosity (Pa s), and ν is the average velocity of the fluid through the porous medium (m sec^{-e}). As mentioned previously, the interfacial forces and wettability-alteration mechanisms drive the fluid's displacement inside porous media, where CEOR processes have great potential to modify it. The CEOR techniques usually require specific chemicals (like polymers or surfactants) at specific dosages that could be expensive and even do not achieve the desired conditions in the whole process. Nonetheless, certain environmental conditions of the reservoir, like high temperature or salinity, can affect the action of chemicals added, increasing the dosages needed, increasing the adsorption onto rock surface, and even affecting the thermal stability of chemicals and their action on viscous properties and reduction of interfacial forces between fluids. So, in this context, nanotechnology emerges as an excellent option to complement and enhance traditional techniques of EOR (Franco et al., 2017), adding additional advantages by nanomaterial inclusion based on their nanometric size, surface energy, volume/area ratio, among others (Saleh, 2020). Particularly for IFT and wettability alteration, the advantages of nanoparticle (NP) addition have been widely reported, based on fluid-fluid and fluid-rock interactions that can be promoted by NPs (Ali et al., 2018). This work presents a wide recompilation of the different studies using NPs for wettability alteration and IFT modification focused on EOR processes based on particular mechanisms influenced by NP inclusion (Ali et al., 2018).

3.2 FUNDAMENTAL CONCEPTS OF RESERVOIR WETTABILITY

Wettability is commonly described as the tendency of a fluid to adhere to or spread on a solid surface in the presence of another fluid. Wetness is vital for the understanding of multiphase flow in the porous media and the subsequent production profiles because the wettability properties of the rock surface tend to control the flow and distribution of reservoir fluids (Goshtap Cheraghian, 2015). Hence, the impact of wettability directly affects the permeability curves, oil- and water-saturation profiles, and capillary pressure, playing an important role in fluid displacement and hydrocarbon production (Panda & Gangawane, 2023a, 2023b).

The interactions between the rock surface and the reservoir fluids establish the wetting property of a reservoir. The rock mineral composition, pore size distribution, fluid composition, and reservoir conditions (pressure and temperature) govern the interactions of the liquid and solid phases in the three regions of contact (Agbalaka et al., 2008). Even though the reservoir rock minerals are generally known as intrinsically hydrophilic (water-wet), the presence of a high fraction of asphaltenes, which are known as natural surface-active agents in crude oil, leads to the adsorption of the oil constituents at the solid-liquid interface (Treiber & Owens, 1972). In this sense, the existence of polar and large fractions of asphaltenes in the oil allows the adsorption onto the solid surface and promotes the formation of an oil film onto the rock surface (Agbalaka et al., 2008). The strength of the oil film depends on three forces, including Van der Waals, hydrogen bonding, and electrostatic mechanism, which control the oil film deposition and spreading capabilities of the oleic phase and hence determine the wetting degree of the rock surface (Panda et al., 2023).

To alter the wettability of porous media, chemicals/additives (like surfactants, low-salinity brine, and selective ions) can be added to reservoir fluids to promote change in surface energy by the addition of specific functional groups onto the rock surface (Ali et al., 2018). The addition of surfactants is a recognized alternative to altering wettability (Standnes & Austad, 2000) based on rock-surface modification with amphiphilic molecules (surfactant) by an adsorptive mechanism. Standnes and Austad (2000) studied the use of surfactants to alter the wettability in chalk from oil-wet to water-wet. They explain the mechanism of wettability alteration based on the formation of an ion-pair between the cationic surfactant and carboxylates negatively charged adsorbed from the oil. More recently, Yao et al. (2021) presented a review of the use of surfactant to modify the wettability in carbonates that usually have strong oil-wet or mixed-wet, making it difficult the mobility of hydrocarbons. The authors highlight the role of surfactants as wettability modifiers, considering their chemistry that could influence the interaction with the rock and promote certain changes (Kumar et al., 2022).

3.3 METHODS OF WETTABILITY MEASUREMENTS

There are numerous methods for the determination of rock-surface wettability, including quantitative and qualitative methods. Quantitative procedures are related to imbibition and forced displacement (Amott), contact angle measurement, and the United States Bureau of Mines (USBM). Otherwise, qualitative methods include relative permeability curves, imbibition rates, flotation, capillary pressure curves, nuclear magnetic resonance, and dye adsorption, among others. Although there are no generally accepted methods, quantitative methods are commonly used for the determination of the reservoir rock surface wettability (Minz & Mishra, 2023; Sahoo & Mishra, 2023).

Quantitative methods can be conducted under static or dynamic conditions. In this sense, static conditions are fulfilled through contact angle measurements as the universal measure of surface wettability. However, contact angle only provides information about a specific portion of the rock surface so it requires an extensive measure through all the surfaces. On the other hand, in the case of dynamic conditions, the Ammot (1959) and the USBM (Donaldson et al., 1969) wettability tests based on oil-water-brine displacement behavior are considered. These two tests are derived from fluid flow conditions by capillary pressure curves, depending on water saturation measurements. The Amott and USBM methods describe the wettability measurement as a function of the displacement properties of the oil-water-rock system (Anderson, 1986).

3.4 CONTACT ANGLE

Conventionally, the wetting state of a solid surface is determined by the measurement of the contact angle for a specific liquid phase. In the case of oil reservoirs where two immiscible fluids co-exist, the wetting state is commonly defined by the measurement of the water contact angle (Agbalaka et al., 2008). In this sense, a rock surface is considered water-wet when the water contact angle is lower than 90°, neutral-wet when the contact angle is equal to 90°, and oil-wet when the contact angle is greater than 90° (Agbalaka et al., 2008; Morrow, 1990; Robin, 2001). However, this is a fragile definition of the wetting state for a reservoir rock surface. Some authors stated that it may be classified into strongly or weakly water-wet and strongly or weakly oil-wet (Agbalaka et al., 2008). Moreover, some researchers introduced the concept of gas-wet, a term commonly used for cases of condensate reservoirs in gas-liquid-rock systems, which usually includes chemical treatments or nanomaterials to promote preferential neutral gas-wet over liquid-wet (Ganie et al., 2019; Panda et al., 2024).

3.5 AMOTT METHOD

The Amott method is derived from the capillary pressure phenomena and the microscopic displacement efficiency, considering that the wetting fluid imbibes spontaneously into the porous media and displaces the nonwetting one (Robin, 2001). The method combines imbibition and forced displacement to obtain the average wettability of the core. In this sense, the core is firstly prepared by centrifugating the sample immersed in water until the residual-oil saturation (Sor) (Donaldson et al., 1969; Ma et al., 1999). In the second step, the rock core is immersed in oil, and the displaced water volume by spontaneous imbibition is measured after 20 hours (V_1). Afterwards, the oil is forced into the core by centrifugation or by forced displacement, and the water volume displaced by the oil is measured (V_2). The third step comprises the immersion of the core into the water to replace the oil by spontaneous imbibition (V_3), and then, the water molecules are forced into the core to displace the oil by forced displacement (V_4). Amott has defined the following two indexes (Amott, 1959):

$$I_o = \frac{V_1}{V_1 + V_2} \tag{3.2}$$

$$I_w = \frac{V_3}{V_3 + V_4} \tag{3.3}$$

Whereas I_o is defined as the displacement-by-oil ratio and I_w is described as the displacement-by-water ratio. When the I_w is closed to 1, oil recovery occurs mainly by spontaneous imbibition, meaning that the rock system is very strongly water-wet. Otherwise, when the I_o is shut to 1, water recovery occurs mainly by spontaneous imbibition, which means that the core is very strongly oil-wet. Later, a unique wettability value was proposed based on the differences between the I_w and I_o, based on the Amott-Harvey relative displacement index (Boneau & Clampitt, 1977; Trantham & Clampitt, 1977).

$$I_{AH} = I_W - I_\alpha \tag{3.4}$$

Where I_{AH} is the Amott-Harvey index that varies from +1 for strongly water-wet to −1 for strongly oil-wet. Cuiec proposed that the rock is water-wet when $0.3 \leq I_{AH} \leq 1$, intermediate-wet when $-0.3 \leq I_{AH} \leq 0.3$, and oil-wet when $-1 \leq I_{AH} \leq -0.3$ (Cuiec, 1984). The main problem related to the application of Amott methods is that the test is not sensitive to near-neutral wettability. It is because when the contact angles vary from approximately 60° to 120°, neither fluid will spontaneously imbibe and place the other one.

3.6 USBM METHOD

It relates to the work needed for an immiscible fluid to displace another immiscible fluid. The required work of a wetting fluid to displace the nonwetting fluid through the porous media is less than that for the opposite displacement. The work is dependent on the area under the capillary pressure curves. When the core sample is water-wet, the area of the water-drive capillary pressure curve is smaller than the area under the oil-drive capillary pressure curve. The USBM method correlates the areas of the two capillary curves for the calculation of the wettability index according to the following expression:

$$W = \log\left(\frac{A_1}{A_2}\right) \tag{3.5}$$

Meanwhile, A_1 and A_2 correspond to the areas under the oil- and brine-drive curves, respectively. When W is higher than zero, the core sample is water-wet. Otherwise, the core is oil-wet when W is less than zero. Finally, when W lies around zero, a neutral-wet condition is displayed. Several authors have proposed modified procedures based on the calculation of the centrifugal capillary pressures (Hassler & Brunner, 1945; Slobod et al., 1951). The significant advantage of the USBM method in contrast to the Amott test is the applicability for near-neutral wettability; however, this test can't determine whether a core system shows mixed or fractional wettability (Anderson, 1986) (Table 3.1).

TABLE 3.1
Relationship between contact angle, wettability, and the Amott and USBM wettability indices (Anderson, 1986)

	Water-wet	Mixed-wet or neutrally wet	Oil-wet
USBM wettability index	W near to 1	W near to 0	W near to −1
Contact angle (°)	0–60	60–105	105–180
Amott wettability index	Positive	Zero	Zero
Amott-Harvey wettability index	$0.3 \leq I \leq 1.0$	$-0.3 < I < 0.3$	$-1.0 \leq I \leq -0.3$

Abbreviation: USBM, United States Bureau of Mines.

3.7 MECHANISMS OF WETTABILITY ALTERATION BY APPLICATION OF NPS

The wettability of rock-oil-water systems can be altered by the dispersion of nanomaterials in aqueous solutions that lead to the creation of wedge-like structures in the three-phase contact region (Wasan & Nikolov, 2003). Wedge-film structures are formed when the NPs in the bulk fluid tend to get confined in the contact region (Wasan et al., 2011). In this sense, the layering and structural organization of the NPs applies an excess amount of pressure film that promotes the structural disjoining pressure and determines the fluid spreading onto the solid surface (Li et al., 2018; Wasan et al., 2011; Wasan & Nikolov, 2003). The fluid spreading process and the formation of the wedge film will separate the deposited formation fluids from the formation surface leading to more oil recovery and hence increasing the fluid production (Kumar & Gangawane, 2022a, 2022b).

Electrostatic repulsion and Brownian motion between the nanomaterials in the bulk solution drive the disjoining pressure and hence the NP wettability alteration (El-Diasty & Ragab, 2013). The spreading process of the water phase depends on the size and polydispersity of the nanomaterials, the bulk volume NP fraction, the salt concentration, as well as the interfacial fluid tension (IFT) between the spreading phase and the drop (Chengara et al., 2004; Kondiparty et al., 2012) (Figure 3.1).

3.8 ADVANTAGES AND EFFICIENCY OF NPS IN WETTABILITY ALTERATION

The use of NPs for wettability alteration has been widely reported because of particular properties of nanomaterials (Franco et al., 2021; Khalil et al., 2017). NPs can interact with the porous media decorating the rock surface leading to changes in surface charge and modifying preferences toward the specific phase. The disjoining pressure effect combined with the NPs is an important mechanism that drives the wettability alteration. Authors such as McElfresh, Holcomb, and Ector (2012) and Chengara, Nikolov, Wasan, Trokhymchuk, Henderson, et al. (2004) studied the effect of NPs on wettability and explained about the formation of an NP thin film onto a rock surface driven by injection pressure of nanofluid injected. The formation of this

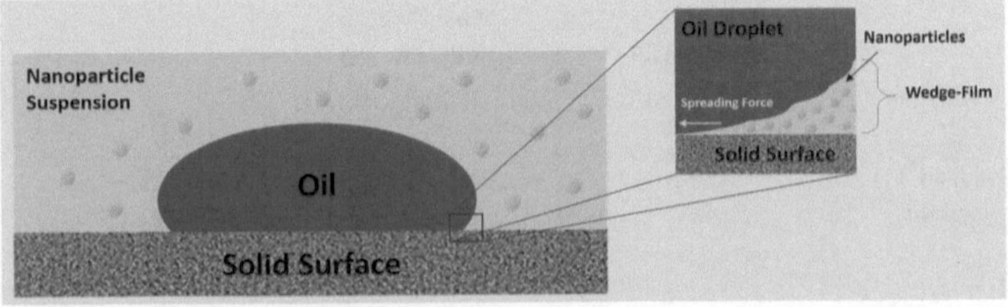

FIGURE 3.1 Representation of structural disjoining pressure gradient by the presence of nanoparticles (Wasan et al., 2011; Wasan & Nikolov, 2003).

thin film can separate and release the reservoir fluids, favoring changes from oil-wet to water-wet promoting a better hydrocarbon displacement.

It is worth mentioning that EOR processes assess both the effects of IFT and wettability alteration as several authors (Iravani et al., 2023; Kazemzadeh et al., 2019) have remarked the importance of oil recovery improvement.

3.9 NPS FOR WETTABILITY ALTERATION ON EOR PROCESSES AND CASE STUDIES

Some authors like Ali et al. (2018) and Goshtasp Cheraghian et al. (2020) present comprehensive reviews of NP use for EOR processes, including particular effects on wettability alteration (Eltoum et al., 2021) highlighting the most chemical nature used as silicon dioxide (SiO_2), aluminum oxide (Al_2O_3), and zirconium dioxide (ZrO_2), followed by magnesium oxide (MgO), titanium dioxide (TiO_2), Fe_2O_3, carbon nanotubes (CNT), among others. These NPs are commonly reported in dosages from 0.05 wt% up to 0.5 wt% and even higher values. These NPs are used in either sandstone or carbonate reservoirs and the use of zirconium oxide NPs is studied for the wettability alteration of carbonate rocks toward a strongly water-wet state. SEM images confirmed that the NP adsorption on the rock surface and the formation of nanostructured ribbons are responsible for the alteration of wetting properties. Additionally, it was observed from free imbibitions that considerable amount of oil was recovered with the addition of nanofluids.

Giraldo et al. (2013) evaluated alumina NPs dispersed in surfactant solutions as wettability modifiers on sandstone cores. They found that concentrations lower than 500 mg/L can improve the efficiency of nonionic surfactants and alter the wettability from oil-wet to strong water-wet, obtaining the best performance at 100 mg/L dosage. Additionally, through displacement tests, a reduction in the residual-water saturation can be related to improved oil recovery and different kinds of NPs were investigated for their influence on wettability alteration, using SiO_2, MgO, Al_2O_3, ZrO_2, calcium carbonate ($CaCO_3$), TiO_2, cerium oxide (CeO_2), and CNT for carbonate reservoirs. The author stated that $CaCO_3$ and SiO_2 reflected the best wettability alteration among others, and the structural disjoining pressure gradient is the main mechanism for wettability alteration, showing an increase of contact angle of decane from 39° to 120° and 122° approximately for $CaCO_3$ and SiO_2 nanofluids, respectively. Besides, the use of nanofluids showed an increase of 8%–9% in oil recovery after water flooding.

Silica NP has been one of the most used NPs on wettability alteration in EOR processes detailed in several works such as Al-Anssari et al. (2016), Azarshin et al. (2017), Hadia et al. (2021), Hendraningrat and Torsæter (2016), Maghzi et al. (2012), Yousefvand and Jafari (2015), among others.

Maghzi et al. (2012) studied the use of silica NP dispersions in distilled water for a water flooding process assessed in a five-spot micromodel. They found an increase in oil recovery efficiency by 8.7% with NP dispersions regarding only water flooding. In addition, when the dosage was increased to 3 wt%, an increase of 26% of oil recovery was obtained. The contact angle was measured, and wettability alteration from oil-wet to water-wet was explained by the hydrophilic property of silica NPs, making strong hydrogen bonds between NPs and water possible. Yousefvand and Jafari (2015) investigated the effect of silica NPs in polymeric solutions, improving the oil recovery by approximately 10% by the addition of NPs, which is related to this result with positive changes in wettability to water-wet conditions.

Al-Anssari et al. (2016) also studied the ability of silica NPs to modify the wettability conditions of calcite surfaces from oil-wet to strongly water-wet, attributing these changes to adsorption phenomenology. Azarshin et al. (2017) evaluated an amine-functionalization onto silica NP surfaces to improve the water flooding in oil-wet reservoirs. They showed changes in water-wet preference with water contact angles from 134° to 20° by the addition of functionalized NPs. Additionally, an increase from 41% to 60% in oil recovery was observed for only water and silica-functionalized NPs, respectively. Recently, Hadia et al. (2021) reported surface modification of silica NPs to improve their performance on wettability alteration at high temperatures and high salinity conditions. The authors showed promising results on thermal stability at 3.5% sodium chloride (NaCl) brine to alter the wettability, remarking on the ability of NPs to adsorb on the rock surface and adjust the wettability toward the water-wet state. They found more significant changes in contact angle for carbonate surfaces in comparison to sandstone surfaces.

Finally, the work realized by Franco et al. (2020) was the first pilot report of nanosurfactant flooding applied in a Colombian Field, designed as a nanofluid surfactant based on the addition of NPs. This work shows the whole experimental set performed, including the IFT and wettability measurements, where the nanofluid presents a better performance, increasing the capillary number corroborated with the core flooding test and the productivity results obtained in the field. In terms of wettability, changes in water contact angle were presented from 56° to 49°, with the addition of NPs to the surfactant solution evidencing changes toward water preference. The pilot presented successful results with a 27.3% production increment.

3.10 ENVIRONMENTAL CHALLENGES

The potential risk of nanomaterials on health, safety, and environment (HSE) depends on the nanostructure type, the surface-area-to-volume ratio, the nanomaterial morphology, and the presence of certain metal oxides (Davoodi et al., 2022; Karkare, 2013). Generally, NPs can be divided into two types of nanostructures such as nanocomposites and free NPs. However, the main HSE concern is related to the application of free NPs because they could be released into the environment or come into contact with human and animal beings (Lau et al., 2017). Most of the NPs used in the petroleum industry are commonly present in powder form or dispersed in a liquid solution and then are added into the aimed fluids. Several regulatory agencies such as the Environmental Protection Agency, The Consumer Protection Directorate of the EU, or The Food and Drug Administration in the US have documented the potential risks associated with nanomaterials. Nevertheless, there is not enough data available on the toxicology of nanomaterials, and the regulatory agencies have not ratified any special protocol regarding their production, handling, or labelling (Lau et al., 2017). Consequently, the toxicology of the NPs to the environment requires extensive research for the development of international regulations (Pal & Sen, 2023a, 2023b).

It is well established that some of the injected NPs will remain in the formations for many years. However, there is insufficient data for this affirmation (Davoodi et al., 2022). Until this time, NPs are being deployed in several field applications (Franco et al., 2021), and no monitoring strategies have been conducted to identify whether the NPs remained in the porous media or whether they tend to move through the reservoir and then can be released during the production stages. This is because the current equipment for routine detection and measurement of nanomaterials in crude oil and water is insufficient (Lau et al., 2017). Hence, it is required an extensive research for the development of monitoring strategies to study the long-term environmental impacts due to the use of NPs in EOR applications, where it is fundamental to monitor the NPs released from the porous media to determine the potential environmental challenges (Sagala et al., 2022). Otherwise, a practical alternative for the environmental concerns raised using NPs in EOR applications is related to the development of highly active and sustainable NPs that can be recoverable and/or recyclable to mitigate the release of nanomaterials to the environment (Davoodi et al., 2022). Therefore, extensive investigation is required to ensure a scalable NP synthesis, a high-performance behavior at EOR operations, and an appropriate recovery strategy at field conditions for the release of nanomaterials coming from the reservoir.

3.11 INTERFACIAL TENSION

The tension that exists in the interface of two different liquids is known as the IFT. This tension prevents the mixing of fluids and keeps them in their present form. If one of the fluids is in the gaseous state, then the tension between them is known as surface tension.

3.12 IMMISCIBLE AND MISCIBLE DISPLACEMENT THROUGH OIL RESERVOIRS

Water injection and chemical injection are two common EOR methods. The existence of high IFT between injected water and in-situ oil avoids suitable displacement and some phenomena like viscous fingering and water breakthrough are inevitable in such cases. While, reducing the IFT during water injection results in miscible displacement, prevents viscous fingering and enhances the sweep efficiency of displacing fluid. Surfactants are common chemical agents that reduce the IFT between water and hydrocarbons (Holm, 1986). In addition, reduction of IFT between displacing and displaced fluids results in more oil production by water injection. NPs are also capable of reducing the IFT. The addition of NPs to the injecting water leads to the enhancement of oil production by two main mechanisms, including wettability alteration and reduction of IFT (Iravani et al., 2023).

3.13 MEASUREMENT OF IFT

Several different methods are derived to measure IFT. Pendant drop, du nouy ring, and spinning drop are the three most used methods for measurement of IFT. For

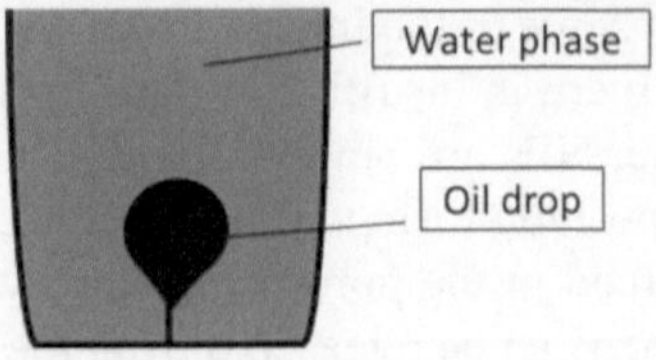

FIGURE 3.2 Measurement of interfacial tension by pendant drop method.

measurement of IFT in an oil-water system by pendant drop method, a drop of oil will be injected into the bulk water phase by a needle, and the state of a drop will be recorded from a view cell. The images will be analyzed, and the IFT will be measured based on their state. This method is applicable for IFT values higher than 1 m N/m (Lee et al., 2008). Figure 3.2 represents a schematic of the pendant drop method.

The governing equation for IFT measurement by pendant drop method is Eq. (3.6):

$$\sigma = \frac{gd_e^2}{t}\left(\rho_w - \rho_o\right) \tag{3.6}$$

Where σ is the IFT, g is the acceleration due to gravity, d_e is the most excellent radius of drop, ρ_w and ρ_o are the densities of water and oil, respectively, and parameter t is the shape factor concerning the curvature of drop.

However, for lower values of IFT, spinning drop is more efficient. During this method, a drop of oil will be located in the middle of a spinning tube that is filled with water. Variation in the shape of oil droplets at different revolutions gives the IFT. This method has a good accuracy in IFT values lower than 1 m N/m. Figure 3.3 illustrates a schematic of a spinning drop device.

Du nouy ring is applicable for the measurement of surface tension. A thin ring will be placed on the surface of the liquid, and then it will be pulled up. The tension for pulling out will be recorded and reported as surface tension. Figure 3.4 represents the schematic of surface tension measurement by the du nouy ring.

3.14 CALCULATION OF IFT

Some correlations could predict IFT in water-hydrocarbon systems. The parachor method is one of the most famous methods for the calculation of IFT in water-hydrocarbon systems. Equation (3.7) represents the Macleod correlation for the prediction of IFT (Schechter & Guo, 1998).

$$\sigma^{1/4} = P_\sigma\left(\rho_M^w - \rho_M^o\right) \tag{3.7}$$

Where σ represents the IFT value, ρ_M^w and ρ_M^o are molar densities of the water and oil phase respectively, and P_σ represents the parachor. The parachor value for some pure paraffinic components is illustrated in Table 3.2 (Danesh, 1998). There are some mixing rules for the calculation of parachor for compositions.

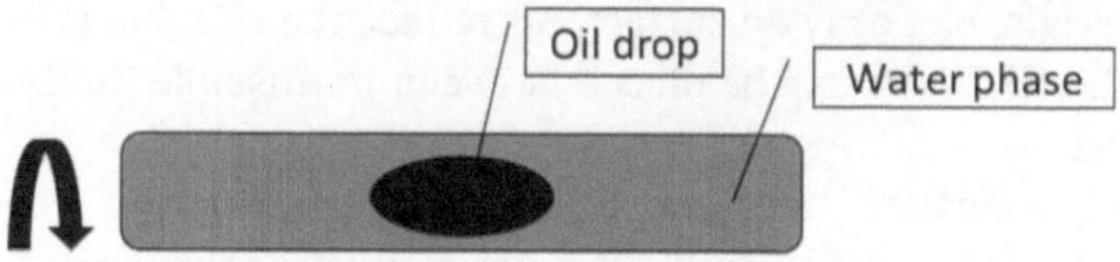

FIGURE 3.3 Measurement of interfacial tension by spinning drop method.

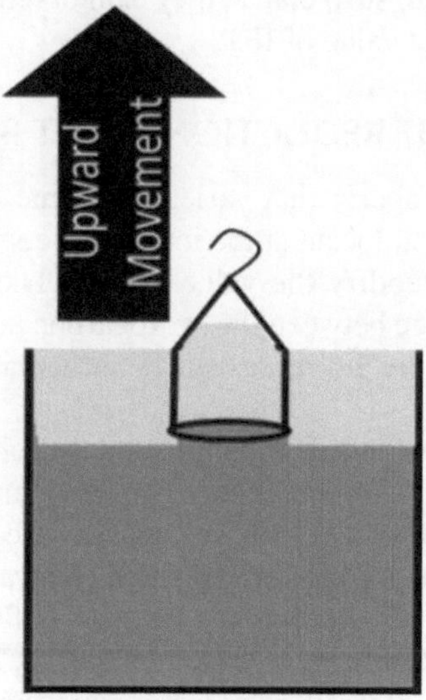

FIGURE 3.4 Measurement of surface tension by du nouy ring.

TABLE 3.2
The parachor values for some paraffinic pure components

Component	Parachor	Component	Parachor
CO_2	78	n-C_8	351.5
C_1	77	n-C_9	393
C_2	108	n-C_{10}	433
C_3	150	n-C_{11}	474.1
n-C_4	189.9	n-C_{12}	514.7
n-C_5	231.5	n-C_{13}	555.2
n-C_6	271	n-C_{14}	595.8
n-C_7	312.5	n-C_{15}	636.4

3.15 REDUCTION OF IFT

As it was mentioned, reducing the IFT is of interest for the EOR process and miscible displacement. IFT reduction can take place through the application of surfactants. Polar tail and organic head at surfactants cause reduction of IFT between contrasting and organic materials. Not only do surfactants reduce the IFT, but NPs are also capable of reducing the IFT. NPs can be placed between immiscible fluids and reduce the IFT between them.

Small size, high surface area, and accessible transportation through a porous medium have made NPs suitable candidates for application in oil reservoirs (Ghaffari et al., 2022). NPs can enhance oil recovery through two main mechanisms: wettability alteration (Rezvani et al., 2018) and IFT reduction (Khalilnejad et al., 2020). Although, in comparison with surfactants, they cannot reduce IFT significantly, they are capable of decreasing the value of IFT.

3.16 MECHANISMS OF REDUCTION OF IFT BY NPS

IFT is a thermodynamic property that varies with time and equilibration between phases. NPs tend to move and locate at the interface between two fluids. This transportation enables them to modify the value of IFT. The placement of NPs at the interface increases the surface between the hydrocarbon and water phases and results in the reduction of IFT. Figure 3.5 represents the schematic of the reduction of IFT by these mechanisms.

Some NPs like surfactants have hydrophilic and hydrophobic parts. This property eases the transportation of NPs from bulk fluids and their attachment to the interface of two immiscible fluids. In addition, NPs have catalytic effects and adsorb asphaltenes. Therefore, they can inhibit asphaltene precipitation (Rezvani et al., 2019). Adsorption of asphaltene and attachment of NPs onto the interface of fluids results in the reduction of IFT (Ahmed et al., 2017; Cheraghian & Hendraningrat, 2016).

Since asphaltenes are contained in polar organic contents, they are known as natural surfactants. Therefore, the movement of asphaltenes to the interface of fluids (which is taken place by NPs) can reduce the value of IFT as a side effect.

Pressure and temperature are high at reservoir conditions. IFT is directly related to pressure. It means increasing the pressure results in higher values of IFT. At the same time, there is an inverse relation between temperature and IFT. In other words, the IFT reduces at higher temperatures. The kinetic energy of the Brownian motion of NPs is increased at high temperatures. The roughness of the surface of NPs can

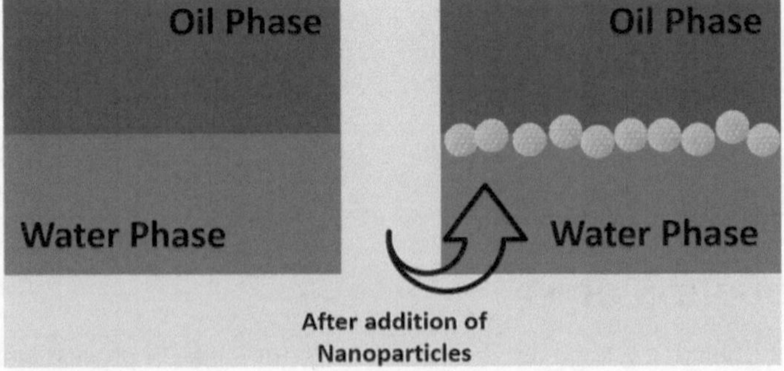

FIGURE 3.5 Varying the interface of fluids after the addition of nanoparticles.

lead to some ruptures in the scale of nano in the interface of liquids. These ruptures reduce the value of IFT (Ahmed et al., 2017).

It should be noted that the properties of NPs almost always change with variations in temperature and pressure. Therefore, the accurate investigation of NPs at desired conditions is vital for the application of NPs for EOR intentions.

There is a critical concentration for NPs in the reduction of IFT. Reduction of IFT in the presence of NPs continues up to a specified amount of NPs. Excess amounts of NPs will not affect the IFT above the identified critical concentration. This phenomenon takes place through the creation of a rigid film of NPs at the interface of fluids, and adding more NPs results in higher IFT values. Therefore, finding this concentration is of high importance in the EOR process. Not only by finding this critical concentration, the efficiency of NPs is optimized, but also the reduction in the consumption of NPs results in lower expenses for the project.

Adsorption of NPs to the interface of fluids needs an adhesion energy. This adhesion energy is calculated by using Eq. (3.8).

$$\Delta E = \pi R^2 \sigma_{12} (1 \pm \cos\theta_{12}) \tag{3.8}$$

Where E is the adhesion energy in $K_B T$, R represents the radius of the NP, σ_{12} is the IFT between fluids, and θ_{12} is the contact angle of particles at the surface of fluids (Tu & Lee, 2014).

3.17 PICKERING EMULSIONS

Concerning the reduction of IFT by NPs, some solid particles in the size range of nano and micron are used to stabilize emulsions. In other words, NPs are used as emulsifiers to stabilize the emulsions. These emulsions are known as pickering emulsions. The tendency of NPs to locate at the interface of two immiscible phases enables them to stabilize emulsions. NPs that are used for the stabilization of emulsions should contain both hydrophilic and hydrophobic sites (Ngai & Bon, 2014).

3.18 LIMITATIONS

Although many benefits are proven for EOR applications of NPs, some limiting parameters exist in reservoirs that restrict the performance of NPs. These parameters should be evaluated carefully to check out the capability of NPs for the considered application (Sun et al., 2017).

The stability of NPs is a challenge at reservoir conditions. The stability of NPs at high temperatures is faced with challenges. Therefore, the optimum concentration of NPs should be specified concerning reservoir temperature and pressure (Ranka et al., 2015). The salinity of formation fluids is another restricting factor for the stabilization of NPs. So the injection scenario should be tuned for the salinity of reservoir fluids, especially at the goal zone (Kazemzadeh et al., 2019).

The pore size distribution of reservoir rock is another determinative factor for the application of NPs in hydrocarbon reservoirs. If the pore size of reservoir rocks is smaller than the radius of NPs, the NPs will not enter the porous medium and also block the entrance. The NPs will have negative effects on production in these cases. To avoid pore blocking, the rock properties should be examined completely (Sameni et al., 2015).

3.19 CASE STUDIES

Silica is the most common NP for EOR intends. For availability, low cost, and efficiency of silica, this NP is investigated widely for application in oil reservoirs. However, metal oxide NPs are of interest for EOR purposes. Some research with a focus on the investigation of the application of NPs to reduce IFT is represented below.

To assess the capability of ZrO_2 NPs to reduce IFT between oil and formation water, Rezvani et al. derived some pendant drop measurements under reservoir conditions. They added various concentrations of NPs to dilutions of formation brine and measured the IFT at wide ranges of pressure and temperature. It is concluded that ZrO_2 NPs can reduce the IFT by 30%. In addition, the critical concentration of NPs is specified as 0.006078 mol/L (Rezvani et al., 2018).

Khalilnezhad et al. investigated the effects of TiO_2 NPs on the reduction of IFT. They found 1,000 ppm as the critical concentration of NPs, at which the lowest value of IFT was achieved. In addition, they observed that TiO_2 NPs can reduce the IFT by around 50%. The catalytic effect of TiO_2 NPs in the adsorption of asphaltenes (at the interface of oil and water) is introduced as the main mechanism for their observation (Khalilnezhad et al., 2019).

In another study, Khalilnezhad et al. examined γ-Al_2O_3 NPs for application in the EOR process. They found that the IFT will be reduced in the presence of γ-Al_2O_3 NPs. The catalytic effects of NPs on the one hand and the porous structure of NPs on the other hand resulted in a significant reduction of IFT. The porosity of NPs enhances the adsorption of asphaltenes and results in intensified activity at the interface (Khalilnezhad et al., 2023).

Sadatshojaei et al. (2019) evaluated the effects of silica NPs on IFT reduction in low-salinity conditions. They have found that increasing the concentration of NPs decreases the value of IFT. Nevertheless, they measured the minimum value of IFT between oil and water (14.5 m N/m) at NP concentration equal to 5,000 ppm (Sadatshojaei et al., 2019).

The application of Janus NPs for EOR purposes has become an interesting topic for scientists. Since Janus NPs have both hydrophilic and lipophilic sections, they will be located at the appropriate interface of oil and water. Giraldo et al. (2018) synthesized Janus NPs of NiO/SiO_2 NPs. In comparison with the bare silica NPs, she found out that Janus NPs can enhance the efficiency of the process significantly. The Janus NPs resulted in the best responses for EOR purposes at a very low concentration (100 ppm) (Giraldo, 2018).

The synergistic effect of the application of NPs with other chemicals is also investigated. It is proved that the simultaneous application of NPs with surfactants improves the efficiency of both agents (NPs and surfactant) for the gradual release of surfactant (Betancur et al., 2019). In other words, the NPs adsorb surfactant micelles and release them slowly. This gradual release avoids surfactant retention and adsorption by porous media.

NOMENCLATURE

EOR:	Enhanced Oil Recovery
CEOR:	Chemical Enhanced Oil Recovery
IFT:	Interfacial Tension
Nc:	Capillary Number
NPs:	Nanoparticles
USBM:	United States Bureau of Mines
HSE:	Health, Safety, and Environment

REFERENCES

Agbalaka, C., Dandekar, A. Y., Patil, S. L., Khataniar, S., & Hemsath, J. R. (2008). The effect of wettability on oil recovery: A review. Paper presented at the *SPE Asia Pacific Oil and Gas Conference and Exhibition*, Perth, Australia, October 2008. doi: https://doi.org/10.2118/114496-MS

Ahmed, A., Saaid, I. M., Tunio, A. H., Pilus, R. M., Mumtaz, M., & Ahmad, I. (2017). Investigation of dispersion stability and IFT reduction using surface modified nanoparticle: Enhanced oil recovery. *Journal of Applied Environmental and Biological Sciences*, 7(4S), 56–62.

Al-Anssari, S., Barifcani, A., Wang, S., Maxim, L., & Iglauer (2016). Wettability alteration of oil-wet carbonate by silica nanofluid. *Journal of Colloid and Interface Science*, 461, 435–442.

Ali, J. A., Kolo, K., Manshad, A. K., & Mohammadi, A. H. (2018). Recent advances in application of nanotechnology in chemical enhanced oil recovery: Effects of nanoparticles on wettability alteration, interfacial tension reduction, and flooding. *Egyptian Journal of Petroleum*, 27(4), 1371–1383.

Amott, E. (1959). Observations relating to the wettability of porous rock. *Transactions of the AIME*, 216(01), 156–162.

Anderson, W. (1986). Wettability literature survey-part 2: Wettability measurement. *Journal of Petroleum Technology*, 38(11), 1246–1262.

Azarshin, S., Moghadasi, J., & Aboosadi, Z. A. (2017). Surface functionalization of silica nanoparticles to improve the performance of water flooding in oil wet reservoirs. *Energy Exploration and Exploitation*, 35(6), 685–697.

Betancur, S., Giraldo, L. J., Carrasco-Marín, F., Riazi, M., Manrique, E. J., Quintero, H., García, H. A., Franco-Ariza, C. A., & Cortés, F. B. (2019). Importance of the nanofluid preparation for ultra-low interfacial tension in enhanced oil recovery based on surfactant–nanoparticle–brine system interaction. *ACS Omega*, 4(14), 16171–16180.

Betancur, S., Olmos, C. M., Pérez, M., Lerner, B., Franco, C. A., Riazi, M., Gallego, J., Carrasco-Marín, F., & Cortés, F. B. (2020). A microfluidic study to investigate the effect of magnetic iron core-carbon shell nanoparticles on displacement mechanisms of crude oil for chemical enhanced oil recovery. *Journal of Petroleum Science and Engineering*, 184, 106589.

Boneau, D., & Clampitt, R. L. (1977). A surfactant system for the oil-wet sandstone of the North Burbank Unit. *Journal of Petroleum Technology*, 29(05), 501–506.

Chengara, A., Nikolov, A. D., Wasan, D. T., Trokhymchuk, A., & Henderson, D. (2004). Spreading of nanofluids driven by the structural disjoining pressure gradient. *Journal of Colloid and Interface Science*, 280(1), 192–201.

Cheraghian, G. (2015). Effects of nanoparticles on wettability: A review on applications of nanotechnology in the enhanced oil recovery. *International Journal of Nano Dimension*, 6(5), 443–452.

Cheraghian, G., & Hendraningrat, L. (2016). A review on applications of nanotechnology in the enhanced oil recovery part A: Effects of nanoparticles on interfacial tension. *International Nano Letters*, 6, 129–138.

Cheraghian, G., Rostami, S., & Afrand, M. J. P. (2020). Nanotechnology in enhanced oil recovery. *Journal of Processes*, 8(9), 1073.

Cuiec, L. (1984). Rock/crude-oil interactions and wettability: An attempt to understand their interrelation. Paper presented at the *SPE Annual Technical Conference and Exhibition? SPE,* pp. 13205–13211.

Danesh, A. (1998). *PVT and Phase Behavior of Petroleum Reservoir Fluids*. Elsevier.

Davoodi, S., Al-Shargabi, M., Wood, D. A., Rukavishnikov, V. S., & Minaev, K. M. (2022). Experimental and field applications of nanotechnology for enhanced oil recovery purposes: A review. *Fuel*, 324, 124669.

Donaldson, E. C., Thomas, R. D., & Lorenz, P. B. (1969). Wettability determination and its effect on recovery efficiency. *Society of Petroleum Engineers Journal*, 9(01), 13–20.

El-Diasty, A. I., & Ragab, A. M. (2013). Applications of nanotechnology in the oil & gas industry: Latest trends worldwide & future challenges in Egypt. Paper presented at the *North Africa Technical Conference and Exhibition*, pp. SPE-164716.

Eltoum, H., Yang, Y.-L., & Hou, J.-R. (2021). The effect of nanoparticles on reservoir wettability alteration: A critical review. *Petroleum Science*, 18, 136–153.

Franco, C. A., Franco, C. A., Zabala, R. D., Bahamón, I. t., Forero, A. n., & Cortés, F. B. (2021). Field applications of nanotechnology in the oil and gas industry: Recent advances and perspectives. *Energy & Fuels*, 35(23), 19266–19287.

Franco, C. A., Giraldo, L. J., Candela, C. H., Bernal, K. M., Villamil, F., Montes, D., Lopera, S. H., Franco, C. A., & Cortés, F. B. (2020). Design and tuning of nanofluids applied to chemical enhanced oil recovery based on the surfactant–nanoparticle–brine interaction: From laboratory experiments to oil field application. *Nanomaterials*, 10(8), 1579.

Franco, C. A., Zabala, R., & Cortés, F. B. (2017). Nanotechnology applied to the enhancement of oil and gas productivity and recovery of Colombian fields. *Journal of Petroleum Science Engineering*, 157, 39–55.

Ganie, K., Idris, A. K., Mohshim, D. F., Sulaiman, W. R. W., Saaid, I. M., Malik, A. A. (2019). A review on the wettability alteration mechanism in condensate banking removal. *Journal of Petroleum Science and Engineering*, 183, 106431.

Ghaffari, Z., Rezvani, H., Khalilnezhad, A., Cortes, F. B., & Riazi, M. (2022). Experimental characterization of colloidal silica gel for water conformance control in oil reservoirs. *Scientific Reports*, 12(1), 1–17.

Giraldo, L. J. (2018). Janus nanoparticles for enhanced oil recovery EOR: Reduction of interfacial tension. In *SPE Annual Technical Conference and Exhibition?* SPE, 194033-STU.

Giraldo, J., Benjumea, P., Lopera, S., Cortés, F. B., Ruiz, M. A. (2013). Wettability alteration of sandstone cores by alumina-based nanofluids. *Energy & Fuels*, 27(7), 3659–3665.

Green, D., & Willhite, G. (1998). *Enhanced Oil Recovery* (vol. 6). SPE Textbook Series.

Guo, H., Song, K., & Hilfer, R. (2020). A critical review of capillary number and its application in enhanced oil recovery. In Paper presented at the *SPE Improved Oil Recovery Conference,* SPE, 200419.

Hadia, N. J., Ng, Y. H., Stubbs, L. P., & Torsæter, O. J. N. (2021). High salinity and high temperature stable colloidal silica nanoparticles with wettability alteration ability for eor applications. *Journal of Nanomaterials*, 11(3), 707.

Hassler, G., & Brunner, E. (1945). Measurement of capillary pressures in small core samples. *Transactions of the AIME*, 160(01), 114–123.

Hendraningrat, L., & Torsæter, O. J. A. N. (2016). A study of water chemistry extends the benefits of using silica-based nanoparticles on enhanced oil recovery. *Journal of Applied Nanoscience*, 6, 83–95.

Holm, L. W. (1986). Miscibility and miscible displacement. *Journal of Petroleum Technology*, 38(08), 817–818.

Iravani, M., Khalilnezhad, Z., & Khalilnezhad, A. (2023). A review on application of nanoparticles for EOR purposes: History and current challenges. *Journal of Petroleum Exploration and Production Technology*, 13(4), 1–36.

Karkare, M. (2013). *Nanotechnology: Fundamentals and Applications*. IK International Pvt Ltd.

Kazemzadeh, Y., Shojaei, S., Riazi, M., & Sharifi, M. (2019). Review on application of nanoparticles for EOR purposes: A critical review of the opportunities and challenges. *Chinese Journal of Chemical Engineering*, 27(2), 237–246.

Khalil, M., Jan, B. M., Tong, C. W., & Berawi, M. A. (2017). Advanced nanomaterials in oil and gas industry: Design, application and challenges. *Applied Energy*, 191, 287–310.

Khalilnejad, A., Lashkari, R., Iravani, M., & Ahmadi, O. (2020). Application of synthesized silver nanofluid for reduction of oil-water interfacial tension. Saint Petersburg 2020, 2020(1), 1–5.

Khalilnezhad, A., Rezvani, H., Ganji, P., & Kazemzadeh, Y. (2019). A complete experimental study of oil/water interfacial properties in the presence of TiO2 nanoparticles and different ions. *Oil & Gas Science and Technology–Revue d'IFP Energies Nouvelles*, 74, 39.

Khalilnezhad, A., Rezvani, H., Talebi, A., Ganji, P., Puntervold, T., & Riazi, M. (2023). Improved oil recovery in carbonate cores using alumina nanoparticles. *Energy & Fuels*, 37(16), 11765–11775.

Kondiparty, K., Nikolov, A. D., Wasan, D., & Liu, K.-L. (2012). Dynamic spreading of nanofluids on solids. Part I: experimental. *Langmuir*, 28(41), 14618–14623.

Kumar, A., & Gangawane, K. M. (2022a). Effect of precipitating agents on the magnetic and structural properties of the synthesized ferrimagnetic nanoparticles by co-precipitation method. *Powder Technology*, 401, 117298.

Kumar, A., & Gangawane, K. M. (2022b). Synthesis and effect on the surface morphology & magnetic properties of ferrimagnetic nanoparticles by different wet chemical synthesis methods. *Powder Technology*, 410(June), 117867.

Kumar, A., & Gangawane, K. M. (2023a). Advanced computational approaches for water treatment. In K. M. Gangawane, M. Dwivedi, & P. Ghodke (Eds.), *Advanced Computational Approaches for Water Treatment* (pp. 1–12). CRC Press.

Kumar, A., & Gangawane, K. M. (2023b). Molecular dynamics for the membrane process. In K. M. Gangawane, M. Dwivedi, & P. Ghodke (Eds.), *Advanced Computational Approaches for Water Treatment* (pp. 13–38). CRC Press.

Kumar, A., Gangawane, K. M., & Ramanjaneyulu, B. (2022). Ferrofluids for waste-water treatment. In *International Conference on Chemical, Bio and Environmental Engineering* (pp. 723–744). Springer International Publishing.

Lake, L. W., & Venuto, P. B. (1990). A niche for enhanced oil recovery in the 1990s. *Oil & Gas Journal*, 88(17), 62–67.

Lau, H. C., Yu, M., & Nguyen, Q. P. (2017). Nanotechnology for oilfield applications: Challenges and impact. *Journal of Petroleum Science and Engineering*, 157, 1160–1169.

Lee, B.-B., Ravindra, P., & Chan, E.-S. (2008). A critical review: surface and interfacial tension measurement by the drop weight method. *Chemical Engineering Communications*, 195(8), 889–924.

Li, K., Wang, D., & Jiang, S. (2018). Review on enhanced oil recovery by nanofluids. *Oil & Gas Science and Technology–Revue d'IFP Energies nouvelles*, 73, 37.

Ma, S., Zhang, X., Morrow, N., & Zhou, X. (1999). Characterization of wettability from spontaneous imbibition measurements. *Journal of Canadian Petroleum Technology*, 38(13), 13–49.

Maghzi, A., Mohammadi, S., Ghazanfari, M. H., Kharrat, R., & Masihi, M. (2012). Monitoring wettability alteration by silica nanoparticles during water flooding to heavy oils in five-spot systems: A pore-level investigation. *Experimental Thermal and Fluid Science*, 40, 168–176.

Minz, P., & Mishra, S. (2023). Jackfruit peel derived ZnCl2-impregnated activated carbon: Optimization, characterization, and application in dye removal. *Biomass Conversion and Biorefinery*, 0123456789. https://doi.org/10.1007/s13399-023-04592-0.

McElfresh, P., Holcomb, D., & Ector, D. (2012). Application of nanofluid technology to improve recovery in oil and gas wells. Paper presented at the *SPE International Oilfield Nanotechnology Conference and Exhibition*. SPE, SPE-1548627.

Morrow, N. R. (1990). Wettability and its effect on oil recovery. *Journal of Petroleum Technology*, 42(12), 1476–1484.

Ngai, T., & Bon, S. A. F. (2014). *Particle-Stabilized Emulsions and Colloids: Formation and Applications*. Royal Society of Chemistry.

Pal, D., & Sen, S. (2023a). Emerging petroleum pollutants and their adverse effects on the environment. In I. D. Behera, & A. P. Das (Eds.), *Impact of Petroleum Waste on Environmental Pollution and Its Sustainable Management through Circular Economy. Environmental Science and Engineering* (pp. 103–137). Springer Nature Switzerland.

Pal, D., & Sen, S. (2023b). In-depth coverage of petroleum waste sources, characteristics, environmental impact, and sustainable remediation process. In I. D. Behera, & A. P. Das (Eds.), *Impact of Petroleum Waste on Environmental Pollution and Its Sustainable Management through Circular Economy* (pp. 1–38). Springer Nature Switzerland.

Panda, D., & Gangawane, K. M. (2023a). Development of superhydrophobic hybrid silica-cellulose aerogel as promising thermal insulation and sound absorption. *International Journal of Materials Research*, 114(4–5), 405–417.

Panda, D., & Gangawane, K. M. (2023b). Recycled cellulose–silica hybrid aerogel for effective oil adsorption: optimization and kinetics study. *Sadhana - Academy Proceedings in Engineering Sciences*, 48(3), 110–132.

Panda, D., Sahu, A. K., & Gangawane, K. M. (2024). Effect of molar concentration and drying methodologies on monodispersed silica sol for synthesis of silica aerogels with temperature-resistant characteristics. *Materials Science and Engineering: B*, 299, 1170x62.

Panda, D., Sharma, S., & Gangawane, K. M. (2023). Synthesis of novel biocatalyst from organic waste protonated with acid treatment for hydrogen production. *International Journal of Hydrogen Energy*, 48(63), 24242–24254.

Ranka, M., Brown, P., & Hatton, T. A. (2015). Responsive stabilization of nanoparticles for extreme salinity and high-temperature reservoir applications. *ACS Applied Materials & Interfaces*, 7(35), 19651–19658.

Rezvani, H., Kazemzadeh, Y., Sharifi, M., Riazi, M., & Shojaei, S. (2019). A new insight into Fe3O4-based nanocomposites for adsorption of asphaltene at the oil/water interface: An experimental interfacial study. *Journal of Petroleum Science and Engineering*, 177, 786–797.

Rezvani, H., Khalilnejad, A., & Sadeghi-Bagherabadi, A. A. (2018). Comparative experimental study of various metal oxide nanoparticles for the wettability alteration of carbonate rocks in EOR processes. *80th EAGE Conference and Exhibition* 2018, 2018(1), 1–5.

Rezvani, H., Khalilnezhad, A., Ganji, P., & Kazemzadeh, Y. (2018). How ZrO2 nanoparticles improve the oil recovery by affecting the interfacial phenomena in the reservoir conditions? *Journal of Molecular Liquids*, 252, 158–168.

Robin, M. (2001). Interfacial phenomena: Reservoir wettability in oil recovery. *Oil & Gas Science and Technology*, 56(1), 55–62.

Sadatshojaei, E., Jamialahmadi, M., Esmaeilzadeh, F., Wood, D. A., & Ghazanfari, M. H. (2019). The impacts of silica nanoparticles coupled with low-salinity water on wettability and interfacial tension: Experiments on a carbonate core. *Journal of Dispersion Science and Technology*, 1159–1173.

Sagala, F., Hethnawi, A., & Nassar, N. N. (2022). Challenges and uncertainties of using nanoparticles in oil and gas applications. In N. N. Nassar, F. B. Cortés, & C. A. Franco (Eds.), *Nanoparticles: An Emerging Technology for Oil Production and Processing Applications* (pp. 497–506). Springer.

Sahoo, A., & Mishra, S. (2023). Synthesis of microporous-activated carbon via direct H3PO4 and KOH activation of Pterospermum acerifolium fruits: Process optimization and characterization. *Biomass Conversion and Biorefinery*, 15, 1–17.

Saleh, T. A. (2020). Nanomaterials: Classification, properties, and environmental toxicities. *Environmental Technology Innovation*, 20, 101067.

Sameni, A., Pourafshary, P., Ghanbarzadeh, M., & Ayatollahi, S. (2015). Effect of nanoparticles on clay swelling and migration. *Egyptian Journal of Petroleum*, 24(4), 429–437.

Schechter, D. S., & Guo, B. (1998). Parachors based on modern physics and their uses in IFT prediction of reservoir fluids. *SPE Reservoir Evaluation & Engineering*, 1(03), 207–217.

Schramm, L. L., Stasiuk, E. N., & Marangoni, D. G. (2003). 2 Surfactants and their applications. *Annual Reports Section "C" (Physical Chemistry)*, 99, 3–48.

Sheng, J. (2010). *Modern Chemical Enhanced Oil Recovery: Theory and Practice*. Gulf Professional Publishing.

Sheng, J. (2013). *Enhanced Oil Recovery Field Case Studies*. Gulf Professional Publishing.

Slobod, R., Chambers, A., & Prehn Jr, W. (1951). Use of centrifuge for determining connate water, residual oil, and capillary pressure curves of small core samples. *Journal of Petroleum Technology*, 3(04), 127–134.

Standnes, D. C., & Austad, T. (2000). Wettability alteration in chalk: 2. Mechanism for wettability alteration from oil-wet to water-wet using surfactants. *Journal of Petroleum Science Engineering*, 28(3), 123–143.

Sun, X., Zhang, Y., Chen, G., & Gai, Z. (2017). Application of nanoparticles in enhanced oil recovery: A critical review of recent progress. *Energies*, 10(3), 345.

Trantham, J., & Clampitt, R. L. (1977). Determination of oil saturation after waterflooding in an oil-wet reservoir the North Burbank Unit, Tract 97 Project. *Journal of Petroleum Technology*, 29(05), 491–500.

Treiber, L., & Owens, W. (1972). A laboratory evaluation of the wettability of fifty oil-producing reservoirs. *Society of Petroleum Engineers Journal*, 12(06), 531–540.

Tu, F., & Lee, D. (2014). One-step encapsulation and triggered release based on Janus particle-stabilized multiple emulsions. *Chemical Communications*, 50(98), 15549–15552.

Wasan, D., Nikolov, A., & Kondiparty, K. (2011). The wetting and spreading of nanofluids on solids: Role of the structural disjoining pressure. *Current Opinion in Colloid & Interface Science*, 16(4), 344–349.

Wasan, D. T., & Nikolov, A. D. (2003). Spreading of nanofluids on solids. *Nature*, 423(6936), 156–159.

Yao, Y., Wei, M., & Kang, W. (2021). A review of wettability alteration using surfactants in carbonate reservoirs. *Advances in Colloid Interface Science*, 294, 102477.

Yousefvand, H., & Jafari, A. J. (2015). Enhanced oil recovery using polymer/nanosilica. *Procedia Materials Science*, 11, 565–570.

4 Factors Affecting Nanofluids by Nanotechnology for Enhanced Oil Recovery

Uma Sankar Behera and Ramanand Singh

4.1 INTRODUCTION

Enhanced oil recovery (EOR) is a vital process in the petroleum industry aimed at maximizing crude oil extraction from hydrocarbon-deposited reservoirs. EOR techniques have become pivotal in extracting more oil from reservoirs in the post-primary and secondary recovery phases by modifying the reservoir rock and fluid properties. Different methods for EOR encompass chemical and gas injection, low-salinity water flooding (LSWF), as well as hybrid approaches that involve a combination of chemicals, nanoparticles (NPs), and low-salinity water (LSW), applicable for light crude oil in conventional reservoirs. On the other hand, in heavy crude oil reservoirs, thermal methods (e.g., steam injection, in situ combustion) are generally employed for heavy and extra-heavy crude oil recovery. However, nano-assisted steam injection is a relatively new technology with a considerable number of investigations (Amanam & Kovscek, 2017; Guo et al., 2015; Medina et al., 2019). The integration of nanotechnology offers a promising pathway to augment EOR processes by manipulating materials at the nanoscale. Recent advancements in nanotechnology have introduced NPs as a promising tool for EOR techniques.

Nevertheless, the success of NP-assisted EOR is significantly affected by the presence of reservoir heterogeneity. The heterogeneity of the reservoirs impacts the NP-assisted EOR by creating non-uniform distribution and retention of NPs within the reservoir. This non-uniformity can affect the overall effectiveness and efficiency of the EOR process. Considerable research is underway regarding the utilization of NPs in EOR; numerous aspects within this domain remain ambiguous and necessitate thorough research and exploration. Specific areas of inquiry include understanding the interactions among NPs in porous media, comprehending their adsorption capacity, and assessing their impact on solid-fluid and fluid-fluid interactions. This work delves into the role played by reservoir heterogeneity in the application of NPs for improved oil recovery. Also, this study explores the multifaceted factors that impact the successful application of nanotechnology in EOR.

DOI: 10.1201/9781032650685-4

4.2 CHARACTERISTICS OF RESERVOIRS

Crude oil reservoirs differ in their rock properties, like how porous and permeable the rock is and how fluids interact with it. These reservoirs are grouped into two types, carbonate and sandstone, based on these properties.

4.2.1 Carbonate Reservoir

Carbonate reservoirs are made up of sedimentary rocks primarily composed of calcium carbonate, originating from marine organisms, skeletons, and algae, as well as chemical precipitation. Unlike sandstone rocks, these carbonates are more chemically active (Jarrahian et al., 2019). Over time, through diagenesis, the rocks undergo various physical and chemical transformations, leading to changes in properties like porosity and permeability. This process results in uneven rock distributions, making it important to note that one EOR technique effective in one carbonate reservoir may not work in another. Carbonate reservoirs can contain minerals like calcite, dolomite, anhydrite (e.g., calcium sulfate), and clay minerals (e.g., illite, kaolinite, smectite). Porosity in carbonate rocks is classified into three categories: connected porosity (occurring between carbonate grains), vugs porosity (unconnected pores formed by calcite dissolution during diagenesis), and fracture porosity (created by geological stress). These distinct porosity types create intricate pathways for fluid flow, directly influencing oil production.

4.2.2 Sandstone Reservoir

Sandstone reservoirs are primarily composed of feldspar, quartz sand, mica, rock fragments, and various minerals like silica. These sediments have undergone extensive transportation and deposition processes, resulting in significant changes to their physical properties. Their mineral composition, geological history, and long transportation distances contribute to this lower chemical reactivity (Jarrahian et al., 2019). In sandstone reservoirs, stable minerals like quartz resist chemical alteration when exposed to environmental fluids. These rocks are formed through the erosion and weathering of source rocks over extended periods, during which less-stable minerals are typically removed. Moreover, the sand grains in sandstone formations travel long distances before deposition, further enhancing their chemical inertness compared to carbonate rocks.

The clay's crystal structure in sandstone reservoirs is typically tetrahedral, with silicon at the center and octahedral aluminum at the center. In terms of physical properties, such as porosity and permeability, sandstone reservoirs generally outperform carbonate reservoirs. However, these properties exhibit heterogeneity, varying from one location to another, often due to the impact of buried sandstone and reservoir heat (Barwis et al., 2012). It's important to note that sandstone reservoirs in different geographical areas can possess unique geological characteristics. Therefore, specific mechanisms may be effective for sandstone reservoirs in one location but not suitable for those in other regions.

4.2.3 Reservoir Heterogeneity

Reservoir heterogeneity can significantly impact the dispersion and mobility of NPs within the reservoir (Esfe et al., 2020). Factors such as variations in pore size distribution, rock permeability, and geological structures must be factored into the design of nanotechnology-based EOR strategies. Reservoir heterogeneity encompasses the inherent non-uniformity present in subsurface geological formations. This non-uniformity contains variations in rock porosity, permeability, mineral composition, and fluid saturations. Reservoir heterogeneity can manifest at various scales, from the macroscopic to the microscopic, and has profound implications on fluid flow patterns and the distribution of injected NPs. Reservoir heterogeneity stands as a critical factor in NP-based EOR applications. An in-depth comprehension of and effective responses to the challenges posed by reservoir heterogeneity through meticulous reservoir characterization, customized NP design, zonal isolation, and the deployment of advanced monitoring techniques can yield significant enhancements in oil recovery. As technology continues to advance, addressing reservoir heterogeneity will remain a pivotal focus for optimizing NP-based EOR methods within the petroleum industry. Some of the successful case studies emphasize the significance of understanding and addressing reservoir heterogeneity. Field trials conducted in heterogeneous carbonate reservoirs showcased improved oil recovery through the injection of NPs and formation damage caused by asphaltene precipitation (Zabala et al., 2014). Integration of intelligent nanosensors enabled continuous monitoring and fine-tuning of NP injection strategies within a highly heterogeneous sandstone reservoir, resulting in EOR. The existence of reservoir heterogeneity presents a range of challenges when implementing NP-based EOR methods, as shown in Figure 4.1.

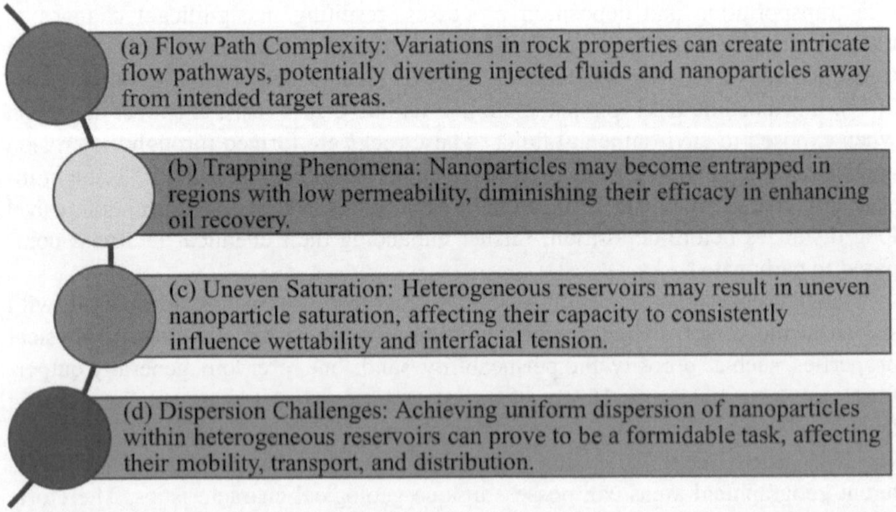

FIGURE 4.1 Range of challenges due to the existence of reservoir heterogeneity when implementing nanoparticle (NP)-assisted enhanced oil recovery (EOR) methods.

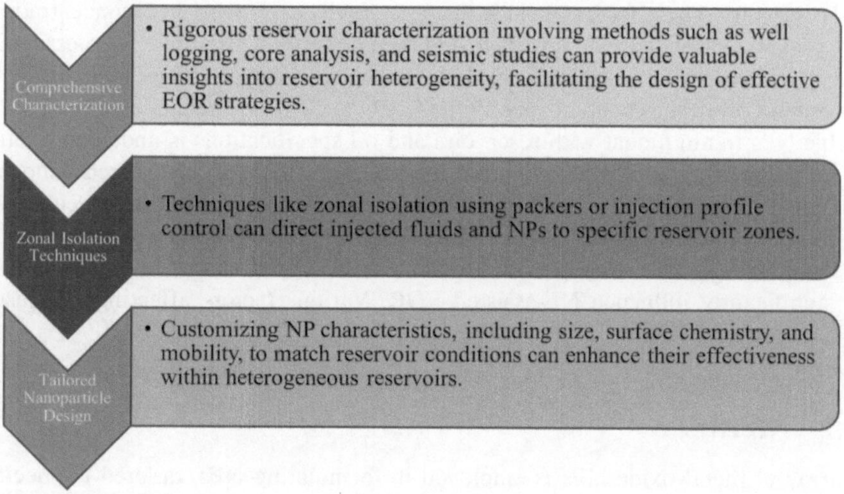

FIGURE 4.2 Various strategies can be adopted to tackle the challenges stemming from reservoir heterogeneity.

Various challenges due to heterogeneity can be neutralized by adopting different strategies. Firstly, a comprehensive reservoir characterization is essential. It involves detailed analysis and mapping of the reservoir's geology and fluid properties, allowing for a more precise understanding of its heterogeneity. Secondly, zonal isolation techniques, such as smart, well completions and selective plugging, can be employed to target specific regions within the reservoir. These techniques help control the movement of injected NPs and fluids, enhancing their distribution in reservoirs. Finally, the modified design of nanofluids (NFs) plays a crucial role. Developing NPs with properties optimized for specific reservoir conditions can significantly improve their mobility and reactivity, thereby increasing the effectiveness of EOR techniques. By combining these strategies, the challenges posed by reservoir heterogeneity can be mitigated. Figure 4.2 represents the various techniques that can help neutralize the obstacles stemming from reservoir heterogeneity.

4.3 NP SUSPENSION OR NF

NFs are a type of engineered fluid that consists of a base liquid (typically a coolant, water, or alcohol) infused with nanometer-sized particles or NPs. These NPs are usually solid materials, such as metal oxides, carbon nanotubes, or other nanomaterials (Panda & Gangawane, 2023). NFs are designed to enhance the desired properties of the base fluid, making them highly efficient for selected applications.

4.3.1 Major Factors Affecting the Performance of NFs

The effectiveness of nanotechnology in EOR critically depends on the careful selection of NFs. Several factors need careful consideration when choosing NFs for EOR in oil reservoirs. These factors include the type of NP of NF, their size, concentration, properties like density and magnetism, and their stability. Different NPs interact differently with reservoir fluids and rock formations, which in turn affect their ability to change fluid properties and enhance EOR (Ooi et al., 2015). The process of selecting suitable NPs in alignment with reservoirs and oil specifications is undeniably intricate. The challenging conditions within the reservoir, marked by elevated temperature, pressure, and salinity, pose a significant risk of compromising and weakening the structural integrity of these NPs). The salinity of both injection water and formation water along with the choice of base fluid for NF formulation are crucial factors that significantly influence NF-assisted EOR. Various factors affecting NF-based EOR are discussed below.

4.3.2 NF Types

An array of metal oxide NPs is employed in formulating NFs, tailored to specific applications in the oil and gas sectors. These NPs serve multiple critical functions, such as enhancing the thermal stability of chemicals used in EOR, altering the wettability characteristics of reservoir rocks, boosting the viscosity of injection fluids, reducing interfacial tension (IFT) in fluid-fluid systems, decreasing the mobility of displacing phases, and improving the desired properties of drilling fluids (Baird and Walz, 2007; Behera and Sangwai, 2021; Naghizadeh et al., 2020).

4.3.3 NP Selection

The integration of nanotechnology into chemical EOR presents a promising solution to address critical limitations within this field and enhance its efficiency. These limitations encompass challenges such as chemical adsorption, retention, degradation, and precipitation, often stemming from reservoir-water brines. The factors contributing to these issues are primarily the high salinity and elevated temperatures within the reservoir, the pH and composition of the reservoir fluids, and the presence of clay minerals in the reservoir rock. Recent research highlights that commonly utilized NPs in chemical EOR applications include AgNPs, silicon dioxide (SiO_2), titanium dioxide (TiO_2), aluminum oxide (Al_2O_3), graphene, and zinc oxide (ZnO). NPs, in combination with surfactant and reservoir fluid, result in the formation of emulsions that enhance conformance efficiency and stable foams that facilitate fluid diversion to low-permeability reservoir locations. NPs play a crucial role in reducing surfactant adsorption onto the reservoir rock surface. Instead, surfactants are adsorbed onto the surface of the NPs, forming surfactant-coated NPs (Gbadamosi et al., 2019). The surfactant-to-NP ratio is a critical factor in this process, with a low ratio yielding a smaller fraction of surfactant-coated NPs, while a high ratio results in a bilayer of surfactant on NP surfaces. Optimal performance is achieved when a single-chain surfactant is utilized, leading to maximal hydrophobicity and flocculation on NPs (Gbadamosi et al., 2019). Table 4.1 represents the summary of the NFs and their impact on various factors relevant for EOR.

TABLE 4.1
Summary of the NFs and their impact on various factors relevant for EOR

	Impact on different oil recovery factors					
NPs	**IFT**	**Wettability**	**Stabilizer**	**Viscosity**	**Recovery factor**	**References**
Al_2O_3	√	√	√	√	√	Ogolo et al. (2012)
SiO_2	√	√	√	–	√	Hendraningrat and Torsæter (2015)
ZnO	–	√	–	–	√	Ogolo et al. (2012)
CuO	–	√	–	√	√	Shah (2009)
Fe_2O_3/Fe_3O_4	–	√	–	√	√	Rezk and Allam (2019)
Ni_2O_3/NiO	–	√	–	–	√	Ogolo et al. (2012)
SnO	√	√	–	–	√ (Low)	
ZrO_2	√	√	–	–	Low	
CNT	–	–	√	–	√	Kadhum et al. (2015)
Clay	√	√	–	√	√	Behera and Sangwai (2020b)

Abbreviations: EOR, enhanced oil recovery; IFT, interfacial tension; NFs, nanofluids; NPs, nanoparticles.

4.3.4 Size of NPs of NFs

The size of the NPs in NFs plays a critical role in modifying the wettability of the reservoir rock surface. This alteration in wettability is predominantly influenced by a phenomenon known as disjoining pressure. Disjoining pressure, which typically arises due to the rearrangement of NPs near the thin film of liquid trapped between the solid surface and the displacing fluid, exhibits a strong size dependence. Specifically, when the NPs size decreases by nearly half, the disjoining pressure increases approximately fourfold (Kondiparty et al., 2011). In contrast, smaller NPs, owing to their higher surface energy, tend to aggregate within the pores, resulting in pore jamming, as indicated by previous research (Hendraningrat et al., 2013). Figure 4.3 visually demonstrates the impact of disjoining pressure on wettability alteration and the subsequent pore jamming caused by NP aggregation. It's important to note that NP aggregation can have a detrimental effect on the EOR process. Therefore, ensuring the stability of the NFs is of paramount importance to optimize the effectiveness of the injection fluid.

4.3.5 NP Concentration in NFs

The concentration of the NPs in the NFs plays a significant role in the formulation of the NFs as it determines the efficacy of the NFs for EOR. The objective of the EOR techniques is to mobilize the trapped oil from the pores of the rocks, and NFs contribute significantly to achieve this. Hence, the concentration of the NPs in the

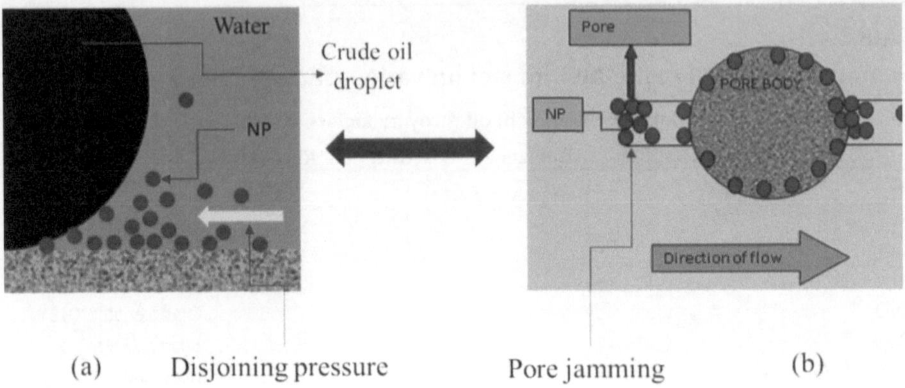

FIGURE 4.3 Rearrangement of NPs, (a) disjoining pressure influence on wettability, (b) pore jamming due to aggregation.

NFs must be carefully calibrated to achieve the desired results. Disjoining pressure increases with the rise of the NPs' concentration in the NFs (Chengara et al., 2004). The concentration of NPs also impacts the other physical properties of the NFs, such as viscosity and thermal conductivity. This, in turn, improves the NPs' ability to modify the reservoir's fluid properties, mitigate the mobility of the displacing phase, and foster greater interaction with the reservoir rock surface.

Nevertheless, there exists an optimal concentration beyond which further augmentation may lead to unfavorable consequences, such as NP aggregation and compromised fluid flow. The behavior of the NFs also changes from Newtonian fluid to non-Newtonian fluid with the rise of the NP concentration in the NFs (Harbottle et al., 2011). However, there is no such theoretical formula available to calculate the NFs' viscosity with a rise in concentration precisely. Therefore, the ideal concentration of the NPs is vital for successful EOR applications.

4.3.6 Water Salinity

The stability of the colloidal solution is influenced by the salinity of the reservoir fluids and the base fluid employed in NF formulation. Elevated salinity levels induce particle aggregation within the colloidal solution, which is normally known from a significant reduction in the absolute value of the zeta potential (McElfresh et al., 2012). Most of the particle interactions happen within the fluid rather than at the rock surface due to the specific charge of the reservoir rock. To keep particles uniformly dispersed and maintain stability, surface modification of the particles is essential. Surfactants help alter particle surfaces, making them more stable and preventing clumping. Hence, the use of surfactants is crucial for keeping the particles well dispersed and ensuring stability. High-salinity conditions enhance surfactant adsorption on the rock's surface due to physiochemical interaction, resulting in an alteration of the rock's wettability from oil-wet to water-wet and promoting oil recovery (Kanj et al., 2020). The reduction in NP aggregation involves a specific salt concentration and surface modification of the particles. This indicates that high salinity aids in altering the rock surface wettability but promotes NP aggregation, which reduces the efficacy of the injection of fluids.

4.3.7 pH of the Reservoir and NFs

The pH of the reservoir has a significant impact on particle surface charge, NP size, and NF stability. Understanding the theory behind zeta potential and the isoelectric point is essential to grasp how pH influences these properties. If the absolute zeta potential value is below 10 mV, particles are deemed highly unstable, between 10 and 20 mV they are relatively stable, between 20 and 30 mV moderately stable, and above 30 mV extremely stable (Yakasai et al., 2021). At the isoelectric point, the zeta potential, net charge, and tendency of particles to aggregate due to weak repulsion forces all equal zero. At pH levels higher or lower than the isoelectric point, electrostatic repulsion becomes stronger, possibly leading to changes in particle size (Yakasai et al., 2021). However, both bare NPs and NPs with modified surfaces can contribute to aggregation depending on their size. Hence, it's crucial to consider the NFs' pH balance in relation to the reservoir's pH, as well as the isoelectric point of both modified and bare NPs in relation to their size. These are fundamental factors for predicting NF behavior in oil reservoirs.

4.3.8 Reservoir Temperature

Temperature impacts both reservoir fluids and injection fluids; hence, the effect on oil recovery is not clearly understood. However, with an increase in the temperature, the IFT of the oil-water system reduces, which is a favorable condition for EOR. In contrast, the wettability of the rock surface remains unaltered when the temperature increases from 25°C to 50°C, regardless of substrate type, brine type, and salt concentration (Lu et al., 2017). However, Al-Anssari et al. (2017) reported a decrease in the water contact angle with increasing temperature in the presence of zirconium oxide NFs, indicating a strong influence of NPs at higher temperatures. Moreover, it has been documented that as the temperature increases, the absolute value of the zeta potential for injection fluids containing various particles, including NPs, surfactants, and ions, decreases. This observation implies that injection fluids, which were initially stable, tend to become unstable at elevated temperatures (Hendraningrat et al., 2013). This change is unfavorable for the oil recovery process. This discussion suggested that temperature has a significant impact on reservoir fluid properties.

4.3.9 Other Factors Impacting NF Application in EOR

Eco-friendly technology represents a crucial factor in environmental protection, ensuring sustainability in production practices. It reduces harmful chemical usage, minimizes resource waste, aligns with strict environmental regulations, and boosts the oil industry's public image. Similarly, economic efficiency and effective technology are considered essential factors in ensuring sustainable production (Kumar & Gangawane, 2022a, 2022b). This section discusses some additional factors for sustainable production.

4.4 ENVIRONMENTALLY FRIENDLY TECHNOLOGY

Emerging as a novel technology in EOR, NF application is gaining attention due to their unique characteristics, including reduction of IFT, wettability alteration, and stable dispersion in injection water and aqueous polymer solution (Behera & Sangwai, 2022; Behera et al., 2022; Joshi et al., 2023). In recent times, EOR methods have garnered significant attention for their potential to recover residual oil from reservoirs that are otherwise inaccessible through secondary recovery techniques (Sun et al., 2017). NFs play a crucial role in improving the thermal stability and rheological properties of brine-based fluids (Singh et al., 2022). The literature highlights a variety of environmentally friendly NPs, such as SiO_2, clay, nitrocellulose, etc., contributing to the increased efficiency of EOR (Agi et al., 2020). With the growing global energy demand, there is an increasing focus on using environmentally friendly EOR agents in the oil and gas industry. Industrial-waste-derived NPs have shown promise due to their enhanced physico-chemical properties, making them suitable for use as an effective EOR agent. These NPs can be incorporated directly into polymer solutions or through grafting onto polymers for their application in EOR. Various industrial and agricultural wastes have been explored as promising candidates for producing NPs (Razak et al., 2022). While numerous methods for synthesizing NPs have been studied in the literature, there is an ongoing need for environmentally friendly and cost-effective processes suitable for large-scale manufacturing (Fang et al., 2020). A comprehensive research is essential to address this need and navigate the complexities of NP production. The studies of cutting-edge, environmentally friendly nano technologies for waste precursor conversion into nanosheets or NPs must be developed, which can then be applied to create highly effective EOR fluids. This approach seeks to contribute to sustainable EOR practices in the oil and gas industry.

4.5 STABILITY OF THE NFS

The stability of NFs is paramount to ensure their efficient transport through the reservoir formations. The propensity for NPs to aggregate or precipitate can undermine their efficacy. Employing appropriate surface modifications and dispersants becomes imperative to maintain NPs' stability under reservoir conditions. This instability of NPs ultimately results in a reduction in their effectiveness. Additionally, a common technical challenge encountered in these applications is the inability to precisely control the dispersion of NPs on a large scale within the reservoir and their inherent properties. In the context of chemical EOR, research indicates that the incorporation of NPs into a polymer solution can significantly improve the stability, performance characteristics, durability under adverse conditions, and thermal properties of the injection fluid (Behera & Sangwai, 2021; Gbadamosi et al., 2018). NPs engage with polymer EOR through various fundamental mechanisms. Collectively, these mechanisms play a pivotal role in shaping the behavior and characteristics of NP polymers for EOR applications. In the context of "physical adsorption," NPs can adhere to polymer surfaces through physical interactions like van der Waals forces or electrostatic attraction. In the process of "steric stabilization," polymers act as a barrier, preventing NPs from clumping together and thus maintaining stability (Worthen et al., 2016).

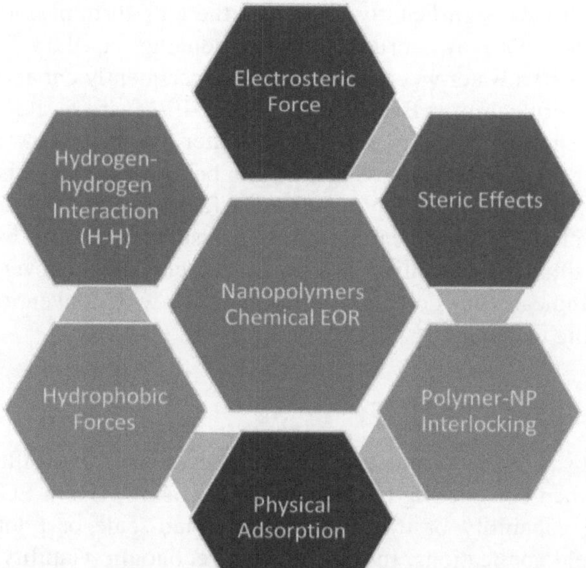

FIGURE 4.4 The mechanism illustrating how NPs interact during chemical EOR.

"Electrostatic interaction" occurs when NPs and polymers have opposite charges; they can form electrostatic bonds that contribute to stability. During the chemical bonding process, NPs and polymers can form covalent or chemical bonds, creating a durable and robust connection. For polymer-NP interlocking (entanglement), polymers can entwine with NPs, effectively trapping them within the polymer matrix. It enhances dispersion and overall stability. The mechanism illustrating how NPs interact during chemical EOR (polymers) is depicted in Figure 4.4.

4.6 COMPATIBILITY WITH RESERVOIR FLUIDS

Harmonization of NPs with reservoir fluids is essential to avoid undesirable reactions or blockages within the reservoir. The selection or modification of NPs should be guided by the goal of minimizing any interactions that might impede their transport or adversely affect fluid properties. Choosing the right NPs in line with well and oil specifications is inherently complex. In specific scenarios, NPs may adhere to the surface of the reservoir rock, forming either a monolayer or a multilayer (Behera & Sangwai, 2020a). The presence of such layers can adversely affect rock properties, including porosity and permeability.

Additionally, the harsh reservoir conditions, characterized by elevated temperature, pressure, and salinity, pose a significant risk of deteriorating and damaging the structural integrity of NPs, ultimately diminishing their efficacy. Furthermore, under reservoir conditions, the attractive forces between NPs may intensify, leading to the aggregation of NPs. This aggregation can consequently obstruct specific pores and pore throats, significantly impacting the long-term performance of NPs in EOR methods. NPs play a crucial role in reducing capillary forces, shifting the reservoir toward a water-wet condition, and consequently enhancing oil mobility and relative permeability (Behera et al., 2024; El-Masry et al., 2023). The oil extraction efficiency is typically superior in water-wet reservoirs as opposed to oil-wet reservoirs. The effectiveness of NFs can be significantly bolstered through the incorporation of NPs into the polymer or surfactant solution. This is attributed to substantial wettability modification of the rock surface induced by the presence of the polymer, ultimately resulting in an enhancement of oil recovery (Gbadamosi et al., 2018). It indicates the compatibility of the NFs with conventional water and chemical flooding technology.

4.7 COST AND SCALABILITY

The potential for field scale implementation and economic feasibility are key factors to assess when considering the practicality of using NFs in EOR. Scalability pertains to the capability of transitioning from lab-scale or pilot-level experiments to full-field applications. In the context of economic viability, conducting a cost-benefit analysis is essential for establishing whether NF-based EOR methods are feasible. Field-scale application allows for a comprehensive cost-benefit analysis, considering factors such as production rates, oil prices, operational expenses, and the project's duration. Developing and producing NPs at a large scale while maintaining cost efficiency is a significant challenge. Rigorous cost-benefit analyzes should be conducted to ascertain the viability of integrating NFs' applications into specific reservoirs.

Figure 4.5 observes the growth and anticipated worth of the worldwide nanomaterial markets up to 2030. Mainly, in the latest year, 2022, the market value for

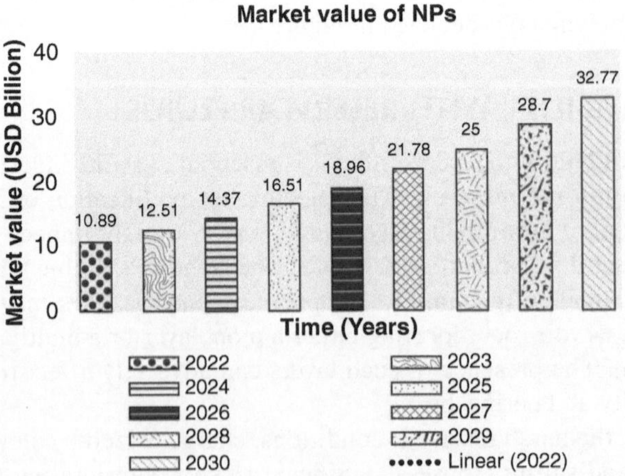

FIGURE 4.5 The figure illustrates the growth trajectory of the global nanomaterial market (starting from the year 2022 on the left), including its anticipated value up to the year 2030. Modified and reproduced from reference (https://www.researchandmarkets.com).

TABLE 4.2
Presents approximate price ranges for the most common NPs per gram, denominated in US dollars ($/25 g) (website: www.us-nano.com). Data are collected on 23rd October 2023

Nanoparticle type	Price in $/25 g, size	Price in $/120 ml, dispersion medium
Clay	46, 18 nm	158, water
Silicon dioxide (SiO_2)	45, 15–20 nm	289, ethylene glycol
Copper oxide (CuO)	45, 80 nm	338, ethylene glycol
Aluminum oxide (Al_2O_3)	45, 80 nm	280, water
Titanium dioxide (TiO_2)	35, 300 nm	380, water
Iron oxide (Fe_2O_3)	79, 30 nm	385, ethanol
Carbon nanotubes	249, short MWCTs, 20–30 nm	155, water

nanomaterials experienced a significant surge, reaching approximately USD 10.89 billion. According to specific assessments, the average yearly growth rate stands at around 14.9%. Projections indicate that by 2030, the value of nanomaterials is expected to soar by an impressive USD 32.77 billion. As global market competition increases, the cost of NPs is expected to decrease, rendering nanotechnology more accessible and affordable for the oil and gas sector.

The present cost of various nanomaterials may be relatively high, making them less cost-effective for widespread use. However, with the increasing competition and growing market size, it's expected that the prices of nanomaterials will decrease, rendering nanotechnology more affordable for various sectors, including oil and gas. Table 4.2 represents the price of multiple NPs per 25 g/120 ml in USD. The pricing of NPs is contingent upon several critical factors, including their physical dimensions, surface modifications such as polymer coatings, the nature of the dispersion medium (e.g., aqueous or alcoholic solvents), and the specific metallic elements incorporated, particularly in the case of alloy NPs.

4.8 MONITORING AND CONTROL

Efficient monitoring and control of NP movement and distribution within the reservoir are essential components of EOR operations. Advanced monitoring techniques, such as nanosensors and tracers, play a significant role in achieving this objective of maximizing oil recovery. Nanosensors are incredibly small-scale sensors capable of detecting and reporting on the presence and behavior of NPs in real time. Tracers, on the other hand, are substances added to the injected fluids that can be tracked and monitored to provide insights into the movement and distribution of NPs. By implementing nanosensors and tracers, technical experts gain the ability to observe closely how NPs disperse and interact within the reservoir. This real-time data is indispensable for making on-the-fly adjustments to injection strategies. For example, suppose the data indicates that NPs are not spreading as desired or are congregating in certain areas; in that case, experts can modify injection rates, pressures, or fluid compositions to optimize the distribution.

4.9 CONCLUSION

Nanotechnology offers a remarkable opportunity for EOR through the manipulation of reservoir properties at the microscale. Its potential to alter the properties of reservoir rock-oil-water systems holds the promise of EOR, enabling precise tuning of reservoir properties at the microscale. The following conclusions are drawn from this study:

- The stability of the NFs in harsh reservoir conditions is a significant issue that may lead to formation damage and should be addressed.
- Hybrid injection fluids, mainly nano-assisted chemical EOR, cause operational problems associated with injectivity (Franco et al., 2021).
- Thermal-based EOR has many opportunities to increase productivity with the integration of nanotechnology that involves nanocatalysts.
- The cost of NPs and optimization of the injection fluid could be major issues of interest for oil production companies for field testing.
- Extensive field studies are needed to gain insights into various issues. Therefore, the active engagement of oil and gas production companies is essential to assessing the practical applicability of nanotechnology in the industry.
- Overcoming these challenges and realizing the full potential of nanotechnology in the petroleum sector requires a holistic approach characterized by interdisciplinary collaboration and sustained research efforts.

REFERENCES

Agi, A., Junin, R., Abdullah, M.O., Jaafar, M.Z., Arsad, A., Wan Sulaiman, W.R., Norddin, M.N.A. M., Abdurrahman, M., Abbas, A., Gbadamosi, A., & Azli, N.B., 2020. Application of polymeric nanofluid in enhancing oil recovery at reservoir condition. *J. Pet. Sci. Eng.* 194, 107476. https://doi.org/10.1016/j.petrol.2020.107476.

Al-Anssari, S., Wang, S., Barifcani, A., Lebedev, M., Iglauer, S., 2017. Effect of temperature and SiO2 nanoparticle size on wettability alteration of oil-wet calcite. *Fuel* 206, 34–42. https://doi.org/10.1016/j.fuel.2017.05.077.

Amanam, U.U., Kovscek, A.R., 2017. Analysis of the effects of copper nanoparticles on in-situ combustion of extra heavy-crude oil. *J. Pet. Sci. Eng.* 152, 406–415. https://doi.org/10.1016/j.petrol.2017.02.018.

Baird, J.C., Walz, J.Y., 2007. The effects of added nanoparticles on aqueous kaolinite suspensions. II. Rheological effects. *J. Colloid Interface Sci.* 306, 411–420. https://doi.org/10.1016/j.jcis.2006.10.066.

Barwis, J.H., McPherson, J.G., Studlick, J.R. eds., 2012. *Sandstone Petroleum Reservoirs*. Springer Science & Business Media.

Behera, U.S., Kumar, G., Sangwai, J.S., 2022. Pore-scale investigation and performance evaluation of smart lowsal flooding for enhanced oil recovery from matured reservoirs using a lab-on-a-chip. *Energy Fuels* 36, 8115–8127. https://doi.org/10.1021/acs.energyfuels.2c01009.

Behera, U.S., Poddar, S., Deshmukh, M.P., Sangwai, J.S., Byun, H.S., 2024. Comprehensive review on the role of nanoparticles and nanofluids in chemical enhanced oil recovery: Interfacial phenomenon, compatibility, scalability, and economic viability. *Energy Fuels* 38(15):13760–13795. https://doi.org/10.1021/acs.energyfuels.4c02248.

Behera, U.S., Sangwai, J.S., 2020a. Interaction of nanoparticles with reservoir fluids and rocks for enhanced oil recovery. In *Green Energy and Technology* (pp. 299–328). Springer International Publishing. https://doi.org/10.1007/978-3-030-33774-2_13.

Behera, U.S., Sangwai, J.S., 2020b. Nanofluids of kaolinite and silica in low saline seawater (lowsal) with and without surfactant: Interfacial tension and wettability alteration of oil–water–rock system for low salinity-enhanced oil recovery. *Ind. Eng. Chem. Res.* 60(1), 291–313. https://doi.org/10.1021/acs.iecr.0c04853.

Behera, U.S., Sangwai, J.S., 2021. Nanofluids of silica nanoparticles in low salinity water with surfactant and polymer (SMART LowSal) for enhanced oil recovery. *J. Mol. Liq.* 342, 117388. https://doi.org/10.1016/j.molliq.2021.117388.

Behera, U.S., Sangwai, J.S., 2022. Silica nanofluid in low salinity seawater containing surfactant and polymer: Oil recovery efficiency, wettability alteration and adsorption studies. *J Pet. Sci. Eng.* 211, 110148. https://doi.org/10.1016/j.petrol.2022.110148.

Chengara, A., Nikolov, A.D., Wasan, D.T., Trokhymchuk, A., Henderson, D., 2004. Spreading of nanofluids driven by the structural disjoining pressure gradient. *J. Colloid Interface Sci.* 280, 192–201. https://doi.org/10.1016/j.jcis.2004.07.005.

El-Masry, J.F., Bou-Hamdan, K.F., Abbas, A.H., Martyushev, D.A., 2023. A comprehensive review on utilizing nanomaterials in enhanced oil recovery applications. *Energies* 16, 691. https://doi.org/10.3390/en16020691.

Esfe, M.H., Esfandeh, S., Hosseinizadeh, E., 2020. Nanofluid flooding in a randomized heterogeneous porous media and investigating the effect of capillary pressure and diffusion on oil recovery factor. *J. Mol. Liq.* 320, 113646. https://doi.org/10.1016/j.molliq.2020.113646.

Fang, Z., Gao, Y., Bolan, N., Shaheen, S. M., Xu, S., Wu, X., Wang, H., 2020. Conversion of biological solid waste to graphene-containing biochar for water remediation: A critical review. *Chem. Eng. J.* 390, 124611. https://doi.org/10.1016/j.cej.2020.124611.

Franco, C.A., Franco, C.A., Zabala, R.D., Bahamón, I., Forero, A., Cortés, F.B., 2021. Field applications of nanotechnology in the oil and gas industry: Recent advances and perspectives. *Energy Fuels* 3, 19266–19287. https://doi.org/10.1021/acs.energyfuels.1c02614.

Gbadamosi, A., Junin, R., Manan, M., Agi, A., Oseh, J., 2019. Nanotechnology application in chemical enhanced oil recovery:current opinion and recent advances. In A. Samsuri (Ed.), *Enhanced Oil Recovery Processes: New Technologies*. Intechopen.

Gbadamosi, A.O., Junin, R., Manan, M.A., Yekeen, N., Agi, A., Oseh, J.O., 2018. Recent advances and prospects in polymeric nanofluids application for enhanced oil recovery. *J. Ind. Eng. Chem.* 66, 1–19. https://doi.org/10.1016/j.jiec.2018.05.020.

Guo, K., Li, H., Yu, Z., 2015. Metallic nanoparticles for enhanced heavy oil recovery: Promises and challenges. *Energy Procedia* 75, 2068–2073. https://doi.org/10.1016/j.egypro.2015.07.294.

Harbottle, D., Bueno, P., Isaksson, R., Kretzschmar, I., 2011. Coalescence of particle-laden drops with a planar oil-water interface. *J. Colloid Interface Sci.* 362, 235–241. https://doi.org/10.1016/j.jcis.2011.05.077.

Hendraningrat, L., Torsæter, O., 2015. Metal oxide-based nanoparticles: Revealing their potential to enhance oil recovery in different wettability systems. *Appl. Nanosci.* 5,181–199. https://doi.org/10.1007/s13204-014-0305-6.

Hendraningrat, L., Li, S., Torsæter, O., 2013. Enhancing oil recovery of low-permeability Berea sandstone through optimized nanofluids concentration. In *SPE Enhanced Oil Recovery Conference*, 2–4 July, Kuala Lumpur, Malaysia. https://doi.org/10.2118/165283-ms.

Jarrahian, K., Boak, L.S., Graham, A.J., Singleton, M.A., Sorbie, K.S., 2019. Experimental investigation of the interaction between a phosphate ester scale inhibitor and carbonate rocks for application in squeeze treatments. *Energy Fuels* 33, 4089–4103. https://doi.org/10.1021/acs.energyfuels.9b00382.

Joshi, D., Maurya, N.K., Mandal, A., 2023. Experimental studies on effectiveness of graphene oxide nanosheets dispersion in water/aqueous PHPA for enhanced oil recovery. *J. Mol. Liq.* 387, 122728. https://doi.org/10.1016/j.molliq.2023.122728.

Kadhum, M.J., Swatske, D.P., Chen, C., Resasco, D.E., Harwell, J.H., Shiau, B., 2015. Propagation of carbon nanotube hybrids through porous media for advancing oil field technology. In *SPE International Symposium on Oilfield Chemistry*, 13–15 April, The Woodlands, TX.

Kanj, M., Sakthivel, S., Giannelis, E., 2020. Wettability alteration in carbonate reservoirs by carbon nanofluids. *Colloids Surfaces A Physicochem. Eng. Asp.* 598, 12481901–12481914. https://doi.org/10.1016/j.colsurfa.2020.124819.

Kondiparty, K., Nikolov, A., Wu, S., Wasan, D., 2011. Wetting and spreading of nanofluids on solid surfaces driven by the structural disjoining pressure: Statics analysis and experiments. *Langmuir* 27, 3324–3335. https://doi.org/10.1021/la104204b.

Kumar, A., Gangawane, K.M., 2022a. Effect of precipitating agents on the magnetic and structural properties of the synthesized ferrimagnetic nanoparticles by co-precipitation method. *Powder Technol.* 401, 117298. https://doi.org/10.1016/j.powtec.2022.117298.

Kumar, A., Gangawane, K.M., 2022b. Synthesis and effect on the surface morphology & magnetic properties of ferrimagnetic nanoparticles by different wet chemical synthesis methods. *Powder Technol.* 410(June), 117867. https://doi.org/10.1016/j.powtec.2022.117867.

Lu, Y., Najafabadi, N.F., Firoozabadi, A., 2017. Effect of temperature on wettability of oil/brine/rock systems. *Energy Fuels*, 31, 4989–4995.

McElfresh, P., Holcomb, D., Ector, D., 2012. Application of nanofluid technology to improve recovery in oil and gas wells. In *SPE International Oilfield Nanotechnology Conference and Exhibition*, 12–14 June, Noordwijk, The Netherlands. https://doi.org/10.2118/154827-ms.

Medina, O.E., Olmos, C., Lopera, S.H., Cortés, F.B., Franco, C.A., 2019. Nanotechnology applied to thermal enhanced oil recovery processes: A review. *Energies* 12(24), 4671. https://doi.org/10.3390/en12244671.

Nanomaterials market size, share & trends analysis report by product (gold, silver, iron, copper, platinum, titanium, nickel, aluminum oxide), by application (aerospace, automotive, medical). *Res. Mark.* 2022, 1–148. https://www.researchandmarkets.com/reports/5505322/nanomaterials-market-size-share-and-trends.

Naghizadeh, A., Azin, R., Osfouri, S., Fatehi, R., 2020. Wettability alteration of calcite and dolomite carbonates using silica nanoparticles coated with fluorine groups, *J. Pet. Sci. Eng.* 188, 10691501–10691514. https://doi.org/10.1016/j.petrol.2020.106915.

Ogolo, N.A., Olafuyi, O.A., Onyekonwu, M.O., 2012. Enhanced oil recovery using nanoparticles. In *SPE Saudi Arabia Section Technical Symposium and Exhibition*, 8–11 April, Al-Khobar, Saudi Arabia. https://doi.org/10.2118/160847-ms.

Ooi, F., Duchene, J.S., Qiu, J., Graham, J.O., Engelhard, M.H., Cao, G., Gai, Z., Wei, W.D., 2015. A facile solvothermal synthesis of octahedral Fe3O4 nanoparticles. *Small* 11(22), 2649–2653. https://doi.org/10.1002/SMLL.201401954.

Panda, D., Gangawane, K.M., 2023. Hybrid NiFe2O4-NiCO4O4 nanoparticles-based eutectic phase change materials for enhancement of thermal efficiency of pin-fin heat sink arrangement. *J. Energy Storage* 60(November 2022), 106644.

Razak, N.A.A., Nur, H.O., Shayuti, M.S.M., Aidah, J., Napisah, S., Woei, J.L., 2022. Agricultural and industrial waste-derived mesoporous silica nanoparticles: A review on chemical synthesis route. *J. Environ. Chem. Eng.* 10, 107322. https://doi.org/10.1016/j.jece.2022.107322.

Rezk, M.Y., Allam, N.K., 2019. Impact of nanotechnology on enhanced oil recovery: A mini-review. *Ind. Eng. Chem. Res.* 58, 16287–16295. https://doi.org/10.1021/acs.iecr.9b03693.

Shah, R.D., 2009. Application of nanoparticle saturated injectant gases for EOR of heavy oils. In *SPE Annual Technical Conference and Exhibition*, 4–7 October, New Orleans, LA. https://doi.org/10.2118/129539-STU.

Singh, R., Sharma, R., Rao, G.R., 2022. Development of a novel high density completion fluid with CuO nanoparticle as promising additive to improve the rheological properties. *Results Eng.*, 15, 100520. https://doi.org/10.1016/j.rineng.2022.100520.

Sun, X., Zhang, Y., Chen, G., Gai, Z., 2017. Application of nanoparticles in enhanced oil recovery: A critical review of recent progress. *Energies* 10, 345. https://doi.org/10.3390/en10030345.

Worthen, A.J., Tran, V., Cornell, K.A., Truskett, T.M., Johnston, K.P., 2016. Steric stabilization of nanoparticles with grafted low molecular weight ligands in highly concentrated brines including divalent ions. *Soft Matter* 12, 2025–2039. https://doi.org/10.1039/C5SM02787J.

Yakasai, F., Jaafar, M.Z., Bandyopadhyay, S., Agi, A., 2021. Current developments and future outlook in nanofluid flooding: A comprehensive review of various parameters influencing oil recovery mechanisms. *J. Ind. Eng. Chem.* 93, 138–162. https://doi.org/10.1016/j.jiec.2020.10.017.

Zabala, R., Mora, E., Botero, O.F., Cespedes, C., Guarin, L., Franco, C.A., Cortes, F.B., Patino, J.E., Ospina, N., 2014, January. Nanotechnology for asphaltenes inhibition in Cupiagua South Wells. In *IPTC 2014: International Petroleum Technology Conference*. European Association of Geoscientists & Engineers. https://doi.org/10.3997/2214-4609-pdb.395.IPTC-17344-MS.

5 Nanoparticle-Assisted Foam for Enhanced Oil Recovery

N.S.V. Dinesh and P. Sivasankar

5.1 INTRODUCTION

Globally, the rise in CO_2 emissions has led to an increase in the Earth's overall temperature, resulting in severe consequences for Earth's environment and climate (Savaresi, 2016). Consequently, the Paris Agreement was established between nations with the goal of reducing CO_2 emissions by 45% by 2030 and achieving net zero CO_2 emissions by 2050 (Hong et al., 2023). However, the International Energy Agency (IEA) in 2022 reported that fossil fuels will remain a predominant source of energy, accounting for approximately 60% of the global energy demand until at least the year 2060 (Peter, 2018). Fossil fuels like coal, gas, and oil are primarily utilized to fulfil this energy requirement, contributing significantly to CO_2 emissions (Gielen et al., 2019; Nikhil et al., 2024; Vishal et al., 2021). Approximately 25%–40% of the oil that is initially present may be extracted by primary and secondary recovery methods (Vishnyakov et al., 2020). Secondary recovery methods, like water flooding, involve maintaining the reservoir pressure by injecting water. However, over time, the efficiency of water flooding decreases due to the increase in water cuts at production wells (Alvarado & Manrique, 2010; Pal & Sen, 2023b). Enhanced Oil Recovery (EOR) techniques such as thermal EOR, gas EOR and chemical EOR are used (Asghari et al., 2006; Kumar et al., 2022). Among these methods, CO_2 EOR techniques have emerged as the best choice for reducing CO_2 emissions and simultaneously increasing oil recovery (Kumar & Gangawane, 2023a; Saraf & Bera, 2023). The CO_2 EOR is generally implemented either in the form of gas or foam (Dai et al., 2014). However, while injecting the CO_2 in gas form, the efficiency of oil recovery and CO_2 sequestration processes decreases due to issues like gravity override, gas channelling and viscous fingering (Telmadarreie & Trivedi, 2020). Sæle et al. (2022) observed that injecting surfactant slugs alternating with the CO_2 gas slugs had resulted in the generation of foam, which increased the CO_2 gas viscosity and subsequently enhanced the oil recovery by mitigating the gas fingering (Kumar & Gangawane, 2023b; Panda & Gangawane, 2023a; Sæle et al., 2022).

It is noteworthy that recent developments in the field of material science have presented the possibility of augmenting extraordinary regeneration (EOR) processes with activated-carbon-based materials obtained from natural sources, including

 DOI: 10.1201/9781032650685-5

jackfruit rind and Pterospermum acerifolium fruits. Research has indicated that activated carbon produced from these precursors can be modified to suit particular purposes, including the elimination of dyes, via procedures that incorporate phosphoric acid (H_3PO_4), potassium hydroxide (KOH) activation, and zinc chloride ($ZnCl_2$) impregnation. The incorporation of these sophisticated materials into EOR methodologies may present innovative avenues for EOR effectiveness while mitigating environmental repercussions (Kumar et al., 2023a; Minz & Mishra, 2023; Pal & Sen, 2024; Sahoo & Mishra, 2023; Santhosh & Sarkar, 2022).

Additionally, Zhang et al. (2015) observed that there is an improvement in oil recovery and sequestration of CO_2 while using it in foam form (Dinesh & Sivasankar, 2024). However, under the harsh conditions of the reservoir, i.e., in the presence of oil, high formation water salinity, pressure and temperature, the foam stability decreases gradually over time (Kalam et al., 2019). This reduction in foam stability causes coalescence and coarsening of gas bubbles and subsequently results in gas breakthroughs (Ab Rasid et al., 2022). To overcome these issues of foam instability in the reservoir, different types of nanoparticles are added to the surfactant solution (Kumar & Gangawane, 2022a; Yekeen et al., 2018a). However, the extent to which nanoparticles increase the stability depends on reservoir properties (i.e., salinity, temperature and the type of crude oil present inside the reservoir) and nanoparticle properties (particle size, concentration of nanoparticles and type of nanoparticles). The main objectives of this chapter are to summarize and review the key findings from the earlier studies and to finally conclude the effect of each parameter on the foam stability.

5.2 FOAM GENERATION MECHANISMS INSIDE THE POROUS MEDIA

Many studies have reported that the main mechanisms that are responsible for in-situ foam generation are (i) lamellae leave-behind mechanism, (ii) snap-off and (iii) lamellae division (Alcorn et al., 2020; Yekeen et al., 2018b). Figure 5.1(a) refers to the generation of foam by the lamella leave-behind mechanism. This mechanism occurs when gas invades the liquid-saturated porous media, where it may not displace the total liquid but rather leave behind the liquid lamella, as shown in Figure 5.1(a) (Hematpur et al., 2018). Figure 5.1(b) refers to the generation of foam by snap-off mechanism. During this mechanism, the giant gas bubble moves through a pore throat and snaps off due to higher capillary pressure at the pore throat (Gugl, 2020). Figure 5.1(c) refers to foam generation by lamella division. This lamella division takes place only when the bubble approaches a "branch point" within the porous media. During this process, the lamella splits and forms two individual bubbles (Al-Shargabi et al., 2022; Farid Ibrahim & Nasr-El-Din, 2020; Kumar & Gangawane, 2022b). It is feasible only if the size of the existing foam bubble is greater than the size of the pore and there is enough pressure to drive the lamellae inside the pore throat. For snap-off to happen, the pressure in front of the gas-liquid interface should be less than the capillary pressure at the pore throat.

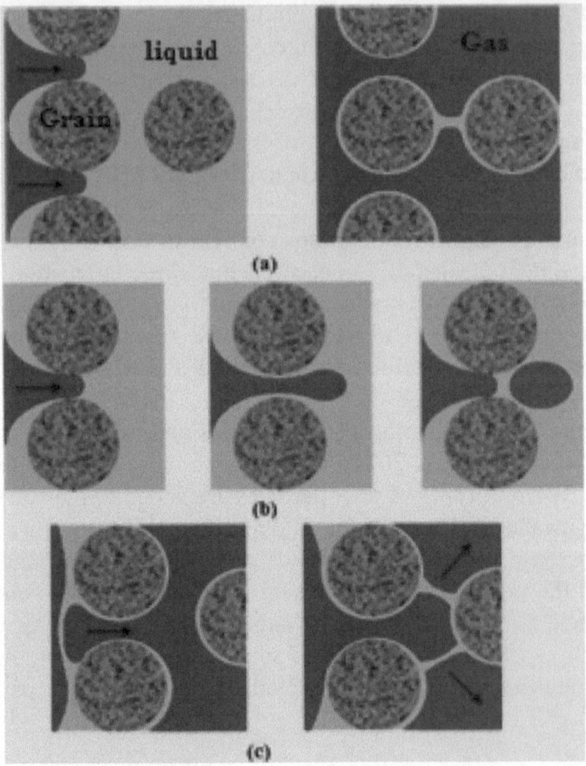

FIGURE 5.1 Mechanisms responsible for in-situ foam generation: (a) lamella left behind; (b) bubble snapped off; (c) division of lamella. This figure was reproduced with permission from Farid Ibrahim and Nasr-El-Din (2020). Copyright@Elsevier, 2018.

5.2.1 Foam Destabilization Mechanisms

The foam structure is thermodynamically unstable and changes over time. This happens due to (i) liquid drainage in foam, (ii) bubble coalescence and (iii) the coarsening of bubbles (Figure 5.2). In Figure 5.2(a), the line indicates the liquid drainage through networks of plateau borders due to capillary forces and gravitational forces (Fameau & Salonen, 2014). This liquid drainage is responsible for the thinning of liquid film. Subsequently, the film between the two bubbles will become unstable and break, resulting in the merging of the two bubbles. This process of merging two bubbles is known as the coalescence of bubbles; refer to Figure 5.2(b) (Yekeen et al., 2018a). Figure 5.2(c) refers to the coarsening of bubbles, in which gas from small bubbles diffuses into more giant bubbles. It happens because of the pressure difference among the bubbles (small bubble has higher pressure). This gas transfer from a smaller bubble to a more considerable bubble results in a further increase in the size of the giant bubble and the disappearance of the smaller bubble. These mechanisms are highly dependent on each other; for instance,

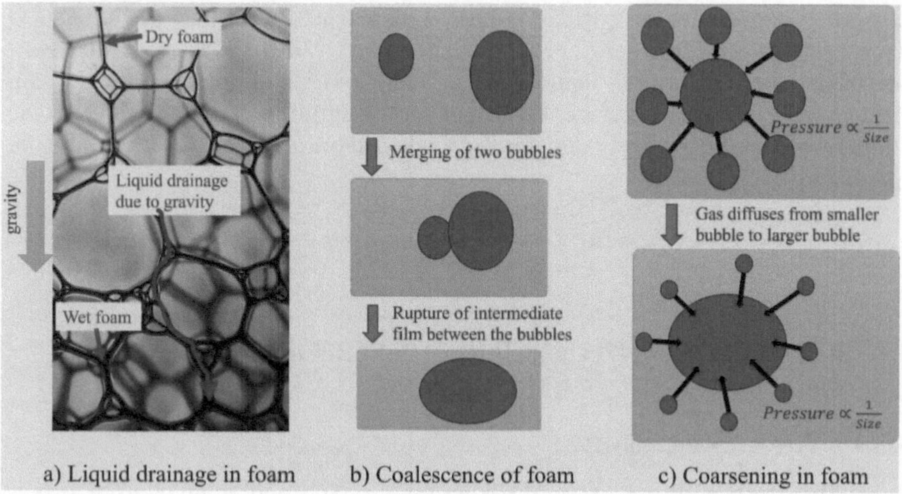

FIGURE 5.2 Foam destabilization mechanisms in porous media: (a) liquid drainage; (b) bubble coalescence; (c) bubble coarsening due to gas diffusion.

the liquid drainage in the foam reduces the lamella thickness, which can result in the breakdown of the liquid film (Carrier & Colin, 2003; Panda et al., 2023; Panda & Gangawane, 2023b).

5.2.2 Foam Stabilization Using Nanoparticles

Earlier, Horozov (2008) discussed three main reasons responsible for foam destabilization phenomena, which are as follows: (i) liquid drainage, (ii) bubble coalescence and (iii) bubble coarsening due to gas diffusion. The first reason is because of the lower viscosity of surfactant, which prompts drainage of liquid from the foam, causing the foam to dry out. The second reason is that less threshold energy is needed to separate the surfactant molecules from the liquid-gas interface. Given this threshold energy, the molecules of surfactant can be easily separated from the interface, which leads to the instability of foam (Binks & Lumsdon, 2000). The third reason is because of the high capillary pressure difference that is present between the liquid-gas interface. The bubble coalescence happens more quickly if the difference in capillary pressure is high. Adding the nanoparticles to the surfactant foam will weaken these three destabilization phenomena and make the foam structure to be more stable. The long-term stability of foam was obtained because of the adsorption of nanoparticles at the liquid-gas interface. However, the desorption of surfactants is nearly irreversible because it requires a high threshold energy (i.e., particle detachment energy), which is not feasible at reservoir conditions (Binks & Lumsdon, 2000; Panda & Gangawane, 2023c). Eq. (5.1) represents the threshold energy (particle detachment energy) that is required to separate the nanoparticles from the liquid-gas interface; W_r denotes it. In Eq. (5.1), the term R represents the particle radius, θ represents the contact angle and $\gamma_{\alpha\beta}$ represents the interface tension (Gbadamosi et al., 2019). Moreover, the addition of nanoparticles in the liquid will act as a barrier and prevent gas diffusion through the liquid-gas interface. Furthermore, the addition of nanoparticles to the liquid increases the viscosity. It helps reduce liquid drainage, coalescence, and the coarsening of bubbles.

$$W_r = \pi R^2 \gamma_{\alpha\beta} (1 \pm \cos\theta)^2 f \tag{5.1}$$

5.3 KEY FACTORS INFLUENCING THE PERFORMANCE OF NANOPARTICLE-BASED FOAMS

5.3.1 Type of Nanoparticle

Researchers have performed several studies on the stability of foam by using different types of nanoparticles. These studies include nanoparticles such as metal oxide nanoparticles, carbon nanotubes, fly ash nanoparticles and silicon dioxide (SiO_2) nanoparticles (Phong et al., 2020; Ramanathan et al., 2021; Rossen & Wang, 1999; Wang et al., 2020; Yekeen et al., 2019). Table 5.1 summarizes the effect of nanoparticle type on the stability of the foam.

In general, all the nanoparticles have shown improved foam stability in comparison to surfactant-based foam. This increased stability is influenced by the nanoparticle properties, which include, size, density, surface charges, wettability and shape, which play a crucial role in the stabilization of foam (Rafati et al., 2016).

5.3.2 Nanoparticle Size

The size of the nanoparticle greatly influences the stability of nanoparticle-surfactant foam. Moreover, Kim et al. (2016) also observed from core flooding and sand pack experiments that with a decrease in nanoparticle size, the apparent foam viscosity increases. Smaller-size nanoparticles are easily adsorbed on the liquid-gas interface due to higher concentration and greater diffusivity (Kim et al., 2016). Table 5.2 summarizes the effect of nanoparticle size on the stability of foam.

5.3.3 Concentration of Nanoparticles

The majority of studies have reported that the increase in nanoparticle concentration results in more foam stability (Karakashev et al., 2011; Yang et al., 2017). Besides the enhancement in the strength of foam and oil recovery, the apparent viscosity of foam and pressure drop also increased with an increment in the concentration of nanoparticles (AlYousef et al., 2018; Eide et al., 2018; Kumar & Mandal, 2017; Yekeen et al., 2017). The surfactant adsorption rate also increased with an increase in the concentration of nanoparticles; as a consequence, zeta potential was reduced (Yang et al., 2017). Table 5.3 summarizes the effect of nanoparticle concentration on the stability of foam.

TABLE 5.1
Influence of nanoparticle type on foam stability

Nanoparticle type	Surfactant used	Authors	Key findings
Fly ash	SDS, AOS, alkylphenol ethoxylates	Wang et al. (2020) and Rossen and Wang (1999)	1. With the increase in surfactant concentration, adsorption of surfactant on nanoparticles also increases. As a result, surface tension was reduced, and the foam stability increased. 2. The viscoelastic nature of liquid film increases with the increase in nanoparticle adsorption. An increase in liquid viscosity results in delayed fluid drainage from the foam network and decreases gas diffusion, resulting in increased foam stability.
TiO_2, ZnO, Al_2O_3, CuO, ZrO_2, SiO_2	SDBS, CTAB, SDS	Yekeen et al. (2019)	1. Foam stability was increased due to electrostatic attraction between surfactant head groups and nanoparticles. Moreover, the IFT between water and n-decane also decreased.
MWCNT, SiO_2	AOS	Ramanathan and Nasr-El-Din (2020)	1. SiO_2 nanoparticles and MWCNT correspondingly increased the foam stability of AOS by 2.9 times and 4.6 times, respectively. 2. Upon adding the salt, MWCNT and SiO_2 nanoparticles started to aggregate due to an increment in Van der Waals forces. At a given salinity, MWCNT was comparatively more stable than the SiO_2 nanoparticle. 3. MWCNT/AOS foam showed better performance than SiO_2/AOS because of rod-like shape, size, partial hydrophobicity and zeta potential.
Fabricated fly ash, fly ash	MFOMAX	Phong et al. (2020)	1. Adsorption of nanoparticles at the interface of liquid–gas system resulted in stability of foam by limiting the contact of fluids, therefore reducing the bubble coalescence, bubble coarsening and liquid drainage. 2. The tendency of aggregation increases with increase in nanoparticle size, which results in increased liquid drainage and decreased foam stability.

Abbreviations: Al_2O_3, aluminium oxide; CuO, copper oxide; IFT, interfacial tension; MWCNT, multi-walled carbon nanotube; SiO_2, silicon dioxide; TiO_2, titanium dioxide; ZnO, zinc oxide; ZrO_2, zirconium dioxide.

TABLE 5.2
Influence of size of nanoparticle on foam stability

Nanoparticle type (nanoparticle size)	Surfactant	Authors	Key findings
SiO_2 A300 (7 nm), SiO_2 R816 (12 nm), ZnO (<100 nm), TiO_2 (25 nm)	MFOMAX	Razali et al. (2018)	1. MFOMAX attained maximum foam stability while using SiO_2 A300 nanoparticles due to small particle size when compared with other nanoparticles.
SiO_2 (20 nm, 100 nm, 500 nm)	CTAB, SDS	Hu et al. (2018)	1. Among the three sizes of SiO_2 nanoparticles (i.e., 20 nm, 100 nm and 500 nm), SiO_2 nanoparticles of 20 nm size exhibit the maximum foam stability. 2. A more compact solid-like layer was created by smaller SiO_2 nanoparticles, which might be effective in preventing coalescence and coarsening of bubbles.
Nano clay (356–419 nm), fumed SiO_2 T30 (179–223 nm)	SDS, LAPB, AOS	Guo and Aryana (2016)	1. The foam stability of nano clay is slightly less when compared to the foam stability of SiO_2. This might be because of the smaller size of SiO_2 particles.
$CaCO_3$ (12,000 nm), $CaSO_4$ (7,600 nm), $BaSO_4$ (1,300 nm), $SrSO_4$ (64,000 nm), Fe_2O_3 (131,000)	AOS	Rafati et al. (2016)	1. $SrSO_4$, having a larger particle size, generated comparatively less-stable foam than $CaCO_3$ and $BaSO_4$. 2. As the particle size increases, it tends to form monolayer bridges. This results in more disjoining pressure at lamellae and correspondingly destabilizes the foam.
SiO_2 (5 nm, 25 nm, 80 nm, 7,600 nm, 13,000 nm, 64,000 nm, 131,000 nm)		Kim et al. (2016)	1. The size of the nanoparticle dramatically influences the stability and apparent viscosity of foam. 2. The smaller nanoparticles have a higher number concentration and greater diffusivity, which increases the adsorption tendency at the brine–oil interface.

Abbreviations: $BaSO_4$, barium sulphate; $CaCO_3$, calcium carbonate; $CaSO_4$, calcium sulphate; Fe_2O_3, ferric oxide; $SrSO_4$, strontium sulphate.

5.3.4 Salinity

It was observed that with an increase in salinity, the performance of foam declines. Meanwhile, Li et al. (2019) reported that the foam stability increased considerably with an increase in salinity between 2 wt% and 10 wt%. This increment or decrement in foam stability is because of the ionic interactions between nanoparticles, surfactants and salts. Some studies reported that the zeta potential reduces with an increase in salinity, which results in the aggregation of particles. The key findings associated with the influence of salinity on the stability of foam are represented in Table 5.4.

TABLE 5.3
Influence of concentration of nanoparticles on the stability of foam

Nanoparticle type	Surfactant	Authors	Key findings
High-purity SiO_2, sepiolite NPs	SDS	Karakashev et al. (2011)	1. With the increase in NP concentration from 1 wt% to 4 wt%, the foamability increased slightly. 2. The apparent viscosity of foam and concentration of NPs are directly proportional to each other.
Alumina solution	SDS	Yekeen et al. (2018a)	1. Higher concentrations of SDS/Alumina dispersions exhibited lower foamability. 2. The optimum concentration ratio of SDS/ Alumina NPs resulted in maximum foam stability.
$CaCO_3$, BN, Al_2O_3, ZrO_2, SiO_2	SDS, CTAB, Polysorbate 80	Kumar and Mandal (2017)	1. Better foam stability was achieved because of the accumulation of surface-active species at the interface of liquid and gas.
NexSil 20	Alcohol ethoxylate	AlYousef et al. (2017)	1. The addition of excessive NPs reduces the foam stability and maximum capillary pressure of coalescence, which results in the formation of large and dense aggregates. 2. Increasing the concentration of NPs does not necessarily increase the stability of foam.
SiO_2, modified Al_2O_3	SDS	Yekeen et al. (2018a)	1. An increase in concentration of NPs from 0.05 wt% to 1 wt% resulted in better foam stability. 2. Further increments in NPs above 1 wt% declined the stability of the foam. The reason for this phenomenon could be because of excessive NP agglomeration in the bulk structure and foam interface.

Abbreviations: BN, boron nitride; NPs, nanoparticles.

5.3.5 Temperature

It was reported that the stability of foam decreased with an increase in temperature. In a temperature range of 25°C to 90°C, Chea et al. (2018) evaluated the foamability and stability of alumina nanoparticles/anionic surfactant and SiO_2 nanoparticles/ anionic surfactant. They found that at the same salinity, surfactant concentration and nanoparticle concentration, the foam's half-life declines with an increase in temperature. This decrease in foam stability with an increase in temperature is attributed to various factors, including the influence of temperature on diffusion rates, adsorption

TABLE 5.4
Effect of salinity on the stability of foam

Nanoparticle type	Surfactant	Salt	Authors	Key findings
SiO_2	AOS	NaCl	Noor et al. (2019)	1. With the increase in NaCl concentration, the foam stability of the SiO_2/AOS system decreases. The foam stability of based foam decreases with an increase in NaCl concentration. 2. An increase in brine concentration reduces the electrostatic repulsions between the particles, which results in agglomeration.
SiO_2 (Ludox CL)	Sulfobetaine (LHSB)	NaCl, $CaCl_2$	Wang et al. (2020)	1. In presence of $CaCl_2$ the adsorption of nanoparticle was reduced. 2. Higher concentration of $CaCl_2$ and NaCl reduces the adsorption of sulfobetaine on nanoparticles.
Alumina-coated SiO_2, SNOWTEX-AK (ST-AK)	AES	NaCl	Qian et al. (2020)	1. The stability of foam generated from surfactant without nanoparticles decreased with increase in salinity. The interaction of nanoparticles and surfactants could be responsible for these phenomena.
SiO_2	ANS	NaCl	Xiao et al. (2016)	1. It was observed that in the presence of NaCl, at minimum foam quality, it had maximum apparent viscosity. 2. The salinity had a significantly negative impact on the apparent viscosity of SiO_2/ANS foam.
Hydrophilic SiO_2	HHSB	NaCl, $CaCl_2$	Li et al. (2019) and Panda et al. (2024)	1. The stability of foam generated from super-critical CO_2 was improved due to the screened electrostatic repulsion between silica nanoparticles. 2. With the increase in salinity, the volume of foam generated from supercritical CO_2 has declined; however, the stability of foam increased.

Abbreviations: $CaCl_2$, calcium chloride; NaCl, sodium chloride.

of surfactant and viscosity of liquid. As the temperature increases, the mobility of surfactant molecules also increases, resulting in less surfactant adsorption at the liquid-gas interface (Kumar et al., 2023b; Pal & Sen, 2023a; Yang et al., 2017; Zhu et al., 2019). Table 5.5 summarizes the influence of temperature on the stability of foam.

TABLE 5.5
Influence of temperature on the stability of foam

Nanoparticle type	Surfactant	Temp (°C)	Authors	Key findings
SiO_2		25, 45, 60	Yu et al. (2012)	1. With the increase in temperature, the height of CO_2-SiO_2 foam decreased. The CO_2 foam started to disappear with an increase in surfactant.
AlOOH	CTAB, EAPB	30, 50, 70	Zhu et al. (2019)	1. The stability of foam under high-temperature conditions was enhanced by adding nanoparticles, although temperature accelerated the liquid drainage and gas diffusion.
SiO_2, Al_2O_3	AOS	25, 40, 60, 90	Chea et al. (2018)	1. At higher temperatures, the nanoparticles did not enhance the foamability. 2. As the temperature increases, the stability of surfactant-based foam declines. However, the presence of nanoparticles in the foam increases the foam stability to an extent. 3. Regardless of the pH and temperature of the system, SiO_2 possessed less stability than Al_2O_3.
Surface-modified SiO_2	CNF	25, 50	AlYousef et al. (2017)	1. During the core flooding studies, it was observed that pressure difference across the core decreases with an increase in temperature. This might be due to the reduction in foam strength and stability.
SiO_2		25, 45, 60	Mo et al. (2014)	1. As the temperature increases from 25°C to 60°C, the corresponding oil recovery obtained from SiO_2/CO_2 foam has reduced from 62.6% to 52.1%.

5.3.6 Type of Oil

Contact of oil with foam is detrimental in the case of conventional surfactant-based foam, which can destabilize the foam instantly. However, by using the nanoparticle along with the surfactant, the stability of foam is increased even in the presence of oil. Table 5.6 summarizes the effect of oil on foam stability.

5.3.7 Synergy Effect of Surfactant Charge vs Nanoparticle Charge

The interaction of charges on surfactant and nanoparticle greatly influences the stability of foam. In conventional surfactant foam, using nanoparticles of similar and opposite charges has enhanced the foam's stability. The synergy effect was observed through the reduction of zeta potential, more stable dispersion of nanoparticles, foam stability and particle aggregation. Using oppositely charged surfactants and

TABLE 5.6
Effect of crude oil on the stability of the foam

Nanoparticle type	Surfactant	Oil	Authors	Key findings
SiO_2 R816 hydrophobic, SiO_2 A300 hydrophilic	MFOMAX	Light crude oil (45° API)	Razali et al. (2018)	1. Hydrophilic SiO_2 foam is two times more stable in the presence of light crude oil than the conventional surfactant foam alone. 2. The SiO_2 nanoparticles have significantly increased the stability of foam in the existence of light crude oil.
Fly-ash-coal	AOS	"Shell Ondina 919", hexadecane	Eftekhari et al. (2015)	1. In the presence of Ondina 919 and hexadecane, bulk nanoparticle foam showed more stability than surfactant foam alone. 2. The mechanism through which the nano ash foam is stabilized was not precise. This phenomenon might occur because nanoparticles at the liquid–gas interface inhibit oil from entering and disrupting the lamellae structure.
Rice husk ash fumed SiO_2	AOS	North Sea oil (54.71 cp)	Bashir et al. (2019)	1. The maximum stability of CO_2 foam in the presence of oil was observed at an optimum concentration of 0.3 wt% xanthan gum and 0.2 wt% SiO_2/RHA nanoparticles.
SiO_2, PEG coating	α-Olefin sulfonate	Crude oil: A (9 cp), B (382 cp), C (36.2 cp)	Singh and Mohanty (2015)	1. The foam volume decreased rapidly when it encountered crude oil A, indicating that the foam is unstable irrespective of the use of nanoparticles. 2. The stability of foam is comparatively higher in the case of crude oils B and C than A. Nanoparticle-surfactant foam is more tolerant and stable than surfactant foam alone.
SNOWTEX-AK (ST-AK), alumina-coated silica	AES	Mineral oil (18 cp), heavy oil (508 cp)	Qian et al. (2019)	1. Surfactant-nanoparticle foam is more stable at partial monolayer coverage and slightly less stable when it contacts the oil phase. 2. Nanoparticles made the foam more stable by not allowing the oil inside the foam structure.

TABLE 5.7
Influence of charge interaction (nanoparticle and surfactant) on foam stability

Nanoparticle type	Surfactant	Authors	Key findings
Nano clay, fumed SiO_2 T30	SDS (anionic), AOS (anionic), LAPB (amphoteric)	Guo and Aryana (2016)	1. It was reported that distribution and particle size affected the stability of nanoparticle-assisted foam. It was also reported that particle aggregation takes place due to the neutralization of nanoparticles by the surface charge. 2. The surface energy of nanoparticles is reduced due to surfactant adsorption on the nanoparticle surface.
SiO_2	CTAB (cationic)	Veyskarami and Ghazanfari (2018)	1. Nanoparticles tend to aggregate and precipitate when mixed with oppositely charged surfactants.
Al_2O_3, CuO, ZrO_2, SiO_2, TiO_2, ZnO	SDS (anionic), SDBS (anionic), CTAB (cationic)	Yekeen et al. (2019)	1. It was reported that the attraction between surfactant and oppositely charged nanoparticles was more dominant than the electrostatic repulsion between similarly charged surfactant molecules and nanoparticles.
SiO_2	n-Amylamine (cosurfactant cationic)	Arriaga et al. (2012)	1. The stability of foam increased because of the electrostatic binding of surfactant molecules (n-amylamine) at the SiO_2 surface.
SiO_2	SDS (anionic), DBSA (anionic)	Vatanparast et al. (2018)	1. The IFT is significantly reduced in the presence of nanoparticles. This behaviour could be due to the repulsion between nanoparticles and the SDS molecules, which increased the surfactant adsorption.

nanoparticles will lead to more surfactant adsorption on the nanoparticle surface. In the case of similarly charged nanoparticles and surfactants, the zeta potential can either increase or remain the same depending on the intensity of electrostatic repulsion. The key findings related to the synergy effect of surfactant charge and nanoparticle charge on the stability of foam are presented in Table 5.7.

5.4 CONCLUSIONS

In every study, it was found that the nanoparticle-foam system was comparatively more stable than the conventional surfactant-based foam without nanoparticles. However, the extent of stability of foam depends on some critical parameters such as reservoir parameters (i.e., the temperature of the reservoir, type of oil in contact and the salinity of the formation water), nanoparticle parameters (i.e., type of nanoparticle, concentration of nanoparticle and size of nanoparticle) and synergy effect of surfactant and nanoparticles.

The key findings of this chapter are:

- *Nanoparticle size*: The foam stability is inversely proportional to the size of the nanoparticle. It was found that nanoparticles of size between 7 nm and 10 μm had improved the foam stability.
- *Nanoparticle concentration*: With the increase in concentration of nanoparticles, the foam stability might or might not improve. This phenomenon is due to the complex synergy effects between the surfactant and nanoparticle charge. So, preliminary screening studies should be performed to determine the optimum nanoparticle concentration.
- *Salinity*: It was found that the stability of foam is inversely related salinity; i.e., lower salinity has higher foam stability. However, in certain cases, depending on the type of nanoparticle and surfactant, the strength of foam has increased due to the ion shielding effect.
- *Temperature and presence of oil*: It was found that surfactant-nanoparticle foam is more stable at high temperatures and in the presence of crude oil when compared to surfactant foam alone.
- *Synergistic effect between the surfactant and nanoparticles*: It was observed that using nanoparticles and surfactants of opposite charge leads to more foam stability. This increase in foam stability is due to the surfactant adsorption on nanoparticles, which results in altering the surface properties of nanoparticles from hydrophilic to partially hydrophobic. However, excessive increments in hydrophobicity may cause the destabilization of foam due to the aggregation of nanoparticles. Therefore, while using oppositely charged nanoparticles, the concentration of surfactant and nanoparticles should be optimized. Moreover, it was also observed that while using similar charged nanoparticles and surfactant, the interfacial tension (IFT) was reduced. In this case, the IFT was decreased because the surfactant molecules tend to move towards the liquid-gas interface due to the electrostatic repulsions.

REFERENCES

Ab Rasid, S. A., Mahmood, S. M., Kechut, N. I., & Akbari, S. (2022). A review on parameters affecting nanoparticles stabilized foam performance based on recent analyses. *Journal of Petroleum Science and Engineering*, *208*(PB), 109475. https://doi.org/10.1016/j.petrol.2021.109475.

Al-Shargabi, M., Davoodi, S., Wood, D. A., Rukavishnikov, V. S., & Minaev, K. M. (2022). Carbon dioxide applications for enhanced oil recovery assisted by nanoparticles: Recent developments. *ACS Omega*, *7*(12), 9984–9994. https://doi.org/10.1021/acsomega.1c07123.

Alcorn, Z. P., Føyen, T., Gauteplass, J., Benali, B., Soyke, A., & Fernø, M. (2020). Pore-and core-scale insights of nanoparticle-stabilized foam for CO2-enhanced oil recovery. *Nanomaterials*, *10*(10), 1–15. https://doi.org/10.3390/nano10101917.

Alvarado, V., & Manrique, E. (2010). Enhanced oil recovery: An update review. *Energies*, *3*(9), 1529–1575. https://doi.org/10.3390/en3091529.

AlYousef, Z. A., Almobarky, M. A., & Schechter, D. S. (2018). The effect of nanoparticle aggregation on surfactant foam stability. *Journal of Colloid and Interface Science, 511*, 365–373. https://doi.org/10.1016/j.jcis.2017.09.051.

AlYousef, Z., Almobarky, M., & Schechter, D. (2017). Enhancing the stability of foam by the use of nanoparticles. *Energy & Fuels, 31*(10), 10620–10627. https://doi.org/10.1021/acs.energyfuels.7b01697.

Arriaga, L. R., Drenckhan, W., Salonen, A., Rodrigues, J. A., Íñiguez-Palomares, R., Rio, E., & Langevin, D. (2012). On the long-term stability of foams stabilised by mixtures of nano-particles and oppositely charged short chain surfactants. *Soft Matter, 8*(43), 11085–11097. https://doi.org/10.1039/c2sm26461g.

Asghari, K., Al-Dliwe, A., & Mahinpey, N. (2006). Effect of operational parameters on carbon dioxide storage capacity in a heterogeneous oil reservoir: A case study. *Indus-trial & Engineering Chemistry Research, 45*(8), 2452–2456. https://doi.org/10.1021/ie050572y.

Bashir, A., Malik, L. A., Ahad, S. et al. (2019). Removal of heavy metal ions from aqueous system by ion-exchange and biosorption methods. *Environ Chem Lett, 17*, 729–754. https://doi.org/10.1007/s10311-018-00828-y.

Binks, B. P., & Lumsdon, S. O. (2000). Influence of particle wettability on the type and stability of surfactant-free emulsions. *Langmuir, 16*(23), 8622–8631. https://doi.org/10.1021/la000189s.

Carrier, V., & Colin, A. (2003). Coalescence in draining foams. *Langmuir, 19*(11), 4535–4538. https://doi.org/10.1021/la026995b.

Chea, T. B., Akhir, N. A. M., & Idris, A. K. (2018). Investigation on the effect of types of nanoparticles and temperature on nanoparticles-foam stability. In *Proceedings of the International Conference on Industrial Engineering and Operations Management, 2018-March* (May) (pp. 365–372).

Dai, Z., Viswanathan, H., Fessenden-Rahn, J., Middleton, R., Pan, F., Jia, W., Lee, S. Y., Mcpherson, B., Ampomah, W., & Grigg, R. (2014). Uncertainty quantification for CO2 sequestration and enhanced oil recovery. *Energy Procedia, 63*, 7685–7693. https://doi.org/10.1016/j.egypro.2014.11.802.

Dinesh, N. S. V., & Sivasankar, P. (2024). Exergy based assessment for decarbonization of CO2 foam flooding enhanced oil recovery process and energy transition with hydrogen. *International Journal of Hydrogen Energy, 52*, 1522–1536. https://doi.org/10.1016/j.ijhydene.2023.08.350.

Eftekhari, S., Salvatore, C. A, Johansson, S., Chen, T. B., Zeng, Z., Edvinsson, L. (2015). Localization of CGRP, CGRP receptor, PACAP and glutamate in trigeminal ganglion. Relation to the blood-brain barrier. Brain Res. 2015 Mar 10;1600:93-109. doi: 10.1016/j.brainres.2014.11.031. Epub 2014 Nov 20. PMID: 25463029.

Eide, Ø., Føyen, T., Skjelsvik, E., Rognmo, A., & Fernø, M. (2018). Nanoparticle stabilized foam in harsh conditions for CO2 EOR. *Day 3 Wed, November 14, 2018*, D031S089R001. https://doi.org/10.2118/193212-MS.

Fameau, A.-L., & Salonen, A. (2014). Effect of particles and aggregated structures on the foam stability and aging. *Comptes Rendus Physique, 15*(8–9), 748–760. https://doi.org/10.1016/j.crhy.2014.09.009.

Farid Ibrahim, A., & Nasr-El-Din, H. A. (2020). CO2 foam for enhanced oil recovery applications. In *Foams - Emerging Technologies*. IntechOpen. https://doi.org/10.5772/intechopen.89301.

Gbadamosi, A. O., Junin, R., Manan, M. A., Agi, A., & Yusuff, A. S. (2019). An overview of chemical enhanced oil recovery: Recent advances and prospects. *International Nano Letters, 9*(3), 171–202. https://doi.org/10.1007/s40089-019-0272-8.

Gielen, D., Boshell, F., Saygin, D., Bazilian, M. D., Wagner, N., & Gorini, R. (2019). The role of renewable energy in the global energy transformation. *Energy Strategy Reviews, 24*(June 2018), 38–50. https://doi.org/10.1016/j.esr.2019.01.006.

Gugl, R. (2020). *Modelling and Simulation of Foam-Assisted Water-Alternating-Gas Injection in Naturally Fractured Carbonate Reservoirs* (Issue September) [Master's Thesis, Montanuniversitaet Leoben (000)].

Guo, F., & Aryana, S. (2016). An experimental investigation of nanoparticle-stabilized CO2 foam used in enhanced oil recovery. *Fuel*, *186*, 430–442. https://doi.org/10.1016/j.fuel.2016.08.058.

Hematpur, H., Mahmood, S. M., Nasr, N. H., & Elraies, K. A. (2018). Foam flow in porous media: Concepts, models and challenges. *Journal of Natural Gas Science and Engineering*, *53*, 163–180. https://doi.org/10.1016/j.jngse.2018.02.017.

Hong, X., Thaore, V. B., Garud, S. S., Karimi, I. A., Farooq, S., Wang, X., Usadi, A. K., & Chapman, B. R. (2023). Decarbonizing singapore via local production of H2 from natural gas. *International Journal of Hydrogen Energy*, *48*(24), 8743–8755. https://doi.org/10.1016/j.ijhydene.2022.12.069.

Horozov, T. S.. (2008). Foams and foam films stabilised by solid particles. *Current Opinion in Colloid & Interface Science*, *13*(3), 134–140. https://doi.org/10.1016/j.cocis.2007.11.009.

Hu, N., Li, Y., Wu, Z., Lu, K., Huang, D., & Liu, W. (2018). Foams stabilization by silica nanoparticle with cationic and anionic surfactants in column flotation: Effects of particle size. *Journal of the Taiwan Institute of Chemical Engineers*, *88*, 62–69. https://doi.org/10.1016/j.jtice.2018.04.008.

Kalam, S., Kamal, M. S., Patil, S., & Hussain, S. M. S. (2019). Role of counterions and nature of spacer on foaming properties of novel polyoxyethylene cationic gemini surfactants. *Processes*, *7*(8), 502. https://doi.org/10.3390/pr7080502.

Karakashev, S. I., Ozdemir, O., Hampton, M. A., & Nguyen, A. V. (2011). Formation and stability of foams stabilized by fine particles with similar size, contact angle and different shapes. *Colloids and Surfaces A: Physicochemical and Engineering Aspects*, *382*(1–3), 132–138. https://doi.org/10.1016/j.colsurfa.2010.09.023.

Kim, I., Worthen, A. J., Johnston, K. P., DiCarlo, D. A., & Huh, C. (2016). Size-dependent properties of silica nanoparticles for Pickering stabilization of emulsions and foams. *Journal of Nanoparticle Research*, *18*(4), 82. https://doi.org/10.1007/s11051-016-3395-0.

Kumar, A., & Gangawane, K. M. (2022a). Effect of precipitating agents on the magnetic and structural properties of the synthesized ferrimagnetic nanoparticles by co-precipitation method. *Powder Technology*, *401*, 117298. https://doi.org/10.1016/j.powtec.2022.117298.

Kumar, A., & Gangawane, K. M. (2022b). Synthesis and effect on the surface morphology & magnetic properties of ferrimagnetic nanoparticles by different wet chemical synthesis methods. *Powder Technology*, *410*(June), 117867. https://doi.org/10.1016/j.powtec.2022.117867.

Kumar, A., & Gangawane, K. M. (2023a). Advanced computational approaches for water treatment. In *Advanced Computational Approaches for Water Treatment* (pp. 1–12). CRC Press. https://doi.org/10.1201/9781003325147-1.

Kumar, A., & Gangawane, K. M. (2023b). Molecular dynamics for the membrane process. In *Advanced Computational Approaches for Water Treatment* (pp. 13–38). CRC Press. https://doi.org/10.1201/9781003325147-2.

Kumar, A., Gangawane, K. M., & Ramanjaneyulu, B. (2022). *Ferrofluids for Waste-Water Treatment BT - Advances in Chemical, Bio and Environmental Engineering* (J. K. Ratan, D. Sahu, N. N. Pandhare, & A. Bhavanam (eds.); pp. 723–744). Springer International Publishing , pp. 723–744.

Kumar, A., Jena, H. M., & Gangawane, K. M. (2023a). Effect of ferrimagnetic nanoparticles as an oxygen absorber on the shelf life of tomatoes. Poster presented at the *9th International Food Convention (IFCoN)*, CSIR-CFTRI, Mysuru, India, 9 December 2023.

Kumar, A., Jena, H. M., & Gangawane, K. M. (2023b). Variations in magnetic behavior of ferrimagnetic nanoparticles with temperature. Poster presented at the *75th Indian Chemical Engineering Congress (CHEMCON 2023)*, Dr. H L Roy Building, Jadavpur University Campus, Kolkata, India, 28 December 2023.

Kumar, S., & Mandal, A. (2017). Investigation on stabilization of CO2 foam by ionic and non-ionic surfactants in presence of different additives for application in enhanced oil recovery. *Applied Surface Science*, *420*, 9–20. https://doi.org/10.1016/j.apsusc.2017.05.126.

Li, W., Wei, F., Xiong, C., Ouyang, J., Shao, L., Dai, M., Liu, P., & Du, D. (2019). A novel supercritical CO2 foam system stabilized with a mixture of zwitterionic surfactant and silica nanoparticles for enhanced oil recovery. *Frontiers in Chemistry*, *7*(October), 1–11. https://doi.org/10.3389/fchem.2019.00718.

Minz, P., & Mishra, S. (2023). Jackfruit peel derived ZnCl2-impregnated activated carbon: Optimization, characterization, and application in dye removal. *Biomass Conversion and Biorefinery*, 0123456789. https://doi.org/10.1007/s13399-023-04592-0.

Mo, C, et al. (2014). Display cellulolytic enzymes on Saccharomyces cerevisiae cell surface by using Flo1p as an anchor protein for cellulosic ethanol production. *Sheng Wu Gong Cheng Xue Bao, 30*(9):1401–13.

Nikhil, P., Prakash, G. B., Kavyasri, K., & Sivasankar, P. (2024). Analysis on CO2 emissions, green hydrogen requirement and geo-storage potential of hydrogen for decarbonization of industrial operations in southeastern coast of India. *International Journal of Hydrogen Energy*, *52*, 1507–1521. https://doi.org/10.1016/j.ijhydene.2023.08.235.

Noor, M. Z. B. M., Teng, W. Y., & Irawan, S. (2019). Stabilization of silicone dioxide nanoparticle foam in tertiary petroleum production. *Indonesian Journal of Chemistry*, *19*(4), 985. https://doi.org/10.22146/ijc.40870.

Pal, D., & Sen, S. (2023a). Emerging petroleum pollutants and their adverse effects on the environment. In I. D. Behera, & A. P. Das (Eds.), *Impact of Petroleum Waste on Environmental Pollution and Its Sustainable Management through Circular Economy. Environmental Science and Engineering* (pp. 103–137). Springer Nature Switzerland. https://doi.org/10.1007/978-3-031-48220-5_5.

Pal, D., & Sen, S. (2023b). In-depth coverage of petroleum waste sources, characteristics, environmental impact, and sustainable remediation process. In I. D. Behera, & A. P. Das (Eds.), *Impact of Petroleum Waste on Environmental Pollution and Its Sustainable Management through Circular Economy* (pp. 1–38). Springer Nature Switzerland. https://doi.org/10.1007/978-3-031-48220-5_1.

Pal, D., & Sen, S. (2024). Optimal synthesis of dolochar derived faujasite zeolite X for highly effective Cd(II) removal. *Environmental Research*, *240*, 117494. https://doi.org/10.1016/j.envres.2023.117494.

Panda, D., & Gangawane, K. M. (2023a). Development of superhydrophobic hybrid silica-cellulose aerogel as promising thermal insulation and sound absorption. *International Journal of Materials Research*, *114*(4–5), 405–417. https://doi.org/10.1515/ijmr-2022-0066.

Panda, D., & Gangawane, K. M. (2023b). Hybrid NiFe2O4-NiCO4O4 nanoparticles-based eutectic phase change materials for enhancement of thermal efficiency of pin-fin heat sink arrangement. *Journal of Energy Storage*, *60*(November 2022), 106644. https://doi.org/10.1016/j.est.2023.106644.

Panda, D., & Gangawane, K. M. (2023c). Recycled cellulose–silica hybrid aerogel for effective oil adsorption: optimization and kinetics study. *Sadhana - Academy Proceedings in Engineering Sciences*, 48(3). https://doi.org/10.1007/s12046-023-02161-9.

Panda, D., Sahu, A. K., & Gangawane, K. M. (2024). Effect of molar concentration and drying methodologies on monodispersed silica sol for synthesis of silica aerogels with temperature-resistant characteristics. *Materials Science and Engineering: B*, *299*, 117062. https://doi.org/10.1016/j.mseb.2023.117062.

Panda, D., Sharma, S., & Gangawane, K. M. (2023). Synthesis of novel biocatalyst from organic waste protonated with acid treatment for hydrogen production. *International Journal of Hydrogen Energy*, *48*(63), 24242–24254. https://doi.org/10.1016/j.ijhydene.2023.03.117.

Peter, S. C. (2018). Reduction of CO2 to chemicals and fuels: A solution to global warming and energy crisis. *ACS Energy Letters*, *3*(7), 1557–1561. https://doi.org/10.1021/acsenergylett.8b00878.

Phong, G. M., M. Pilus, R., Mustaffa, A., Thangavel, L., & Mohamed, N. M. (2020). Relationship between fly ash nanoparticle-stabilized-foam and oil production in core displacement and simulation studies. *Fuel*, *266*(January), 117033. https://doi.org/10.1016/j.fuel.2020.117033.

Qian, X., Song, H., Ming, G. L. (2019). Brain organoids: advances, applications and challenges. Development. 2019 Apr 16;146(8):dev166074. doi: 10.1242/dev.166074. PMID: 30992274; PMCID: PMC6503989.

Qian, C., Telmadarreie, A., Dong, M., & Bryant, S. (2020). Synergistic effect between surfactant and nanoparticles on the stability of foam in EOR processes. *SPE Journal*, *25*(02), 883–894. https://doi.org/10.2118/195310-PA.

Rafati, R., Haddad, A. S., & Hamidi, H. (2016). Experimental study on stability and rheological properties of aqueous foam in the presence of reservoir natural solid particles. *Colloids and Surfaces A: Physicochemical and Engineering Aspects*, *509*, 19–31. https://doi.org/10.1016/j.colsurfa.2016.08.087.

Ramanathan, R., and Hisham A. Nasr-El-Din. (2020). Chelating Agents for Iron Sulfide Scale Removal at 300°F (149°C). Paper presented at the CORROSION 2020, physical event cancelled, June 2020.

Ramanathan, R., Abdelwahab, O., & Nasr-El-Din, H. A. (2021). A new effective multiwalled carbon nanotube-foam system for mobility control. *SPE Journal*, *26*(05), 3205–3221. https://doi.org/10.2118/203210-PA.

Razali, N., Hsia, I. C. C., Zainal, S., & Manap, A. A. A. (2018). Nano-sized particle as foam stabiliser designed for application at high temperature and light crude oil condition for enhanced oil recovery - A fluid-fluid case study. *Day 3 Thu, March 22, 2018*, D032S003R007. https://doi.org/10.4043/28431-MS.

Rossen, W. R., & Wang, M. W. (1999). Modeling foams for acid diversion. *SPE Journal*, *4*(02), 92–100. https://doi.org/10.2118/56396-PA.

Sæle, A., Graue, A., & Alcorn, Z. P. (2022). Unsteady-state CO2 foam injection for increasing enhanced oil recovery and carbon storage potential. *Advances in Geo-Energy Research*, *6*(6), 472–481. https://doi.org/10.46690/AGER.2022.06.04.

Sahoo, A., & Mishra, S. (2023). Synthesis of microporous-activated carbon via direct H3PO4 and KOH activation of Pterospermum acerifolium fruits: process optimization and characterization. *Biomass Conversion and Biorefinery*, 0123456789. https://doi.org/10.1007/s13399-023-05072-1.

Santhosh, R., & Sarkar, P. (2022). Jackfruit seed starch/tamarind kernel xyloglucan/zinc oxide nanoparticles-based composite films: Preparation, characterization, and application on tomato (Solanum lycopersicum) fruits. *Food Hydrocolloids*, *133*(April), 107917. https://doi.org/10.1016/j.foodhyd.2022.107917.

Saraf, S., & Bera, A. (2023). Carbon dioxide injection for enhanced oil recovery and underground storage to reduce greenhouse gas. Dr. Inamuddin and Tariq Altalhi (eds). In *Green Sustainable Process for Chemical and Environmental Engineering and Science* (pp. 149–176). Elsevier.

Savaresi, A. (2016). The Paris Agreement: A new beginning. *Journal of Energy and Natural Resources Law*, *34*(1), 16–26. https://doi.org/10.1080/02646811.2016.1133983.

Singh, R. and Mohanty, K. K. (2015). Synergy between Nanoparticles and Surfactants in Stabilizing Foams for Oil Recovery. *Energy & Fuels*, *29*, 467–479. https://doi.org/10.1021/ef5015007.

Telmadarreie, A., & Trivedi, J. J. (2020). CO2 foam and CO2 polymer enhanced foam for heavy oil recovery and CO2 storage. *Energies*, *13*(21), 5735. https://doi.org/10.3390/en13215735.

Vatanparast, H., Shahabi, F., Bahramian, A., Javadi, A., & Miller, R. (2018). The role of electrostatic repulsion on increasing surface activity of anionic surfactants in the presence of hydrophilic silica nanoparticles. *Scientific Reports*, *8*(1), 1–11. https://doi.org/10.1038/s41598-018-25493-7.

Veyskarami, M., & Ghazanfari, M. H. (2018). Synergistic effect of like and opposite charged nanoparticle and surfactant on foam stability and mobility in the absence and presence of hydrocarbon: A comparative study. *Journal of Petroleum Science and Engineering*, *166*(January), 433–444. https://doi.org/10.1016/j.petrol.2018.03.076.

Vishal, V., Chandra, D., Singh, U., & Verma, Y. (2021). Understanding initial opportunities and key challenges for CCUS deployment in India at scale. *Resources, Conservation and Recycling*, *175*(August), 105829. https://doi.org/10.1016/j.resconrec.2021.105829.

Vishnyakov, V., Suleimanov, B., Salmanov, A., & Zeynalov, E. (2020). Oil recovery stages and methods. In *Primer on Enhanced Oil Recovery* (pp. 53–63). Elsevier. https://doi.org/10.1016/B978-0-12-817632-0.00007-4.

Wang, T., Fan, H., Yang, W., & Meng, Z. (2020). Stabilization mechanism of fly ash three-phase foam and its sealing capacity on fractured reservoirs. *Fuel*, *264*(December 2019), 116832. https://doi.org/10.1016/j.fuel.2019.116832.

Xiao, C., Balasubramanian, S. N., & Clapp, L. W. (2016). Rheology of supercritical CO2 foam stabilized by nanoparticles. In *SPE - DOE Improved Oil Recovery Symposium Proceedings*, *2016-Janua*, SPE-179621. https://doi.org/10.2118/179621-ms.

Yang, W., Wang, T., Fan, Z., Miao, Q., Deng, Z., & Zhu, Y. (2017). Foams stabilized by in situ - modified nanoparticles and anionic surfactants for enhanced oil recovery. *Energy & Fuels*, *31*(5), 4721–4730. https://doi.org/10.1021/acs.energyfuels.6b03217.

Yekeen, N., Idris, A. K., Manan, M. A., Samin, A. M., Risal, A. R., & Kun, T. X. (2017). Bulk and bubble-scale experimental studies of influence of nanoparticles on foam stability. *Chinese Journal of Chemical Engineering*, *25*(3), 347–357. https://doi.org/10.1016/j.cjche.2016.08.012.

Yekeen, N., Manan, M. A., Idris, A. K., Padmanabhan, E., Junin, R., Samin, A. M., Gbadamosi, A. O., & Oguamah, I. (2018a). A comprehensive review of experimental studies of nanoparticles-stabilized foam for enhanced oil recovery. *Journal of Petroleum Science and Engineering*, *164*(January), 43–74. https://doi.org/10.1016/j.petrol.2018.01.035.

Yekeen, N., Manan, M. A., Idris, A. K., Samin, A. M., & Risal, A. R. (2018b). Mechanistic study of nanoparticles–surfactant foam flow in etched glass micro-models. *Journal of Dispersion Science and Technology*, *39*(5), 623–633. https://doi.org/10.1080/01932691.2017.1378581.

Yekeen, N., Padmanabhan, E., & Idris, A. K. (2019). Synergistic effects of nanoparticles and surfactants on n-decane-water interfacial tension and bulk foam stability at high temperature. *Journal of Petroleum Science and Engineering*, *179*(May), 814–830. https://doi.org/10.1016/j.petrol.2019.04.109.

Yu, J., Liu, N., Li, L., & Lee, R. (2012). Generation of nanoparticle-stabilized supercritical CO2 foams. *Society of Petroleum Engineers - Carbon Management Technology Conference 2012*, *1*, 189–199. https://doi.org/10.7122/150849-ms.

Zhang, Y., Wang, Y., Xue, F., Wang, Y., Ren, B., Zhang, L., & Ren, S. (2015). CO2 foam flooding for improved oil recovery: Reservoir simulation models and influencing factors. *Journal of Petroleum Science and Engineering*, *133*, 838–850. https://doi.org/10.1016/j.petrol.2015.04.003.

Zhu, J., Yang, Z., Li, X., Hou, L., & Xie, S. (2019). Experimental study on the microscopic characteristics of foams stabilized by viscoelastic surfactant and nanoparticles. *Colloids and Surfaces A: Physicochemical and Engineering Aspects*, *572*(March), 88–96. https://doi.org/10.1016/j.colsurfa.2019.03.087.

6 Functional Groups and Functionalized Nanoparticles in EOR Techniques

Krishna Raghav Chaturvedi, Abhishek Kumar and Anand Raghav Chaubey

6.1 INTRODUCTION

The quest for sustainable and efficient methods of oil extraction has led to a paradigm shift in the oil and gas industry, with nanotechnology emerging as a transformative force (Christian et al., 2008; Jiang et al., 2003). Within this dynamic landscape, the utilization of functional groups and functionalized nanoparticles (NPs) has garnered significant attention, offering tailored solutions to the challenges inherent in Enhanced Oil Recovery (EOR) techniques. This chapter embarks on an in-depth exploration of the multifaceted role played by functional groups and their integration into NPs, providing a comprehensive understanding of their mechanisms, applications, and prospects in the context of EOR. As global energy demands escalate and conventional oil recovery methods approach their limits, the need for innovative approaches becomes increasingly apparent. Functional groups, with their distinct chemical characteristics, and NPs, with their unique physical properties, present a synergistic avenue for addressing the intricacies of reservoir conditions (Fazelabdolabadi et al., 2015; Kumar et al., 2023a; Panda et al., 2024). The chapter begins by elucidating the fundamental principles of functional groups, their interactions, and the strategic functionalization of NPs, setting the stage for a nuanced exploration of their applications in EOR.

Surface chemistry forms the cornerstone of the discussion, as the modification of NPs through functionalization plays a pivotal role in tailoring their behavior in reservoir environments (Kumar & Gangawane, 2022a; Murshed et al., 2009; Wu et al., 2008). Understanding the intricate interplay between functional groups and NPs is crucial for optimizing their performance, a theme that permeates throughout the chapter. Recent research has provided evidence for the viability of producing activated-carbon-based materials from jackfruit peel and Pterospermum acerifolium fruits. This has been accomplished by optimizing the synthesis processes with phosphoric acid (H_3PO_4), potassium hydroxide (KOH) activation, and zinc chloride

 DOI: 10.1201/9781032650685-6

($ZnCl_2$) impregnation (Minz & Mishra, 2023; Pal & Sen, 2024; Sahoo & Mishra, 2023; Santhosh & Sarkar, 2022). Superior materials with customized characteristics that are well suited for adsorption purposes may present innovative strategies to improve EOR processes. The incorporation of these materials into EOR strategies may reduce environmental impact and increase oil recovery efficiency, thereby harmonizing with the industry's objectives of efficiency and sustainability (Pal & Sen, 2023a, 2023b). A systematic breakdown of the mechanisms through which functionalized NPs influence EOR processes follows, with a focus on their impact on interfacial tension (IFT), wettability alteration, and capillary forces within porous rock formations. Real-world applications are brought to the forefront through case studies and experimental findings, showcasing the adaptability and efficacy of these advanced materials in diverse oil recovery scenarios. However, the practical implementation of functionalized NPs in EOR is not without its challenges. Regulatory considerations, environmental implications, and economic feasibility are pivotal aspects that demand careful examination to navigate the path toward field-scale application successfully. Looking ahead, the chapter peers into the future of EOR techniques with a discussion on emerging trends and potential breakthroughs. Responsive NPs, capable of adapting to the dynamic conditions of reservoirs, represent one such frontier, offering innovative strategies to overcome current limitations and further enhance the efficiency of oil recovery (Behzadi & Mohammadi, 2016; Kumar & Gangawane, 2022b). In summation, this chapter serves as a gateway into the intricate world of functional groups and functionalized NPs in the realm of EOR. By unraveling the complexities of nanoscale interactions within reservoirs, the chapter contributes to the ongoing pursuit of sustainable and efficient energy solutions, marking a significant stride forward in the evolution of oil recovery practices.

6.2 NEED FOR NANOMATERIAL FUNCTIONALIZATION

There is broad consensus that the conduct of NPs exerts a substantial influence on a multitude of physicochemical phenomena. In specific applications, for the sake of convenience, the geometry of NPs has been altered to faceted or to resemble a traditional sphere. However, the concept of NPs has evolved to encompass a range of dimensionalities, including zero, one, two, or three dimensions. The nanostructures above consist of nanodiamonds, nanotubes, nanocapsules, nanospheres, nanostars, nanorods, nanodisks, nanopyramids, and undoubtedly many others (Christian et al., 2008; Kumar & Gangawane, 2023a; Yang & Han, 2006; Zhang et al., 2018). These are all distinguished by NPs on their surfaces that resemble those observed on the nanospheres. The continuous advancements in the investigation of NPs, which include their characterization and synthesis (with a particular focus on controlling their size and structure), have been seamlessly incorporated into a multitude of scientific and technological developments. The shape of an NP helps ensure the widespread use of NPs in biomedical applications is due to their dimensions being similar to those of biomolecules and biological phenomena. Therefore, the exploration of functionalizing NP applications has been widely proposed as a modern strategy to improve their efficacy and subsurface utilization potential.

The intentional adsorption of diverse species onto the surface of NPs (e.g., metals, ions, molecules, polymers, biomolecules, and so forth) is the conventional definition of functionalization (Kumar & Gangawane, 2023b; Nahri et al., 2020; Panda & Gangawane, 2023a; Wu et al., 2008). Consequently, physical deposition or chemical junction of those specific species has been utilized to functionalize the surface of NPs, in accordance with universally acknowledged fundamental laws and principles of physics, chemistry, biology, and other interrelated scientific fields. It is imperative to underscore the fact that functionalized NPs generate functional systems characterized by superior performance and a strong hierarchical structure when compared to non-functionalized systems. Upon realization of the projected biomedical applications, in addition to mitigating toxicity effects, a number of benefits would accrue, depending on the particular system and application. In most cases, physical functionalization, also known as "decoration," is carried out after the synthesis of NPs, barring any limiting constraints. A delicate interplay emerges between the stabilized agent and the NPs, which has the potential to deteriorate or become unstable in extreme conditions. On the contrary, chemical functionalization results in covalent bonding, which typically denotes a stable interaction between the molecular attachment agents and the NP surfaces.

6.3 STEPS TO FUNCTIONALIZE NANOMATERIALS

NP functionalization involves modifying the surface of NPs with various chemical groups or molecules to tailor their properties for specific applications. The functionalization process is crucial for enhancing the stability, solubility, reactivity, and compatibility of NPs in a given environment. Here is an overview of how NP functionalization takes place:

- *Selection of NP*: Choose the appropriate NP based on the desired properties and application. Standard NPs include metals (e.g., gold, silver), metal oxides (e.g., iron oxide), quantum dots, and carbon-based materials (e.g., graphene, carbon nanotubes). Usually, for oil and gas applications, silica nanomaterials are most widely preferred due to their low cost, superior subsurface compatibility, and other oil recovery properties (Al-Anssari et al., 2017; Dabhade et al., 2023; Hendraningrat & Torsæter, 2014; Panda & Gangawane, 2023c).
- *Surface Modification*: The surface of NPs typically contains functional groups or ligands from the synthesis process. However, additional modifications may be necessary for specific applications. The surface modification involves replacing or introducing new functional groups to enhance the properties of the NPs.
- *Functional Group Selection*: Choose functional groups based on the targeted application. Common functional groups include amino groups (–NH_2), carboxyl groups (–COOH), hydroxyl groups (–OH), thiol groups (–SH), and various organic molecules. The selection depends on the desired interactions with other molecules, solubility requirements, and reactivity.

- *Chemical Conjugation*: Functionalization is achieved through chemical reactions that form covalent bonds between the selected functional groups and the NP surface. Various methods, such as co-precipitation, ligand exchange, or direct chemical reactions, can be employed (Corredor-Rojas et al., 2018).
- *Surface Ligand Exchange*: In some cases, the existing ligands on the NP surface may be exchanged with the desired functional groups. This ligand exchange process is common for colloidal NPs and involves introducing a new ligand that displaces the original ones.
- *Surface Activation*: Prior to functionalization, the NP surface may require activation to make it more reactive and amenable to chemical bonding. Activation methods include treatment with acids, bases, or other reactive agents to create suitable binding sites for functional groups (Lee et al., 2020).

Several techniques can be employed for NP functionalization (Corredor-Rojas et al., 2018; Kumar et al., 2014; Lee et al., 2020), including:

- *Chemical Conjugation*: Involves direct chemical reactions between the NP surface and functional groups.
- *Click Chemistry*: A specific type of chemical reaction known for its efficiency and selectivity in joining molecular components (Figure 6.1).

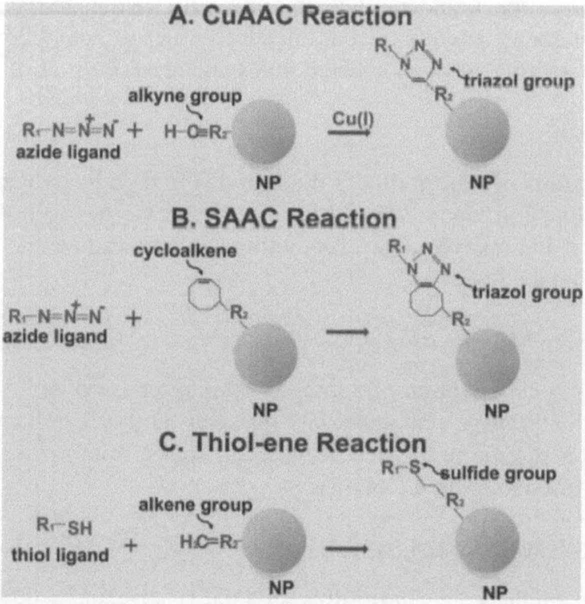

FIGURE 6.1 Observations of lick chemistry reactions onto nanoparticle surfaces. (A) CuAAC, (B) SPAAC, and (C) thiol-ene coupling. Adapted with permission from Tovar et al. (2023). 10.1016/B978-0-12–821637-8.00004-3

- *Coating or Encapsulation*: Surrounding the NP with a protective layer containing the desired functional groups.
- *Adsorption*: Adsorbing molecules with specific functional groups onto the NP surface.

After functionalization, the NPs are characterized to confirm the success of the modification. Techniques such as Fourier-transform infrared spectroscopy (FTIR), X-ray photoelectron spectroscopy (XPS), and nuclear magnetic resonance (NMR) are commonly used to analyze surface functional groups.

Next, there is a need to tailor the functionalization process based on the specific requirements of the intended application. For instance, in biomedical applications, functionalization may focus on improving biocompatibility and targeting specific biological interactions. In oil field applications, nanomaterials are often modified and tailored to meet specific requirements related to EOR and reservoir engineering. These application-specific modifications are essential for optimizing the performance of nanomaterials in the challenging and dynamic conditions of oil reservoirs (Kumar et al., 2023b; Panda & Gangawane, 2023b). Here are some key application-specific modifications for nanomaterials in oil field applications:

Surface Wettability Modification

- *Objective*: Alter the wetting properties of reservoir rocks to enhance fluid displacement.
- *Modification*: Functionalize nanomaterials with specific surface groups to induce wettability alteration from oil-wet to water-wet conditions, improving the interaction between injected fluids and reservoir rocks.

IFT Reduction

- *Objective*: Enhance the ability of fluids to displace oil by reducing IFT.
- *Modification*: Introduce NPs with surfactant properties to lower the IFT between oil and injected fluids, promoting better spreading and penetration of fluids into the reservoir rock.

Capillary Force Modification

- *Objective*: Reduce capillary trapping of oil in reservoir pores.
- *Modification*: Design nanomaterials that can modify capillary forces by altering the surface properties of reservoir rocks, minimizing oil entrapment and improving fluid mobility.

Improved Reservoir Sweep Efficiency

- *Objective*: Enhance the uniformity of fluid displacement throughout the reservoir.
- *Modification*: Develop nanomaterials that mitigate fingering effects and improve the overall sweep efficiency, ensuring injected fluids reach and displace oil from previously unrecovered regions.

Enhanced Mobility in Heavy-Oil Reservoirs

- *Objective*: Improve the flow of heavy oil by reducing viscosity.
- *Modification*: Introduce NPs that interact with heavy-oil components, reducing viscosity and enhancing the mobility of heavy oil for improved recovery.

Smart NPs for Responsive Behavior

- *Objective*: Enable adaptive responses to changing reservoir conditions.
- *Modification*: Design smart NPs that can respond to variations in temperature, pressure, or chemical composition, optimizing their behavior in real-time based on reservoir dynamics.

In-Situ Gelling and Plugging

- *Objective*: Control fluid flow and divert injected fluids to specific zones.
- *Modification*: Develop nanostructured materials that can form in-situ gels or plugs, managing fluid flow and directing injected fluids to desired regions within the reservoir.

Reservoir Conformance Control

- *Objective*: Improve the uniformity of fluid distribution in heterogeneous reservoirs.
- *Modification*: Utilize NPs to control the conformance of injected fluids, ensuring even distribution and minimizing bypassed zones.

Monitoring with Nanosensors

- *Objective*: Real-time monitoring of EOR processes.
- *Modification*: Integrate nanosensors into nanomaterials to provide continuous monitoring of key parameters such as fluid displacement, pressure, and temperature within the reservoir.

Environmental Compatibility

- *Objective*: Mitigate environmental impact and ensure compatibility with reservoir conditions.
- *Modification*: Choose nanomaterials and functionalization methods that are environmentally benign and suitable for the specific geological and chemical environment of the reservoir.

These application-specific modifications highlight the versatility of nanomaterials in addressing the unique challenges of oil field applications. Tailoring nanomaterial properties to meet the specific demands of EOR processes is essential for realizing their full potential in optimizing oil extraction from reservoirs. From these observations, it can be identified that NP functionalization is a versatile and critical step in customizing NPs for a wide range of applications, from drug delivery and diagnostics to catalysis and EOR. The choice of functional groups and the method of functionalization depend on the targeted functionality and the surrounding environment in which the NPs will be employed.

6.4 CHALLENGES IN WORKING WITH FUNCTIONALIZED NANOMATERIALS

Working with functionalized NPs in various applications, including EOR, presents a set of challenges that researchers and engineers must carefully navigate (Estellé et al., 2018; Hunter et al., 2008; Kumar et al., 2022; Li et al., 2018; Panda et al., 2023). These challenges encompass various aspects of synthesis, characterization, stability, toxicity, scalability, and field implementation. Here's a detailed exploration of the key challenges associated with functionalized NPs.

6.4.1 Synthesis and Functionalization

- *Precise Control*: Achieving precise control over the size, shape, and surface properties of functionalized NPs during synthesis is challenging. Variability in these parameters can significantly impact the NPs' performance and behavior in specific applications.
- *Reproducibility*: Ensuring the reproducibility of functionalized NPs is crucial for large-scale applications. Small variations in synthesis conditions or functionalization processes can lead to inconsistencies in NP properties.

6.4.2 Characterization

- *Characterization Techniques*: The characterization of functionalized NPs requires sophisticated techniques such as Transmission Electron Microscopy (TEM), XPS, and NMR. However, these techniques may have limitations in providing comprehensive information about the size distribution, surface charge, and functional group density.
- *In-Situ Characterization*: Real-time and in-situ characterization of functionalized NPs under relevant operating conditions can be challenging but is essential for understanding their behavior in dynamic environments.

6.4.3 Stability and Agglomeration

- *Agglomeration*: Functionalized NPs may tend to agglomerate, affecting their dispersibility and performance. Strategies to prevent or control agglomeration, such as appropriate surface modifications, are essential for maintaining stability.
- *Long-Term Stability*: Ensuring the long-term stability of functionalized NPs, especially in complex environments like reservoirs, is critical for sustained performance over the desired application period.

6.4.4 Toxicity and Environmental Impact

- *Biocompatibility*: Functionalized NPs may have potential toxicity concerns, especially when applied in biomedical or environmental settings. Evaluating the biocompatibility of these materials is crucial for ensuring safety.
- *Environmental Impact*: The environmental impact of functionalized NPs, including their potential accumulation and persistence, needs to be carefully studied to avoid unintended consequences in ecosystems.

6.4.5 Scalability

- *Production Scalability*: Transitioning from laboratory-scale synthesis to large-scale production of functionalized NPs can be challenging. The scalability of synthesis methods is essential for practical and cost-effective applications.
- *Economic Feasibility*: The cost of producing functionalized NPs, including the expense of functionalization agents and precursor materials, may pose economic challenges, particularly in industries with cost-sensitive applications.

6.4.6 Field Implementation

- *Reservoir Compatibility*: Adapting functionalized NPs for use in specific reservoir conditions presents challenges. Reservoir heterogeneity, varying rock properties, and fluid compositions necessitate tailoring NPs to specific geological settings.
- *Regulatory Compliance*: Complying with regulatory requirements and obtaining approvals for the field application of functionalized NPs is a complex process that requires addressing safety, environmental, and health concerns.

6.4.7 Dynamic Conditions and Responsiveness

- *Adaptability*: Functionalized NPs may need to adapt to dynamic conditions within reservoirs. Developing NPs that respond to changes in temperature, pressure, or chemical composition is an ongoing challenge.
- *Real-Time Monitoring*: Incorporating real-time monitoring systems to track the behavior of functionalized NPs during EOR operations is challenging but crucial for optimizing their performance.

Addressing these challenges requires interdisciplinary collaboration, innovative technologies, and a thorough understanding of both nanomaterial science and the specific requirements of the target application. Ongoing research aims to overcome these challenges and unlock the full potential of functionalized NPs in various fields, including EOR.

6.5 CONCLUSION

In conclusion, this chapter of the book has provided an in-depth exploration of the multifaceted challenges and opportunities associated with working with functionalized NPs in the context of EOR techniques. As we navigate the intricacies of nanotechnology applications, particularly within the dynamic and complex environments of reservoirs, it becomes evident that harnessing the potential of functionalized NPs requires a holistic understanding and strategic approach. The challenges encountered in the synthesis and functionalization of NPs underscore the need for precise control, reproducibility, and scalability in manufacturing processes. Overcoming these hurdles is essential to ensure the reliability and consistency of NPs for large-scale applications. Characterization of functionalized NPs presents its own set of challenges, necessitating advanced techniques and methodologies that allow for real-time and in-situ analysis. Bridging the gap between laboratory characterization and real-world conditions is crucial for the effective deployment of these nanomaterials. Stability concerns, including aggregation and long-term performance, highlight the importance of engineering strategies that maintain the stability of functionalized NPs in diverse environments. Addressing toxicity and environmental impact concerns is imperative for the responsible and sustainable application of these materials. The scalability of production processes and the economic feasibility of functionalized NPs demand innovative solutions to ensure cost-effective implementation on a field scale. Furthermore, considerations related to reservoir compatibility, regulatory compliance, and adaptability to dynamic conditions emphasize the need for comprehensive approaches that encompass geological, engineering, and regulatory perspectives. Despite these challenges, the potential benefits of functionalized NPs in EOR and other applications are profound. Real-world case studies and experimental findings presented in this chapter demonstrate the adaptability and efficacy of these advanced materials across diverse scenarios. Looking forward, the future of functionalized NPs holds exciting possibilities. Responsive NPs that can adapt to changing reservoir conditions, coupled with ongoing advancements in nanotechnology, herald a new era in EOR. Strategic research initiatives, interdisciplinary collaboration, and innovative technologies will be pivotal in overcoming current limitations and realizing the full potential of functionalized NPs in enhancing oil recovery processes.

REFERENCES

Al-Anssari, S., Arif, M., Wang, S., Barifcani, A., Lebedev, M., & Iglauer, S. (2017). Wettability of nano-treated calcite/CO2/brine systems: Implication for enhanced CO2 storage potential. *International Journal of Greenhouse Gas Control*, *66*, 97–105. https://doi.org/10.1016/j.ijggc.2017.09.008.

Behzadi, A., & Mohammadi, A. (2016). Environmentally responsive surface-modified silica nanoparticles for enhanced oil recovery. *Journal of Nanoparticle Research*, *18*(9), 1–19. https://doi.org/10.1007/s11051-016-3580-1.

Christian, P., Von Der Kammer, F., Baalousha, M., & Hofmann, T. (2008). Nanoparticles: Structure, properties, preparation and behaviour in environmental media. *Ecotoxicology*, *17*(5), 326–343. https://doi.org/10.1007/s10646-008-0213-1.

Corredor-Rojas, L. M., Hemmati-Sarapardeh, A., Husein, M. M., Dong, M., & Maini, B. B. (2018). Rheological behavior of surface modified silica nanoparticles dispersed in partially hydrolyzed polyacrylamide and xanthan gum solutions: Experimental measurements, mechanistic understanding, and model development. *Energy and Fuels*, *32*(10), 10628–10638. https://doi.org/10.1021/acs.energyfuels.8b02658.

Dabhade, A. H., Verma, R. P., Paramasivan, B., Kumawat, A., & Saha, B. (2023). Development of silver nanoparticles and aptamer conjugated biosensor for rapid detection of E. coli in a water sample. *3 Biotech*, *13*(7), 244. https://doi.org/10.1007/s13205-023-03663-3.

Estellé, P., Cabaleiro, D., Żyła, G., Lugo, L., & Murshed, S. M. S. (2018). Current trends in surface tension and wetting behavior of nanofluids. In *Renewable and Sustainable Energy Reviews* (Vol. 94, pp. 931–944). Elsevier Ltd. https://doi.org/10.1016/j.rser.2018.07.006.

Fazelabdolabadi, B., Khodadadi, A. A., & Sedaghatzadeh, M. (2015). Thermal and rheological properties improvement of drilling fluids using functionalized carbon nanotubes. *Applied Nanoscience (Switzerland)*, *5*(6), 651–659. https://doi.org/10.1007/s13204-014-0359-5.

Hendraningrat, L., & Torsæter, O. (2014, October). Understanding fluid-fluid and fluid-rock interactions in the presence of hydrophilic nanoparticles at various conditions. In *SPE Asia Pacific Oil & Gas Conference and Exhibition*. https://doi.org/10.2118/171407-MS.

Hunter, T. N., Pugh, R. J., Franks, G. V., & Jameson, G. J. (2008). The role of particles in stabilising foams and emulsions. *Advances in Colloid and Interface Science*, *137*(2), 57–81. https://doi.org/10.1016/J.CIS.2007.07.007.

Jiang, L., Gao, L., & Sun, J. (2003). Production of aqueous colloidal dispersions of carbon nanotubes. *Journal of Colloid and Interface Science*, *260*(1), 89–94. https://doi.org/10.1016/S0021-9797(02)00176-5.

Kumar, A., & Gangawane, K. M. (2022a). Effect of precipitating agents on the magnetic and structural properties of the synthesized ferrimagnetic nanoparticles by co-precipitation method. *Powder Technology*, *401*, 117298. https://doi.org/10.1016/j.powtec.2022.117298.

Kumar, A., & Gangawane, K. M. (2022b). Synthesis and effect on the surface morphology & magnetic properties of ferrimagnetic nanoparticles by different wet chemical synthesis methods. *Powder Technology*, *410*(June), 117867. https://doi.org/10.1016/j.powtec.2022.117867.

Kumar, A., & Gangawane, K. M. (2023a). Advanced computational approaches for water treatment. In *Advanced Computational Approaches for Water Treatment* (pp. 1–12). CRC Press. https://doi.org/10.1201/9781003325147-1.

Kumar, A., & Gangawane, K. M. (2023b). Molecular dynamics for the membrane process. In *Advanced Computational Approaches for Water Treatment* (pp. 13–38). CRC Press. https://doi.org/10.1201/9781003325147-2.

Kumar, A., Gangawane, K. M., & Ramanjaneyulu, B. (2022). Ferrofluids for waste-water treatment. In *International Conference on Chemical, Bio and Environmental Engineering* (pp. 723–744). https://link.springer.com/chapter/10.1007/978-3-030-96554-9_48.

Kumar, A., Jena, H. M., & Gangawane, K. M. (2023a). Effect of ferrimagnetic nanoparticles as an oxygen absorber on the shelf life of tomatoes. Poster presented at the *9th International Food Convention (IFCoN)*, CSIR-CFTRI, Mysuru, India, 9 December 2023.

Kumar, A., Jena, H. M., & Gangawane, K. M. (2023b). Variations in magnetic behavior of ferrimagnetic nanoparticles with temperature. Poster presented at the *75th Indian Chemical Engineering Congress (CHEMCON 2023)*, Dr. H L Roy Building, Jadavpur University Campus, Kolkata, India, 28 December 2023.

Kumar, A., Ujjwal, R. R., Mittal, A., Bansal, A., & Ojha, U. (2014). Polyacryloyl hydrazide: An efficient, simple, and cost effective precursor to a range of functional materials through hydrazide based click reactions. *ACS Applied Materials and Interfaces*, *6*(3), 1855–1865. https://doi.org/10.1021/AM404837F

Lee, J. W., Kim, S., Torres Pineda, I., & Kang, Y. T. (2020). Review of nanoabsorbents for capture enhancement of CO2 and its industrial applications with design criteria. *Renewable and Sustainable Energy Reviews*, *138*, 110524. https://doi.org/10.1016/j.rser.2020.110524.

Li, K., Wang, D., & Jiang, S. (2018). Review on enhanced oil recovery by nanofluids. *Oil & Gas Science and Technology – Revue d'IFP Energies Nouvelles*, *73*, 37. https://doi.org/10.2516/ogst/2018025.

Minz, P., & Mishra, S. (2023). Jackfruit peel derived ZnCl2-impregnated activated carbon: Optimization, characterization, and application in dye removal. *Biomass Conversion and Biorefinery*, 0123456789. https://doi.org/10.1007/s13399-023-04592-0.

Murshed, S. M. S., Leong, K. C., & Yang, C. (2009). A combined model for the effective thermal conductivity of nanofluids. *Applied Thermal Engineering*, *29*(11–12), 2477–2483. https://doi.org/10.1016/j.applthermaleng.2008.12.018.

Nahri, S. Y., Nielsen, J. L., & Chen, Y. (2020). Effect of nanoparticles on the nucleation and agglomeration rates of hydrate growth using THF–water clathrates. *Petroleum Science*, *17*(2), 467–476. https://doi.org/10.1007/s12182-019-00390-4.

Pal, D., & Sen, S. (2023a). Emerging petroleum pollutants and their adverse effects on the environment. In I. D. Behera, & A. P. Das (Eds.), *Impact of Petroleum Waste on Environmental Pollution and Its Sustainable Management through Circular Economy. Environmental Science and Engineering* (pp. 103–137). Springer Nature Switzerland. https://doi.org/10.1007/978-3-031-48220-5_5.

Pal, D., & Sen, S. (2023b). In-depth coverage of petroleum waste sources, characteristics, environmental impact, and sustainable remediation process. In I. D. Behera, & A. P. Das (Eds.), *Impact of Petroleum Waste on Environmental Pollution and Its Sustainable Management through Circular Economy* (pp. 1–38). Springer Nature Switzerland. https://doi.org/10.1007/978-3-031-48220-5_1.

Pal, D., & Sen, S. (2024). Optimal synthesis of dolochar derived faujasite zeolite X for highly effective Cd(II) removal. *Environmental Research*, *240*, 117494. https://doi.org/10.1016/j.envres.2023.117494.

Panda, D., & Gangawane, K. M. (2023a). Development of superhydrophobic hybrid silica-cellulose aerogel as promising thermal insulation and sound absorption. *International Journal of Materials Research*, *114*(4–5), 405–417. https://doi.org/10.1515/ijmr-2022-0066.

Panda, D., & Gangawane, K. M. (2023b). Hybrid NiFe2O4-NiCO4O4 nanoparticles-based eutectic phase change materials for enhancement of thermal efficiency of pin-fin heat sink arrangement. *Journal of Energy Storage*, *60*(November 2022), 106644. https://doi.org/10.1016/j.est.2023.106644.

Panda, D., & Gangawane, K. M. (2023c). Recycled cellulose–silica hybrid aerogel for effective oil adsorption: optimization and kinetics study. *Sadhana - Academy Proceedings in Engineering Sciences,* 48(3). https://doi.org/10.1007/s12046-023-02161-9.

Panda, D., Sahu, A. K., & Gangawane, K. M. (2024). Effect of molar concentration and drying methodologies on monodispersed silica sol for synthesis of silica aerogels with temperature-resistant characteristics. *Materials Science and Engineering: B*, *299*, 117062. https://doi.org/10.1016/j.mseb.2023.117062.

Panda, D., Sharma, S., & Gangawane, K. M. (2023). Synthesis of novel biocatalyst from organic waste protonated with acid treatment for hydrogen production. *International Journal of Hydrogen Energy*, *48*(63), 24242–24254. https://doi.org/10.1016/j.ijhydene.2023.03.117.

Sahoo, A., & Mishra, S. (2023). Synthesis of microporous-activated carbon via direct H3PO4 and KOH activation of Pterospermum acerifolium fruits: process optimization and characterization. *Biomass Conversion and Biorefinery*, 0123456789. https://doi.org/10.1007/s13399-023-05072-1.

Santhosh, R., & Sarkar, P. (2022). Jackfruit seed starch/tamarind kernel xyloglucan/zinc oxide nanoparticles-based composite films: Preparation, characterization, and application on tomato (Solanum lycopersicum) fruits. *Food Hydrocolloids*, *133*(April), 107917. https://doi.org/10.1016/j.foodhyd.2022.107917.

Wu, W., He, Q., & Jiang, C. (2008). Magnetic iron oxide nanoparticles: Synthesis and surface functionalization strategies. *Nanoscale Research Letters*, *3*(11), 397–415. https://doi.org/10.1007/s11671-008-9174-9.

Yang, B., & Han, Z. H. (2006). Temperature-dependent thermal conductivity of nanorod-based nanofluids. *Applied Physics Letters*, *89*(8). https://doi.org/10.1063/1.2338424.

Zhang, Z., Cai, J., Chen, F., Li, H., Zhang, W., & Qi, W. (2018). Progress in enhancement of CO2 absorption by nanofluids: A mini review of mechanisms and current status. *Renewable Energy*, *118*, 527–535. https://doi.org/10.1016/j.renene.2017.11.031.

7 Effects of Silica Nanoparticles Coupled with Low-Salinity Water on Oil Recovery in a Part of Upper Assam Basin, India

Nayan Medhi, Uttaron Goswami and Pushpa Sharma

7.1 INTRODUCTION

The interest in Enhanced Oil Recovery (EOR) techniques has remained at the forefront of the oil and gas sector as a result of the increasing trend in crude oil prices, which is the result of the growing realization that significant new hydrocarbon discoveries are unlikely to emerge anytime soon. This is because many accessible and easily discovered reserves have already been exploited (Schwartz et al., 2011). The remaining undiscovered resources are in more remote or technically challenging areas, making exploration costly and complex. While technology has improved, the geological formations containing large hydrocarbon reserves are becoming difficult to detect and produce. Therefore, more advanced exploration techniques are required to find and access these reserves.

Moreover, the increasing negative impacts on the environment make it challenging to explore and exploit sensitive areas (Mudumba et al., 2023). The rising global demand for hydrocarbon has pushed the petroleum industry to increase the production rate as well as develop advanced recovery techniques. Therefore, the primary objective for any oil and gas company is to maximize the hydrocarbon recovery over the field life (Engeset, 2012; Pal & Sen, 2023a).

The use of nanoparticles for increasing oil recovery is a novel alternative method to be applied as a recent EOR technique that has solved residual oil saturation problems in the last decade (Aliabadian et al., 2020; Binshan et al., 2006; Khalilinezhad et al., 2017; Miranda et al., 2012; Roustaei et al., 2013). This technique has advanced significantly during the past ten years and revealed its potential EOR applications

 DOI: 10.1201/9781032650685-7

(Engeset, 2012; Hendraningrat et al., 2013). In the last decade, numerous research has shown that nanoparticles and nanofluids have improved EOR processes (Cheraghian, 2016; Cheraghian & Tardasti, 2012). Nanoparticles are materials that are made up of a thin shell and a core that ranges in size from 1 nm to 100 nm (Kanj et al., 2009; Krishnamoorti, 2006). Earlier studies have shown that the enhanced properties of nanoparticles, such as a particular surface area, outstanding thermal features, and chemical capabilities, can enhance oil recovery through the modification of rock wettability. It is observed that nanoparticles have the ability to lower water cut, improve sweep efficiency, and lower water relative permeability & capillary force, which can enhance the recovery of oil in the porous media (Li et al., 2012; Tian et al., 2012; Yuan et al., 2010). Moreover, nanoparticles can change the water viscosity and oil-nanofluid Interfacial Tension (IFT) favourably, allowing for greater mobility of the nanofluids to release the trapped oil (Negin et al., 2016). Different nanoparticles such as silicon dioxide (SiO_2), iron oxide (Fe_2O_3/Fe_3O_4), aluminium oxide (Al_2O_3), copper oxide (CuO), nickel(III) oxide (Ni_2O_3), zinc oxide (ZnO), titanium dioxide (TiO_2), magnesium oxide (MgO), zirconium oxide (ZrO_2) etc. have shown positive outcomes in improving the oil recovery (Kumar & Gangawane, 2022a, 2022b; Mohammadi et al., 2011). However, out of the different nanoparticles, 'Silica Nanoparticle' offers the highest possibility due to its affordability, environment friendliness, and availability. The application of Silica Nanoparticles shows improved recovery of oil through oil-nanofluid IFT reduction and rock wettability alteration (Negin et al., 2016; Panda et al., 2023, 2024; Panda & Gangawane, 2023b).

The different suggested mechanisms of nanofluid flooding that increase the oil recovery in a sandstone reservoir are the reduction of oil-brine IFT, wettability alteration of rock, improvement of mobility ratio, formation of structural disjoining pressure, plugging of pore channels, and prevention of asphaltene precipitation (Sun et al., 2017).

Numerous pieces of data demonstrate that nanoparticles can lower the IFT between water and oil. Mass transfer between oil and water controls the IFT changes. As the mass transfer increases, the IFT decreases. Some nanoparticles feature hydrophilic and hydrophobic parts, which help them to move in the bulk fluid and adhere to the fluid interfaces. The primary reasons behind the creation of a nanoparticle layer in the oil-water interface are the propensity of nanoparticles to comply with the interface and their catalytic activity in asphaltene adsorption (Iravani et al., 2023; Minz & Mishra, 2023). When nanoparticles are present, mass transfer between fluids increases due to their propensity to adsorb asphaltenes, and eventually, IFT decreases. It is observed that an optimum concentration of nanoparticles can lower oil-water IFT to about 50% (Hosseini et al., 2019; Kumar et al., 2022).

According to Karimi et al. (2012), the properties of the nanoparticles are regulated by their adsorption on the solid surfaces and surface energies, which alters the solid wettability. This rock wettability plays a crucial role in fluid saturation and relative permeability in the Crude Oil/Brine/Rock (COBR) system. The mineralogy of the reservoir rock surface and the characteristics of the formation fluid have a significant role in wettability. According to earlier research, nanoparticles can increase the water-wetness of the rock, which improves oil recovery (Dehghan Monfared et al., 2016; Kumar & Gangawane, 2023a, 2023b).

Earlier research has observed that the incorporation of nanoparticles into brine can increase the effective viscosity of the nanofluid (Ebrahim et al., 2019). Several factors, such as temperature, nanoparticle concentration, and share rate, affect the nanofluid viscosity (Kumar et al., 2023b; Sun et al., 2017). For example, as the shear rate decreases, the silica nanofluid's viscosity increases. Study shows that when nanoparticle concentration rises, so does the viscosity of the nanofluids. It has also been found that certain nanoparticles ([CuO, Al_2O_3, Fe_2O_3, and NiO]) can reduce the viscosity of light oil as well as heavy oil (Ayatollahi & Zerafat, 2012; Hasçakir et al., 2010; Samba et al., 2019; Shah, 2009; Shokrlu & Babadagli, 2014; Srinivasan & Shah, 2014). Nanoparticles can lower the residual oil saturation by improving the mobility ratio by reducing the oil viscosity and raising the nanofluid viscosity.

Structural disjoining pressure is the pressure that occurs as a result of two surface layers overlaying each other reciprocally. This effect happens due to the combined influence of forces with different natural properties. It is thought that structural disjoining pressure improves the dispersing and wetting of nanofluids across solid surfaces, which leads to the enhancement of oil recovery (Wasan et al., 2011). It is observed that an oil droplet is detached from a solid surface because of structural disjoining pressure applied to a discontinuous phase by a wedge-shaped film (Moon, 2010). According to Birkeland (2013), in nanoparticle EOR, the structural disjoining pressure occurs due to the consolidation of millions of nanoparticles at the oil-aqueous phase interface. Previous studies have shown that the different components of structural disjoining pressure are van der Waals forces or combining forces, that is, electrostatic forces, van der Waals forces, and hydration forces.

In nanoparticle EOR, pore channel plugging may occur due to the log-jamming effect. In this mechanism, pore channels of the reservoir rock are plugged by the flowing nanoparticles. When nanofluids flow through the pore channels, a differential pressure is developed in the pore throats, which causes an increase in nanofluid velocity. As a result, water molecules flow faster, whereas the nanoparticles get accumulated and blocked in the pore throats. As a result, high pressure is developed in the adjacent pore throats, which removes the oil that has become trapped in the throats (Skauge et al., 2010). However, once the oil is forced out from the pore throats, the surrounding pressure drops, and the log-jamming disappears. The different factors that mainly govern pore channel plugging are the diameter of the pore throats, the size and concentration of nanoparticles, and the nanofluid's flow rate.

The process of separating asphalt molecules from crude oil into a solid particle is called asphaltene precipitation, which gets deposited on the tubings, surface facilities, and reservoir rocks. The alteration of pressure, temperature, and crude oil composition may affect the initial equilibrium state of the COBR system, which can cause asphaltene precipitation. Precipitation and deposition of asphaltene decreases the effective permeability of oil, alters rock wettability to more oil wetness, causes pore channel plugging, reduces oil recovery, and sometimes causes formation damage (Doryani et al., 2016; Fakher et al., 2020; Shen & Sheng, 2018; Taheri-Shakib et al., 2018). As a result, residual oil saturation increases in the reservoir. Earlier studies have observed that some nanoparticles (SiO_2, TiO_2, Fe_3O_4, and ZrO_2) can stabilize asphaltene molecules in crude oil and improve the recovery (Kazemzadeh et al., 2015; Kumar et al., 2023a; Mohammadi et al., 2011; Nassar et al., 2011).

On the other hand, another novel EOR method called Low-Salinity Waterflooding (LSW) has been proposed to increase oil recovery in sandstone reservoirs where the salinity of the injection brine is lower than the formation brine and generally below 5,000 ppm (Hilner et al., 2015; Piñerez Torrijos et al., 2016; Qiao et al., 2016; Robbana et al., 2012; Skauge et al., 2010). This technique may be very efficient if the sandstones contain clay minerals, crude oil contains polar compounds, and formation brine contains divalent ions like Ca^{2+} and Mg^{2+} (Alotaibi et al., 2010; Austad et al., 2010; Lager et al., 2008b; Vledder et al., 2010). The mechanisms proposed for LSW include Multi-Component Ion Exchange (MIE) (Lager et al., 2020), Electrical Double Layer (EDL) expansion between fine particles and limited fine release (Lager et al., 2008a; Tang & Morrow, 1999a; Tang & Morrow, 1999b), EDL expansion between rock–oil interfaces, migration of fines, and pH increase of bulk fluid. These mechanisms essentially change the rock's wettability toward more incredible water wetness, improving oil recovery (McGuire et al., 2005).

Recent studies have shown very encouraging oil recovery results during the application of LSW in the study area in both the tertiary and secondary recovery modes (Medhi & Das, 2022a; Medhi & Das, 2022b). Studies have shown that the stability of the nanoparticles in nanofluid is increased in a low-salinity environment (El-sayed et al., 2012). According to earlier studies, nanoparticle-assisted LSW uses both ions and nanoparticles as EOR agents in the injection brine. Recent laboratory experiment shows that the usage of Silica Nanoparticles along with LSW improves oil recovery by modifying rock wettability and mobility ratio (Ebrahim et al., 2019). The experiment shows a 4% additional oil recovery from low-salinity nanofluid flooding over LSW alone. Thus, the low-salinity nanofluid is more effective on EOR than the nanofluid or LSW alone, which is a new hybrid EOR technique in recent times. In this work, a study has been made on the effects of Silica Nanoparticles coupled with low-salinity water on oil recovery in the Tipam Reservoir Sandstone of the Upper Assam Basin.

7.2 MATERIALS AND METHODS

As mentioned above, the study will be conducted in the Tipam Reservoir Sandstone of the Upper Assam Basin. Therefore, for this study, crude oil and formation brine samples were collected from the study area. Additionally, 12 numbers of reservoir rock samples were taken from the Tipam Reservoir Sandstone, which is available at a depth between 2,894 and 2,966.5 m. For preparing the nanofluid solutions, the SiO_2 nanoparticles were purchased from the Sisco Research Laboratories Pvt. Ltd., Maharashtra, India. In contrast, the magnesium chloride ($MgCl_2.6H_2O$), calcium chloride ($CaCl_2.2H_2O$), and sodium chloride (NaCl) were purchased from Molychem, Mumbai. The distilled water (15 l) was prepared in the laboratory for composing the nanofluids.

7.3 RESERVOIR ROCK ANALYSIS

7.3.1 Thin-Section Analysis

A thin rock was cut out from the collected subsurface rock sample and optically flattened for this study. After mounting on a glass slide, the sample was smoothed out using progressively finer abrasive grit until it was just around 30 μm thick. The

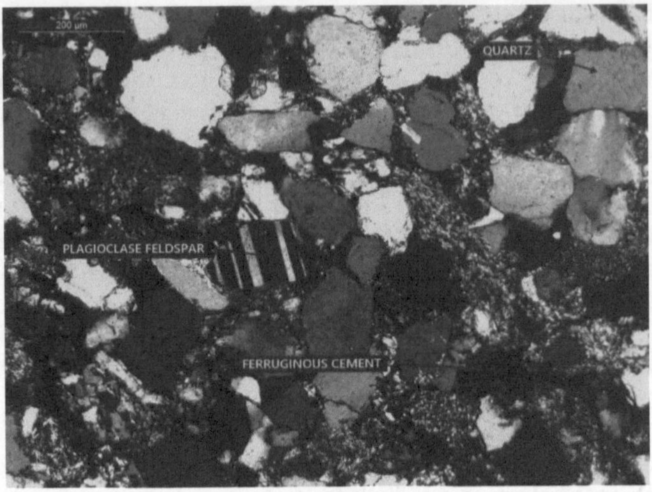

FIGURE 7.1 Photomicrograph showing plagioclase feldspar, ferruginous cement, and quartz.

FIGURE 7.2 Photomicrograph showing ferruginous cement, plagioclase feldspar, and quartz.

section was then examined with a polarized light microscope. The study shows the presence of plagioclase feldspar, ferruginous cement, and quartz in the rock of the study area (Figures 7.1 and 7.2).

7.3.2 Scanning Electron Microscopic Analysis

The optimal size of the rock samples prepared for the Scanning Electron Microscopic (SEM) analysis was around (5 by 10 by 10) [length/width/breadth] mm. Using a

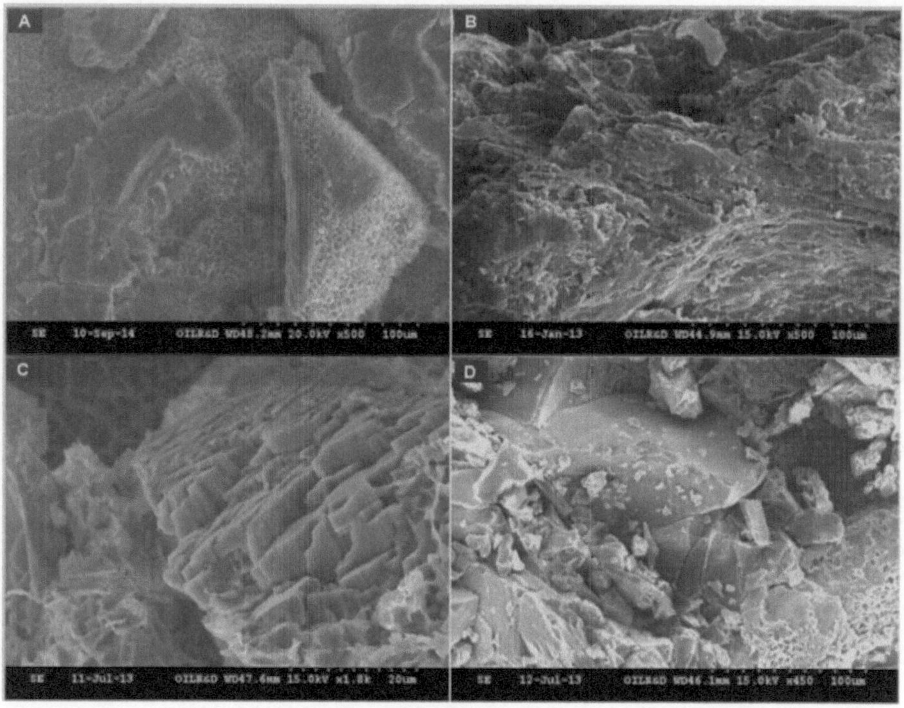

FIGURE 7.3 SEM images showing smectite (A), illite (B), feldspar (C), and quartz (D).

Feron duster, the small debris was removed from the sample surfaces. Carbon tape was used to attach the cut samples to a 15 mm SEM specimen stub so that an electrical ground could be established between the sample and the specimen stub. The splutter coating system (model BIO-RAD SC-502) was then used to apply a gold coating to the samples. The coated samples were put in the multi-holder, placed in the sample chamber, and evacuated to a high vacuum. A SE detector and 20 kV accelerating voltage were used for the analysis of the samples. SEM analysis reveals that smectite, illite, feldspar, and quartz are present in the reservoir rock (Figure 7.3).

7.3.3 X-Ray Diffraction Analysis

Here, the original grain size was decreased to less than 2-micron size for X-Ray Diffraction (XRD) analysis to identify minerals. X-ray diffractometer was used to conduct the study in the 2θ range of 4° to 40°. For this study, less than a 2-micron fraction of the samples was employed after sieving and pipetting. The study demonstrates that the reservoir rock contains smectite, illite, kaolinite, quartz, and feldspar (Figures 7.4 and 7.5).

7.4 ANALYSIS OF THE CRUDE OIL AND RESERVOIR BRINE

Determining the acid number and the presence of polar compounds (asphaltene and resin) is the primary goal of the crude oil analysis because these are essential

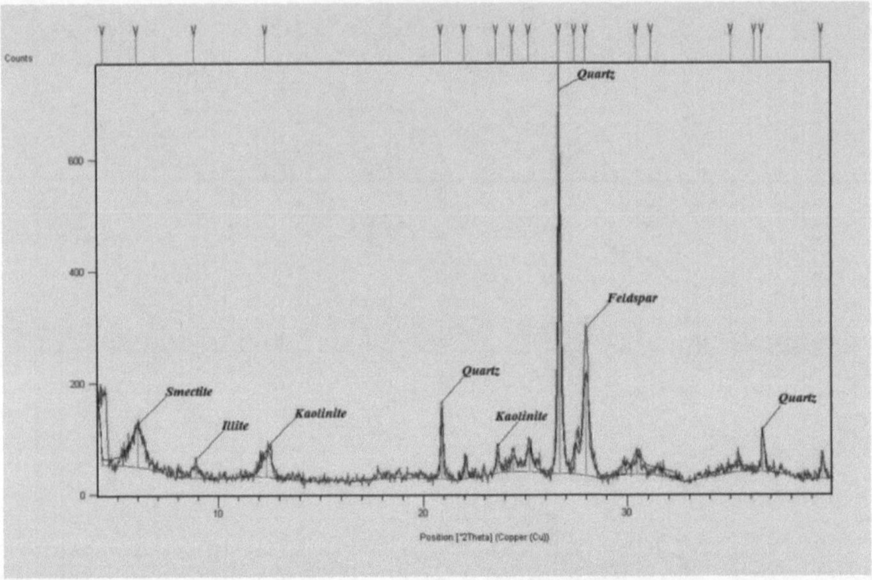

FIGURE 7.4 XRD patterns showing smectite, illite, kaolinite, quartz, and feldspar (Medhi & Das, 2022b).

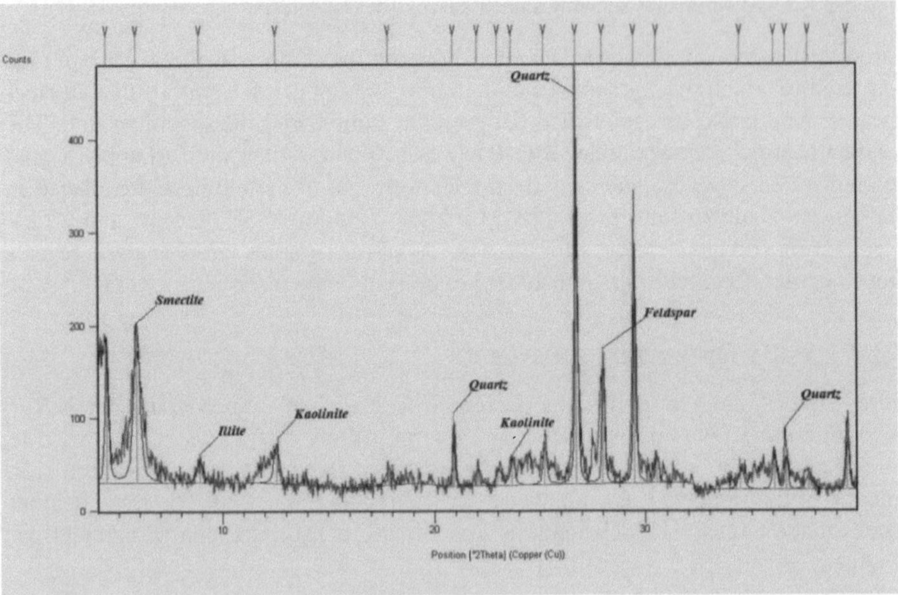

FIGURE 7.5 XRD patterns showing smectite, illite, kaolinite, quartz, and feldspar (Medhi, 2018).

to low-salinity nanofluid flooding. The acid number was determined as per the IP standard (IP 001) for Petroleum and its Products (Part-I). In contrast, the resin and asphaltene were determined based on IOC-AOD and IP143 standards, respectively.

To determine the acid number of oil, a 250 ml conical flask was filled with 10 g of the oil sample. Additionally, 50 ml of alcohol and 1 ml of phenolphthalein solution were mixed in a different flask. After heating the mixture to 40–50°C, a potassium hydroxide solution was added to neutralize it. After adding the neutralized alcohol to the oil sample, the mixture was heated for five minutes and swirled. Then, 1 ml of phenolphthalein solution was added, and it was promptly titrated using a potassium hydroxide solution. The following formula was then used to determine the acid number (Medhi, 2018; Medhi & Das, 2022b).

$$AN = 56.1NV/W \tag{7.1}$$

Where

AN = acid number

N = normality of the potassium hydroxide solution

V = volume of the potassium hydroxide solution (ml)

W = weight of the oil sample (g)

To determine the amount of asphaltene present in the crude oil, 2 ml crude oil was mixed with 50 ml N-heptane. The mixture was heated to 50–60°C and kept for two hours. A filter paper was used to separate the mixture of N-heptane+resin and asphaltene (gravity filtration). After collecting the N-heptane+resin mixture in a beaker, benzene was mixed with the residue, which dissolved the asphaltene. The benzene+asphaltene mixture was then collected in a beaker and heated to about 80°C. As a result, benzene boiled out, leaving behind the asphaltene. Next, the weight of the asphaltene was calculated. Further, silica gel was added with the N-heptane+resin mixture, where the silica gel absorbed resin. The solid layer of silica+resin deposited at the bottom of the beaker was then separated from the N-heptane and kept in a filter paper where benzene was mixed. Benzene dissolved the resin and filtered it out, leaving behind silica on the filter paper. The filtrate (benzene+resin) was then heated to about 80°C. Benzene boiled out, leaving behind the resin in the beaker. The weight of the resin was then measured (Medhi, 2020).

According to the analysis, the crude oil has a 0.55% (w/w) asphaltene and a 15.49% (w/w) resin content. The acid number of the oil is 0.58 (Medhi & Das, 2022b). Earlier studies have observed that a high acid number (>0.2) of crude oil containing resin and asphaltene is crucial for wettability alteration in the direction of more water wetness during LSW that results in improved recovery of oil (Ehrlich et al., 1974; Ehrlich & Wygal, 1977; Jensen & Radke, 1988; Lager et al., 2006).

The analysis of the formation brine was done to determine mainly the pH, salinity, and the presence of divalent cations. The salinity of the brine was determined by 'Titration with Silver Nitrate', whereas the divalent cation concentrations were determined by 'Complexometric Titration (Using EDTA)'. The pH of the brine was measured using a Metrohm pH meter. The analysis shows that the pH, salinity (as NaCl), calcium ion concentration, and magnesium ion concentration in the formation brine of the study area are 8.10, 1,404 ppm, 6 ppm, and 8 ppm, respectively (Minati & Nayan, 2014). Previous studies have observed that salinity, pH, and the divalent cations of brine control the COBR system stability, which affects oil production (Boussour et al., 2009).

7.5 STUDIES ON ROCK WETTABILITY AND CRUDE OIL-NANOFLUID IFT

As mentioned earlier, wettability and oil-nanofluid IFT alteration are some of the significant suggested mechanisms of nanofluid flooding that result in better oil recovery. Here, the effects of the Silica Nanoparticle concentrations (0–0.4 wt%) on rock wettability were studied by determining the contact angles of the rock. The values of the contact angle determine the wettability state of the rock, which has a range of values between 0° and 180°. If the contact angle is less than 90°, the surface is called water-wet, and if it is more significant, it is called oil-wet. A neutral system exhibits a 90° contact angle. For this study, 0.4-inch-thick rock samples were taken, which were first cleaned in the Soxhlet extractor using a mixture of methanol (20%), toluene (40%), and chloroform (40%), and then in ultrasonic cleaner using acetone. The samples were then properly dried in the humidity cabinet at 63°C dry bulb temperature and 40% relative humidity. It is already observed that a low-salinity brine of 800 ppm (as NaCl) salinity is able to provide the highest oil recovery efficiency in the study area (Medhi & Das, 2022b). Therefore, for this study, the brine salinity was taken at 800 ppm (as NaCl) to prepare the low-salinity nanofluid. After saturating the cleaned and dried samples in the vacuum desiccator using the prepared nanofluids and crude oil, the contact angles were determined using 'Drop Shape Analyzer - DSA25' at atmospheric pressure and 70°C. Also, the effects of the Silica Nanoparticle concentrations (0–0.4 wt%) on crude oil-nanofluid IFT were studied using the same Drop Shape Analyzer - DSA25 at 70°C. The study's findings are shown in Table 7.1.

TABLE 7.1
The alteration of contact angle and crude oil-nanofluid IFT with various concentrations of SiO_2 nanoparticles

Sl. no.	NP concentration (wt%)	Contact angle (degree)	Crude oil-nanofluid IFT (mN/m)
1	0	44.56	36.78
2	0.01	42.34	31.56
3	0.04	40.86	28.16
4	0.08	38.78	25.98
5	0.15	37.82	24.14
6	0.2	35.24	20.66
7	0.3	33.76	20.24
8	0.4	33.12	19.98

Abbreviations: IFT, Interfacial Tension; NP, Nanoparticle.

7.6 RESULTS AND DISCUSSION

It is found that thermal, chemical, and miscible methods are the three major classifications of EOR (Ayatollahi & Zerafat, 2012; Kong & Ohadi, 2010; Viebahn et al., 2015). The wettability alteration, mobility control, IFT reduction, gravity drainage, and change of physical properties of the reservoir are the reasons behind the improved recovery of oil by the mentioned-above EOR techniques (Alvarado & Manrique, 2010; De Souza et al., 2005; Gonzalez Da Silva et al., 2007; Pal & Sen, 2024). However, the drawbacks of these EOR methods are the loss of chemicals in formation (Seright et al., 2006), high mobility ratio of oil and gas (Sun et al., 2015), high cost of chemicals, and lower volumetric sweep efficiency (Fokker, 1996). Moreover, some of these methods have detrimental effects on the reservoir's health. However, low-salinity nanofluid flooding is a newly developed, inexpensive, and green hybrid EOR method. Of all the nanoparticles used in EOR, 'Silica Nanoparticle' exhibits the most significant potential due to its affordability, accessibility, and eco-friendliness. As previously stated, Silica Nanoparticles in EOR show better recovery of oil through rock wettability alternation and decrease oil-aqueous solution IFT (Negin et al., 2016; Panda & Gangawane, 2023a).

The application of nanoparticles coupled with LSW is a new hybrid EOR technique where it is possible to combine the strengths of both nanoparticles and LSW to enhance the recovery of oil. However, the prerequisites for implementing this hybrid EOR in a sandstone reservoir are the existence of certain clay minerals, divalent cations, and polar organic components in the reservoir COBR system as mentioned above. Moreover, the effectiveness of environmentally friendly nanoparticles, like 'Silica Nanoparticles', in lowering residual oil saturation has been observed in sandstone reservoirs (Pal & Sen, 2023b).

The XRD and petrographic analysis of the reservoir rock samples from the study area reveal that smectite, illite, kaolinite, quartz, plagioclase feldspar, and ferruginous cement are present (Figures 7.1–7.5). It is observed that smectite, illite, and kaolinite clay minerals can recover more oil during LSW through the migration of clays (mainly kaolinite), oil-brine IFT reduction, and sweep efficiency improvement (de Bruin, 2012; Rezaeidoust et al., 2009). Additionally, the presence of plagioclase feldspar in the rock strongly suggests that the research area is appropriate for the application of low-salinity nanofluid EOR (Reinholdtsen et al., 2011).

The analysis of the crude oil of the study area shows that it contains 0.55% (w/w) asphaltene and 15.49% (w/w) resin. Also, the acid number of the crude oil is 0.58. As previously stated, crude oil with an acid number higher than 0.2 and containing polar components is suitable for changes in rock wettability in the direction of more water wetness during LSW, which results in improved recovery of oil. The reservoir brine analysis reveals that the pH, salinity (as NaCl), calcium ion concentration, and magnesium ion concentration of the brine are 8.10, 1,404 ppm, 6 ppm, and 8 ppm, respectively. It is observed that the salinity, pH, and divalent cations of brine control the stability of the reservoir COBR system, which influences oil production. These divalent cations (Ca^{2+} and Mg^{2+}) can improve the oil recovery through the MIE mechanism, where the organic components are replaced from the clay surfaces with simple cations, leading to changes in wettability to a more water-wet state (Anderson, 1987).

An experimental investigation was done on the impact of the low-salinity silica nanofluids on rock wettability and crude oil-nanofluid IFT. As mentioned earlier, the concentration of the SiO_2 nanoparticles was taken from 0 wt% to 0.4 wt% in 800 ppm (as NaCl) brine. The study shows that both the contact angle and IFT decrease with the increase in nanoparticle concentration (Table 7.1). The contact angle decreases from 44.56° (in the absence of nanoparticles) to 33.12° (with 0.4 wt% nanoparticle concentration). This suggests that when the concentration of nanoparticles increases, the wettability of the rock shifts to a more water-wet state. According to earlier research, this alteration of rock wettability is due to the nanoparticle adsorption on solid surfaces, which are regulated by their adsorption affinity and surface energies. The 'Structural Disjoining Pressure' can improve the dispersing and wetting of nanofluids on the surfaces of rocks, which leads to the improvement of oil recovery.

Moreover, pH increase, MIE, EDL expansion, and acceptable migration are the other mechanisms of low-salinity nanofluid flooding that can favourably alter rock wettability, as mentioned above. Earlier studies have already proved that fine migration and pH increase occur in the core plugs of the study area during the injection of low-salinity water. Also, the crude oil-nanofluid IFT decreases from 36.78 mN/m (in the absence of nanoparticles) to 19.98 mN/m (with 0.4 wt% nanoparticle concentration). This reduction of IFT is a crucial element that can improve the recovery of oil. The reasons behind the decrease of the IFT may be the migration of fines, an increase in the pH of the bulk fluid, and the propensity of nanoparticles to adhere to the oil-brine interface.

This hybrid EOR technique shows promise in improving oil recovery by leveraging the combined effects of nanoparticles and LSW. Its ability to shift the rock wettability towards a more water-wet state can facilitate better oil recovery from the reservoir. Certain clay minerals present in the reservoir, such as kaolinite, can aid in oil recovery by reducing IFT during LSW injection. Moreover, kaolinite, illite, and smectite can improve the sweep efficiency during low-salinity brine injection. As observed in the experimental studies, lowering the IFT between crude oil and nanofluid can significantly enhance oil production. Using eco-friendly nanoparticles like Silica Nanoparticles enhances the technique's appeal in the context of sustainable practices.

However, the success of this technique heavily relies on the presence of certain clay minerals, divalent cations, and polar organic components in the reservoir. Not all reservoirs might possess these characteristics, limiting their applicability. The effectiveness of the method is contingent upon the chemistry of the reservoir brine, including pH, salinity, and ion concentrations. Deviations from ideal conditions might diminish the technique's efficacy. While the IFT and wettability changes are observed with increasing nanoparticle concentration, finding an optimal concentration without inducing negative effects on the reservoir's health is crucial. The full-scale efficiency and long-term effects of this technique need further validation through extensive field trials and prolonged studies to ensure its practical viability.

7.7 CONCLUSION

The prospect of combining LSW with Silica Nanoparticles to improve oil recovery in the Upper Assam Basin's Tipam Sandstone has been examined through a series of laboratory studies. According to the study, low-salinity silica nanofluid flooding might favourably enhance oil recovery through the modification of rock wettability and oil-nanofluid IFT. Therefore, it is feasible to apply this hybrid EOR technique in the study area, which contains clay minerals, multivalent cations, and polar components in its COBR system. In order to determine their impact on oil recovery, more research is required on a number of laboratory core flooding studies utilizing various concentrations of Silica Nanoparticles combined with low-salinity brine.

ACKNOWLEDGEMENT

The authors are thankful to the Oil and Natural Gas Corporation Limited (ONGCL), Jorhat & Nazira, Assam, India, for providing the required samples and data for this work. The authors also thank Oil India Limited (OIL), Duliajan, and Assam, India, for providing laboratory facilities for analyzing oil, rock, and brine samples. The authors are also grateful to Dr. Dhrubajyoti Neog, Assistant Professor, Department of Petroleum Technology, Dibrugarh University, for providing the laboratory facility for the determination of contact angle and IFT.

REFERENCES

Aliabadian, E., Sadeghi, S., Rezvani Moghaddam, A., Maini, B., Chen, Z., & Sundararaj, U. (2020). Application of graphene oxide nanosheets and HPAM aqueous dispersion for improving heavy oil recovery: Effect of localized functionalization. *Fuel*, *265*(December 2019), 116918. https://doi.org/10.1016/j.fuel.2019.116918.

Alotaibi, M. B., Azmy, R. M., & Nasr-El-Din, H. A. (2010). A comprehensive EOR study using low salinity water in sandstone reservoirs. *SPE - DOE Improved Oil Recovery Symposium Proceedings*, *2*, 1454–1473. https://doi.org/10.2118/129976-ms.

Alvarado, V., & Manrique, E. (2010). Enhanced oil recovery: An update review. *Energies*, *3*(9), 1529–1575. https://doi.org/10.3390/en3091529.

Anderson, W. G. (1987). Wettability literature survey - Part 5: The effects of wettability on relative permeability. *JPT, Journal of Petroleum Technology*, *39*(11), 1453–1468. https://doi.org/10.2118/16323-PA.

Austad, T., RezaeiDoust, A., & Puntervold, T. (2010). Chemical mechanism of low salinity water flooding in sandstone reservoirs. *All Days*, *1*, 679–695. https://doi.org/10.2118/129767-MS.

Ayatollahi, S., & Zerafat, M. M. (2012). Nanotechnology-assisted EOR techniques: New solutions to old challenges. In *Society of Petroleum Engineers - SPE International Oilfield Nanotechnology Conference* 2012 (pp. 382–396). https://doi.org/10.2118/157094-ms.

Binshan, J., Tailiang, F., & Mingxue, M. (2006). Enhanced oil recovery by flooding with hydrophilic nanoparticles. *China Particuology*, *4*(1), 41–46.

Birkeland, M. Å. (2013). Investigation of nanoparticle effect on wettability and interfacial tension [Norwegian University of Science and Technology]. In *Petroleum Geoscience* (Issue June). https://hdl.handle.net/11250/240096.

Boussour, S., Cissokho, M., Cordier, P., Bertin, H., & Hamon, G. (2009). Oil recovery by low salinity brine injection: Laboratory results outcrop and reservoir cores. *Proceedings - SPE Annual Technical Conference and Exhibition*, *3*, 1595–1606. https://doi.org/10.2118/124277-ms.

Cheraghian, G. (2016). A new thermal method concept for ior from oil reservoir using optimized in-situ combustion. *78th EAGE Conference and Exhibition 2016: Efficient Use of Technology - Unlocking Potential, 2016*(1), 1–3. https://doi.org/10.3997/2214-4609.201600923.

Cheraghian, G., & Tardasti, S. (2012). Improved oil recovery by the efficiency of nanoparticle in imbibition mechanism. In *74th European Association of Geoscientists and Engineers Conference and Exhibition 2012 Incorporating SPE EUROPEC 2012: Responsibly Securing Natural Resources* (pp. 5486–5488). https://doi.org/10.3997/2214-4609.20142967.

de Bruin, W. J. (2012). Simulation of geochemical processes during low salinity water flooding by coupling multiphase Buckley-Leverett flow to the geochemical package PHREEQC. In *Delpht Msc Thesis* (Issue August). Delft University of Technology.

De Souza, J. C., Da Cursino, D. F. S., & De Pádua, K. G. O. (2005). Twenty years of steam injection in heavy-oil fields. In *SPE Latin American and Caribbean Petroleum Engineering Conference Proceedings, 2005-June*. https://doi.org/10.2118/94808-ms.

Dehghan Monfared, A., Ghazanfari, M. H., Jamialahmadi, M., & Helalizadeh, A. (2016). Potential application of silica nanoparticles for wettability alteration of oil–wet calcite: A mechanistic study. *Energy & Fuels*, *30*(5), 3947–3961. https://doi.org/10.1021/acs.energyfuels.6b00477.

Doryani, H., Malayeri, M. R., & Riazi, M. (2016). Visualization of asphaltene precipitation and deposition in a uniformly patterned glass micromodel. *Fuel*, *182*, 613–622. https://doi.org/10.1016/j.fuel.2016.06.004.

Ebrahim, T., Mohsen, V. S., Mahdi, S. M., Esmaeel, K. T., & Saeb, A. (2019). Performance of low-salinity water flooding for enhanced oil recovery improved by SiO2 nanoparticles. *Petroleum Science*, *16*(2), 357–365. https://doi.org/10.1007/s12182-018-0295-1.

Ehrlich, R., Hasiba, H. H., & Raimondi, P. (1974). Alkaline waterflooding for wettability alteration - Evaluating a potential field application. *JPT, Journal of Petroleum Technology*, *26*(12), 1335–1343. https://doi.org/10.2118/4905-PA.

Ehrlich, R., & Wygal, R. J. (1977). Interrelation of crude oil and rock properties with the recovery of oil by caustic waterflooding. *Society of Petroleum Engineers of AIME Journal*, *17*(4), 263–270. https://doi.org/10.2118/5830-pa.

El-sayed, G. M., Kamel, M. M., Morsy, N. S., & Taher, F. A. (2012). Encapsulation of nano Disperse Red 60 via modified miniemulsion polymerization. I. Preparation and characterization. *Journal of Applied Polymer Science*, *125*(2), 1318–1329. https://doi.org/10.1002/app.35102.

Engeset, B. (2012). The potential of hydrophilic silica nanoparticles for EOR purposes. In *MSC Thesis* (Issue May). Norwegian University of Science and Technology.

Fakher, S., Ahdaya, M., Elturki, M., & Imqam, A. (2020). Critical review of asphaltene properties and factors impacting its stability in crude oil. *Journal of Petroleum Exploration and Production Technology*, *10*(3), 1183–1200. https://doi.org/10.1007/s13202-019-00811-5.

Fokker, P. (1996). Principles of petroleum reservoir engineering, volume 2. *Earth-Science Reviews*, *40*(1–2), 169–170. https://doi.org/10.1016/0012-8252(96)90067-7.

Gonzalez Da Silva, I. P., De Melo, M. A., Luvizotto, J. M., & Lucas, E. F. (2007). Polymer flooding: A sustainable enhanced oil recovery in the current scenery. *Proceedings of the SPE Latin American and Caribbean Petroleum Engineering Conference*, *2*, 999–1007. https://doi.org/10.2118/107727-ms.

Hasçakir, B., Babadagli, T., & Akin, S. (2010). Field-scale analysis of heavy-oil recovery by electrical heating. *SPE Reservoir Evaluation and Engineering*, *13*(1), 131–142. https://doi.org/10.2118/117669-PA.

Hendraningrat, L., Li, S., & Torsæter, O. (2013). Enhancing oil recovery of low-permeability Berea sandstone through optimized nanofluids concentration. In *Society of Petroleum Engineers - SPE Enhanced Oil Recovery Conference, EORC 2013: Delivering the Promise NOW!* (pp. 713–722). https://doi.org/10.2118/165283-ms.

Hilner, E., Andersson, M. P., Hassenkam, T., Matthiesen, J., Salino, P. A., & Stipp, S. L. S. (2015). The effect of ionic strength on oil adhesion in sandstone - The search for the low salinity mechanism. *Scientific Reports*, *5*, 1–9. https://doi.org/10.1038/srep09933.

Hosseini, E., Hajivand, F., Yaghodous, A., & Soltani, R. (2019). Experimental investigation of the effect of dispersed silica and alumina nanoparticles on oil-aqueous phase interfacial tension. *Petroleum Science and Technology*, *37*(13), 1485–1494. https://doi.org/10.1080/10916466.2018.1476530.

Iravani, M., Khalilnezhad, Z., & Khalilnezhad, A. (2023). A review on application of nanoparticles for EOR purposes: History and current challenges. *Journal of Petroleum Exploration and Production Technology*, *13*(4). https://doi.org/10.1007/s13202-022-01606-x.

Jensen, J. A., & Radke, C. J. (1988). Chromatographic transport of alkaline buffers through reservoir rock. *SPE Reservoir Engineering (Society of Petroleum Engineers)*, *3*(3), 849–856. https://doi.org/10.2118/14295-PA.

Kanj, M. Y., Funk, J. J., & Al-Yousif, Z. (2009). Nanofluid coreflood experiments in the ARAB-D. *All Days*, SPE-126161. https://doi.org/10.2118/126161-MS.

Karimi, A., Fakhroueian, Z., Bahramian, A., Pour Khiabani, N., Darabad, J. B., Azin, R., & Arya, S. (2012). Wettability alteration in carbonates using zirconium oxide nanofluids: EOR implications. *Energy & Fuels*, *26*(2), 1028–1036. https://doi.org/10.1021/ef201475u.

Kazemzadeh, Y., Eshraghi, S. E., Kazemi, K., Sourani, S., Mehrabi, M., & Ahmadi, Y. (2015). Behavior of asphaltene adsorption onto the metal oxide nanoparticle surface and its effect on heavy oil recovery. *Industrial and Engineering Chemistry Research*, *54*(1), 233–239. https://doi.org/10.1021/ie503797g.

Khalilinezhad, S. S., Cheraghian, G., Roayaei, E., Tabatabaee, H., & Karambeigi, M. S. (2017). Improving heavy oil recovery in the polymer flooding process by utilizing hydrophilic silica nanoparticles. *Energy Sources, Part A: Recovery, Utilization and Environmental Effects*, *12*, 1–10. https://doi.org/10.1080/15567036.2017.1302521.

Kong, X., & Ohadi, M. M. (2010). Applications of micro and nano technologies in the oil and gas industry-an overview of the recent progress. *Society of Petroleum Engineers -14th Abu Dhabi International Petroleum Exhibition and Conference 2010, ADIPEC 2010, 3*, 1703–1713. https://doi.org/10.2118/138241-ms.

Krishnamoorti, R. (2006). Technology tomorrow: Extracting the benefits of nanotechnology for the oil industry. *Journal of Petroleum Technology*, *58*(11), 24–26. https://doi.org/10.2118/1106-0024-JPT.

Kumar, A., & Gangawane, K. M. (2022a). Effect of precipitating agents on the magnetic and structural properties of the synthesized ferrimagnetic nanoparticles by co-precipitation method. *PowderTechnology*, *401*, 117298. https://doi.org/10.1016/j.powtec.2022.117298.

Kumar, A., & Gangawane, K. M. (2022b). Synthesis and effect on the surface morphology & magnetic properties of ferrimagnetic nanoparticles by different wet chemical synthesis methods. *Powder Technology*, *410*(June), 117867. https://doi.org/10.1016/j.powtec.2022.117867.

Kumar, A., & Gangawane, K. M. (2023a). Advanced computational approaches for water treatment. In *Advanced Computational Approaches for Water Treatment* (pp. 1–12). CRC Press. https://doi.org/10.1201/9781003325147-1.

Kumar, A., & Gangawane, K. M. (2023b). Molecular dynamics for the membrane process. In *Advanced Computational Approaches for Water Treatment* (pp. 13–38). CRC Press. https://doi.org/10.1201/9781003325147-2.

Kumar, A., Gangawane, K. M., & Ramanjaneyulu, B. (2022). *Ferrofluids for Waste-Water Treatment BT - Advances in Chemical, Bio and Environmental Engineering* (J. K. Ratan, D. Sahu, N. N. Pandhare, & A. Bhavanam (eds.); pp. 723–744). Springer International Publishing.

Kumar, A., Jena, H. M., & Gangawane, K. M. (2023a). Effect of ferrimagnetic nanoparticles as an oxygen absorber on the shelf life of tomatoes. In *9th International Food Convention (IFCoN)*, "TRIMSAFE – Technology Re-engineering for Innovation and Mitigating the risk for a Safe, sustainable Affordable & secure Food Eco-system", CSIR-CFTRI campus, Mysuru, 7–10 December 2023. http://hdl.handle.net/2080/4172.

Kumar, A., Jena, H. M., & Gangawane, K. M. (2023b). Variations in magnetic behavior of ferrimagnetic nanoparticles with temperature. In *76th Indian Chemical Engineering Congress (CHEMCON)*, Kolkata, India, 27–30 December 2023. http://hdl.handle.net/2080/4265.

Lager, A., Webb, K. J., & Black, C. J. J. (2020). Impact of brine chemistry on oil recovery. In *IOR 2007–14th European Symposium on Improved Oil Recovery*, cp–24. https://doi.org/10.3997/2214-4609-pdb.24.a24.

Lager, A., Webb, K. J., Black, C. J. J., Singleton, M., & Sorbie, K. S. (2008a). Low salinity oil recovery - An experimental investigation. *Petrophysics*, *49*(1), 28–35. https://onepetro.org/petrophysics/article-abstract/171146/Low-Salinity-Oil-Recovery-An-Experimental.

Lager, A., Webb, K. J., Collins, I. R., & Richmond, D. M. (2008b). LoSalTM enhanced oil recovery: Evidence of enhanced oil recovery at the reservoir scale. *Proceedings - SPE Symposium on Improved Oil Recovery*, *3*, 1313–1324. https://doi.org/10.2118/113976-ms.

Lager, A., Collins, I. R., Jerauld, G. R., Mcguire, P. L., & Webb, K. (2006). *BP Low Salinity Patent Application (Hydrocarbon Recovery Process). PCT/GB2007/003337*-WO 2008/029124 *A1*. PCT/GB2007/003337-WO 2008/029124.

Li, X., Ying, Z., Jia, Y., Liu, X., Yang, T., & Ma, L. (2012). Application of nanosphere deep profile control and displacement technology in Chanqing oilfield. *Oilfield Chem*, *29*(4), 419–422.

McGuire, P., Chatham, J., Paskvan, F., Sommer, D., & Carini, F. (2005). Low salinity oil recovery: An exciting new EOR opportunity for Alaska's north slope. In *Proceedings of SPE Western Regional Meeting* (pp. 439–453). https://doi.org/10.2523/93903-MS.

Medhi, N. (2018). A study on Fine'S migration during low salinity water flooding in Geleki oilfield of Upper Assam Basin. *Journal of Applied Geochemistry*, *20*(3), 344–351.

Medhi, N. (2020). Feasibility study of low salinity waterflooding in a part of Cambay Basin, India. *Research Journal of Chemistry and Environment*, *24*(12), 125–133.

Medhi, N., & Das, M. (2022a). Experimental analysis of low salinity waterflooding in tertiary recovery mode in Geleki oil field of Upper Assam Basin, India. *Advances in Science, Technology and Innovation*, 413–415. https://doi.org/10.1007/978-3-030-73026-0_95.

Medhi, N., & Das, M. (2022b). An experimental study of low salinity waterflooding in secondary recovery mode in Geleki oil field of Assam-Arakan Basin, India. *Indian Journal of Natural Sciences*, *13*(71), 39633–39644.

Minati, D., & Nayan, M. (2014). A study on the effects of crude oil/brine/rock interaction on oil recovery during low saline waterflooding in a part of Geleki oil field of Upper Assam Basin. *Indian Journal of Applied Geochemistry*, *16*(3), 267–276. https://www.appliedgeochemistsindia.org.in/pdf/vol16/no3/5.pdf.

Minz, P., & Mishra, S. (2023). Jackfruit peel derived ZnCl2-impregnated activated carbon: Optimization, characterization, and application in dye removal. *Biomass Conversion and Biorefinery*, 0123456789. https://doi.org/10.1007/s13399-023-04592-0.

Miranda, C. R., de Lara, L. S., & Tonetto, B. C. (2012). Stability and mobility of functionalized silica nanoparticles for enhanced oil recovery applications. *All Days*, 311–321. https://doi.org/10.2118/157033-MS.

Mohammadi, M., Akbari, M., Fakhroueian, Z., Bahramian, A., Azin, R., & Arya, S. (2011). Inhibition of asphaltene precipitation by TiO2, SiO2, and ZrO2 nanofluids. *Energy and Fuels*, *25*(7), 3150–3156. https://doi.org/10.1021/ef2001635.

Moon, T. (2010). Nanofluid technology promises large-scale performance gains from tight reservoirs. *Journal of Petroleum Technology*.

Mudumba, T., Stimpson, B., Jingo, S., & Montgomery, R. A. (2023). The implications of global oil exploration for the conservation of terrestrial wildlife. *Environmental Challenges*, *11*(March), 100710. https://doi.org/10.1016/j.envc.2023.100710.

Nassar, N. N., Hassan, A., & Pereira-Almao, P. (2011). Metal oxide nanoparticles for asphaltene adsorption and oxidation. *Energy & Fuels*, *25*(3), 1017–1023. https://doi.org/10.1021/ef101230g.

Negin, C., Ali, S., & Xie, Q. (2016). Application of nanotechnology for enhancing oil recovery – A review. *Petroleum*, *2*(4), 324–333. https://doi.org/10.1016/j.petlm.2016.10.002.

Pal, D., & Sen, S. (2023a). Emerging petroleum pollutants and their adverse effects on the environment. In I. D. Behera, & A. P. Das (Eds.), *Impact of Petroleum Waste on Environmental Pollution and Its Sustainable Management through Circular Economy. Environmental Science and Engineering* (pp. 103–137). Springer Nature Switzerland. https://doi.org/10.1007/978-3-031-48220-5_5.

Pal, D., & Sen, S. (2023b). In-depth coverage of petroleum waste sources, characteristics, environmental impact, and sustainable remediation process. In I. D. Behera, & A. P. Das (Eds.), *Impact of Petroleum Waste on Environmental Pollution and Its Sustainable Management through Circular Economy* (pp. 1–38). Springer Nature Switzerland. https://doi.org/10.1007/978-3-031-48220-5_1.

Pal, D., & Sen, S. (2024). Optimal synthesis of dolochar derived faujasite zeolite X for highly effective Cd(II) removal. *Environmental Research*, *240*, 117494. https://doi.org/10.1016/j.envres.2023.117494.

Panda, D., & Gangawane, K. M. (2023a). Hybrid NiFe2O4-NiCO4O4 nanoparticles-based eutectic phase change materials for enhancement of thermal efficiency of pin-fin heat sink arrangement. *Journal of Energy Storage*, *60*(November 2022), 106644. https://doi.org/10.1016/j.est.2023.106644.

Panda, D., & Gangawane, K. M. (2023b). Recycled cellulose–silica hybrid aerogel for effective oil adsorption: Optimization and kinetics study. *Sadhana - Academy Proceedings in Engineering Sciences*, *48*(3). https://doi.org/10.1007/s12046-023-02161-9.

Panda, D., Sahu, A. K., & Gangawane, K. M. (2024). Effect of molar concentration and drying methodologies on monodispersed silica sol for synthesis of silica aerogels with temperature-resistant characteristics. *Materials Science and Engineering: B*, *299*, 117062. https://doi.org/10.1016/j.mseb.2023.117062.

Panda, D., Sharma, S., & Gangawane, K. M. (2023). Synthesis of novel biocatalyst from organic waste protonated with acid treatment for hydrogen production. *International Journal of Hydrogen Energy*, *48*(63), 24242–24254. https://doi.org/10.1016/j.ijhydene.2023.03.117.

Piñerez Torrijos, I. D., Puntervold, T., Strand, S., Austad, T., Abdullah, H. I., & Olsen, K. (2016). Experimental study of the response time of the low-salinity enhanced oil recovery effect during secondary and tertiary low-salinity waterflooding. *Energy & Fuels*, *30*(6), 4733–4739. https://doi.org/10.1021/acs.energyfuels.6b00641.

Qiao, C., Johns, R., & Li, L. (2016). Modeling low-salinity waterflooding in chalk and limestone reservoirs. *Energy & Fuels*, *30*(2). https://doi.org/10.1021/acs.energyfuels.5b02456.

Reinholdtsen, A. J., RezaeiDoust, A. R., Strand, S., & Austad, T. (2011). Why such a small low salinity EOR – Potential from the Snorre formation? In *16th European Symposium on Improved Oil Recovery 2011* (pp. 748–757). https://doi.org/10.3997/2214-4609.201404796.

Rezaeidoust, A., Puntervold, T., Strand, S., & Austad, T. (2009). Smart water as wettability modifier in carbonate and sandstone: A discussion of similarities/differences in the chemical mechanisms. *Energy and Fuels*, *23*(9), 4479–4485. https://doi.org/10.1021/ef900185q.

Robbana, E., Buikema, T., Mair, C., Williams, D., Mercer, D., Webb, K., Hewson, A., & Reddick, C. (2012). Low salinity enhanced oil recovery - Laboratory to day one field implementation - LoSal EOR into the Clair Ridge Project. *All Days*, *4*, 2582–2596. https://doi.org/10.2118/161750-MS.

Roustaei, A., Saffarzadeh, S., & Mohammadi, M. (2013). An evaluation of modified silica nanoparticles' efficiency in enhancing oil recovery of light and intermediate oil reservoirs. *Egyptian Journal of Petroleum*, *22*(3), 427–433. https://doi.org/10.1016/j.ejpe.2013.06.010.

Samba, M. A., Hassan, H. A., Munayr, M. S., Yusef, M., Eschweido, A., Burkan, H., & Elsharafi, M. O. (2019). Nanoparticles EOR aluminum oxide (Al2O3) used as a spontaneous imbibition test for sandstone core. *ASME International Mechanical Engineering Congress and Exposition, Proceedings (IMECE)*, *6*, V006T06A054. https://doi.org/10.1115/IMECE2019-10283.

Schwartz, B. S., Parker, C. L., Hess, J., & Frumkin, H. (2011). Public health and medicine in an age of energy scarcity: The case of petroleum. *American Journal of Public Health*, *101*(9), 1560–1567. https://doi.org/10.2105/AJPH.2010.205187.

Seright, R. S., Chang, H. L., Zhang, Z. Q., Wang, Q. M., Xu, Z. S., Guo, Z. D., Sun, H. Q., Cao, X. L., & Qi, Q. (2006). Advances in polymer flooding and alkaline/surfactant/polymer processes as developed and applied in the People's Republic of China. *JPT, Journal of Petroleum Technology*, *58*(5), 80–81. https://doi.org/10.2118/0506-0080-jpt.

Shah, R. D. (2009). Appl ication of nanoparticle saturated injectant gases for EOR of heavy oils. *Proceedings - SPE Annual Technical Conference and Exhibition*, *7*, 4784–4795. https://doi.org/10.2118/129539-stu.

Shen, Z., & Sheng, J. J. (2018). Experimental and numerical study of permeability reduction caused by asphaltene precipitation and deposition during CO2 huff and puff injection in Eagle Ford shale. *Fuel*, 211(September 2017), 432–445. https://doi.org/10.1016/j.fuel.2017.09.047.

Shokrlu, Y. H., & Babadagli, T. (2014). Viscosity reduction of heavy oil/bitumen using micro- and nano-metal particles during aqueous and non-aqueous thermal applications. *Journal of Petroleum Science and Engineering*, *119*, 210–220. https://doi.org/10.1016/j.petrol.2014.05.012.

Skauge, T., Spildo, K., & Skauge, A. (2010). Nano-sized particles For EOR. *Proceedings of SPE Improved Oil Recovery Symposium*, *2*, 1281–1290. https://doi.org/10.2523/129933-MS.

Srinivasan, A., & Shah, S. N. (2014). Surfactant-based fluids containing copper-oxide nanoparticles for heavy oil viscosity reduction. *All Days*, *4*, 2994–3007. https://doi.org/10.2118/170800-MS.

Sun, X., Dong, M., Zhang, Y., & Maini, B. B. (2015). Enhanced heavy oil recovery in thin reservoirs using foamy oil-assisted methane huff-n-puff method. *Fuel*, *159*, 962–973. https://doi.org/10.1016/j.fuel.2015.07.056.

Sun, X., Zhang, Y., Chen, G., & Gai, Z. (2017). Application of nanoparticles in enhanced oil recovery: A critical review of recent progress. *Energies*, *10*(3). https://doi.org/10.3390/en10030345.

Taheri-Shakib, J., Shekarifard, A., & Naderi, H. (2018). Experimental investigation of the asphaltene deposition in porous media: Accounting for the microwave and ultrasonic effects. *Journal of Petroleum Science and Engineering*, *163*(August 2017), 453–462. https://doi.org/10.1016/j.petrol.2018.01.017.

Tang, G.-Q., & Morrow, N. R. (1999a). Influence of brine composition and fines migration on crude oil/brine/rock interactions and oil recovery. *Journal of Petroleum Science and Engineering*, *24*(2–4), 99–111. https://doi.org/10.1016/S0920-4105(99)00034-0.

Tang, G., & Morrow, N. R. (1999b). Oil recovery by waterflooding and imbibition – invading brine cation valency and salinity. Society of Core Analysts, Paper SCA 9911.

Tian, Y., Wang, L., Tang, Y., Liu, C., Ma, C., & Wang, T. (2012). Research and application of nano polymer microspheres diversion technique of deep fluid. *All Days*, 261–266. https://doi.org/10.2118/156999-MS.

Viebahn, P., Vallentin, D., & Höller, S. (2015). Integrated assessment of carbon capture and storage (CCS) in South Africa's power sector. *Energies*, *8*(12), 14380–14406. https://doi.org/10.3390/en81212432.

Vledder, P., Fonseca, J. C., Wells, T., Gonzalez, I., & Ligthelm, D. (2010). Low salinity water flooding: Proof of wettability alteration on a field wide scale. *All Days*, *1*, 200–209. https://doi.org/10.2118/129564-MS.

Wasan, D., Nikolov, A., & Kondiparty, K. (2011). The wetting and spreading of nanofluids on solids: Role of the structural disjoining pressure. *Current Opinion in Colloid & Interface Science*, *16*(4), 344–349. https://doi.org/10.1016/j.cocis.2011.02.001.

Yuan, W., Liu, X., Wei, H., Liu, J., Yang, H., Hu, S., Li, Y., & Wang, D. (2010). Research and application effect of polymeric microsphere in Wen-10 of Sinopec Zhongyuan oil field. *Inner Mongolia Petrochemical Industry*, *12*, 122–126.

8 Case Studies of EOR by Using Nanoparticles

Krishna Raghav Chaturvedi, Abhishek Kumar and Anand Raghav Chaubey

8.1 INTRODUCTION

The pursuit of sustainable energy resources has intensified in recent years, prompting a reevaluation of traditional methods in the oil and gas industry. In this context, the utilization of nanofluids for enhanced oil recovery (EOR) emerges as a promising avenue, holding the potential to redefine the landscape of hydrocarbon extraction. With the global demand for energy continually escalating, there is a pressing need to enhance the efficiency of oil recovery processes while minimizing environmental impact. Nanomaterials and nanofluids, i.e., engineered suspensions of nanoparticles in base fluids, exhibit unique thermal and rheological properties that can significantly augment the recovery of hydrocarbons from reservoirs. This chapter explores the transformative role of nanomaterials/nanofluids in addressing this imperative, focusing on their application in field scenarios. Initially, the various challenges before the field application of nanofluids/nanomaterials have been recapitulated for the readers of this work. Next, the importance of favorable project economics for the successful implementation of the proposed methodology in the field is discussed. Finally, some successful field studies and pilots have been discussed to understand what worked in those cases to ensure the success of the project. By delving into pilot studies and envisioning future implementations, this chapter seeks to unravel the potential benefits of nanofluid-EOR and the role it may play in fostering a sustainable trajectory for the industry by critically examining the inherent challenges in translating laboratory success to field-scale applications, emphasizing the necessity of overcoming these hurdles to ensure the viability and ecological sustainability of nanofluid-based EOR (Kumar et al., 2022).

8.2 CHALLENGES BEFORE NANOMATERIAL APPLICATIONS IN EOR

As discussed in the previous chapter and as observed in several experimental studies that aimed to replicate field conditions, several challenges hinder the successful Implementation of nanofluid flooding in the field. These include:

- *Saline Effect*: Nanofluids may face stability issues in high-salinity environments commonly found in oil reservoirs. The presence of salts can induce

DOI: 10.1201/9781032650685-8

aggregation or precipitation of nanoparticles, altering their performance and hindering their effectiveness in EOR processes.

- *Thermal Effects*: Nanofluids often exhibit unique thermal properties, such as enhanced heat transfer capabilities. However, managing these thermal effects in the field can be challenging. Temperature variations within reservoirs may impact nanofluid stability and alter rheological properties, necessitating careful consideration of thermal dynamics for successful implementation.
- *High Cost*: The production and integration of nanofluids can be costly, posing an economic barrier to their widespread use in field applications. The expense associated with synthesizing nanoparticles, ensuring stability, and scaling up production processes raises concerns about the overall cost effectiveness of nanofluid-EOR.
- *Scale-Up Challenges*: While promising results are often observed in laboratory and pilot studies, scaling up nanofluid applications to field levels presents significant engineering and logistical challenges. Achieving consistent performance and maintaining nanoparticle dispersion over large reservoir volumes require innovative solutions and optimized processes.
- *Environmental Impact*: The environmental implications of nanofluid use, including the potential toxicity of nanoparticles and their long-term effects on ecosystems, must be thoroughly assessed. Establishing clear regulations and guidelines for nanofluid deployment is essential to mitigate any adverse environmental consequences.

Addressing these challenges demands interdisciplinary collaboration between researchers, engineers, and industry experts. Comprehensive studies, technological innovations, and ongoing research efforts are essential to overcoming these obstacles and realizing the full potential of nanofluids in enhancing the sustainability of oil recovery processes (Kumar & Gangawane, 2022a, 2022b).

8.3 IMPACT OF ECONOMICS ON THE IMPLEMENTATION OF THE PROJECT

The economics of field applications, including the use of chemicals in EOR, play a pivotal role in decision-making within the oil and gas industry. Several key economic factors influence the feasibility and adoption of chemical applications in oilfield operations:

- *Cost of Chemicals*: The upfront and ongoing costs associated with acquiring and deploying chemicals, such as nanofluids or other EOR agents, directly impact economic viability. High material costs may be a barrier to adoption, necessitating a careful cost-benefit analysis to justify their use.
- *Production Efficiency Gains*: The economic rationale for employing chemicals in EOR lies in the potential efficiency gains in oil production. Assessing the impact on recovery rates and overall production output is crucial to determine whether the associated costs are justified by increased yields.
- *Reservoir Conditions*: The characteristics of the reservoir, such as its size, depth, and geological complexity, influence the choice of chemicals and their

application. Tailoring chemical solutions to specific reservoir conditions is essential for optimizing economic returns.

- *Scale-Up Costs*: Transitioning from laboratory or pilot-scale applications to full-field implementation involves additional engineering and logistical challenges. The costs associated with scaling up operations, ensuring consistency, and addressing unforeseen issues impact the economic feasibility of large-scale chemical deployment.
- *Operational Costs*: Beyond the direct costs of chemicals, factors like transportation, storage, and handling contribute to operational expenses. Efficient logistical planning is crucial to minimize these costs and ensure cost-effectiveness.
- *Oil Price Dynamics*: Fluctuations in global oil prices directly influence the economic landscape for EOR. In periods of low oil prices, operators may be more cautious about investing in additional technologies or chemicals, while higher prices may make such investments more attractive.
- *Regulatory Compliance*: Adherence to environmental and safety regulations is a non-negotiable aspect of oilfield operations. Meeting these standards may incur additional costs, but it is essential for ensuring sustainable and responsible practices.
- *Technological Innovations*: Ongoing advancements in technology can influence the economic viability of chemical applications. Innovations that improve efficiency, reduce costs, or address specific challenges can enhance the attractiveness of using chemicals in EOR.

In essence, the economic considerations for chemical applications in oilfield operations involve a delicate balance between upfront costs, potential gains in production efficiency, and the prevailing market conditions. One factor that has limited the wider exploration of nanomaterials for oilfield applications is their high cost. Table 8.1 indicates the tentative cost of some nanomaterials identified in previous work (Sergis & Hardalupas, 2011).

TABLE 8.1
Indicates the cost of nanomaterials identified in \$/kg by authors of previous work (Sergis & Hardalupas, 2011)

Nanoparticle type	Price in \$/kg
Aluminum oxide (Al_2O_3)	700
Aluminum (Al)	3,800
Carbon nanotubes	9,300–12,500
Copper oxide (CuO)	750
Copper (Cu)	500
Silica dioxide (SiO_2)	700
Silver (Ag)	400
Gold (Au)	5,500
Titanium dioxide (TiO_2)	800

From Table 8.1, it can be observed that the cost of nanomaterials is significantly high. In field applications, even though a minimal amount of nanomaterials is used, the price would reach several hundreds of thousands of dollars. Thus, rigorous economic assessments by the leadership and continuous monitoring of technological advancements are essential for informed decision-making in the dynamic landscape of EOR.

8.4 SUBSURFACE APPLICATIONS OF NANOMATERIALS: TOWARD EOR

The practical use of nanomaterials in the context of EOR is still somewhat restricted, with just a handful of recorded cases seen globally. An exemplary instance is a field application carried out by Aramco. However, it is crucial to emphasize that the main objective of this undertaking, which started in 2014, was not primarily directed at improving oil recovery mechanisms or regulating mobility. However, the main objective of the research was to clarify the routes by which oil moves throughout the reservoir. The National Iranian Oil Company (NIOC) carried out an experiment in the Belal Oil Field, located in the Hormozgan Province, where they injected 18,000 barrels of nanofluids into a well. This project consisted of many steps, starting with the creation of stable emulsions on a small scale in the laboratory. This was followed by thorough experimental and modeling research. Following that, during the third stage, the chosen nanofluids were manufactured on a semi-industrial level and then injected into the Surmeh reservoir of the Belal field. Ultimately, during the fourth stage, the nanofluids that were created were introduced into a well in order to facilitate the retrieval of oil. Nevertheless, it is crucial to recognize that there is a lack of comprehensive data available about this study. As a result, it is prudent to use care when interpreting its conclusions. Authors should take caution in deriving conclusions from this study until peer-reviewed papers or data from NIOC are accessible.

8.5 CONCLUSION

In conclusion, the exploration of nanofluids for EOR has shown promising results in pilot studies, demonstrating their potential to revolutionize the field. However, the transition from laboratory success to full-scale field applications presents several challenges that demand careful consideration. Issues such as nanofluid stability, cost-effectiveness, and potential environmental impacts need to be meticulously addressed before widespread implementation.

Navigating these challenges requires collaborative efforts between researchers, industry professionals, and regulatory bodies. Rigorous testing, optimization of nanofluid formulations, and the development of scalable production processes are imperative steps toward ensuring the feasibility of nanofluids in real-world oil recovery operations. Additionally, establishing clear guidelines and regulations will play a crucial role in addressing environmental concerns and ensuring the responsible deployment of nanofluid technologies.

As the journey from pilot studies to field applications continues, the industry must remain proactive in overcoming these challenges. By addressing these issues head-on, we can pave the way for the successful integration of nanofluids into EOR practices, ultimately contributing to increased efficiency, sustainability, and resilience in the dynamic landscape of oil extraction.

REFERENCES

Kumar, A., & Gangawane, K. M. (2022a). Effect of precipitating agents on the magnetic and structural properties of the synthesized ferrimagnetic nanoparticles by co-precipitation method. *Powder Technology*, *401*, 117298. https://doi.org/10.1016/j.powtec.2022.117298.

Kumar, A., & Gangawane, K. M. (2022b). Synthesis and effect on the surface morphology & magnetic properties of ferrimagnetic nanoparticles by different wet chemical synthesis methods. *Powder Technology*, *410*(June), 117867. https://doi.org/10.1016/j.powtec.2022.117867.

Kumar, A., Gangawane, K. M., & Ramanjaneyulu, B. (2022). *Ferrofluids for Waste-Water Treatment BT - Advances in Chemical, Bio and Environmental Engineering* (J. K. Ratan, D. Sahu, N. N. Pandhare, & A. Bhavanam (eds.); pp. 723–744). Springer International Publishing.

News Article. Link accessed on 8th Jan, 2024. https://www.tehrantimes.com/news/461182/IOOC-implements-new-high-tech-EOR-pilot-project.

Sergis, A., & Hardalupas, Y. (2011). Anomalous heat transfer modes of nanofluids: A review based on statistical analysis. *Nanoscale Research Letters*, *6*, 391.

9 Synthesis of Iron Oxide Nanoparticles for Enhanced Oil Recovery

Aditi Pradeep Dahake and Nirmalya Bachhar

9.1 INTRODUCTION

In recent years, there has been a substantial surge in research, development, and practical implementation of nanotechnology across diverse domains, including biology, the environment, and technology. This phenomenon is intrinsically linked to the distinctive properties exhibited by nanoparticles (NPs) and their adaptability to specific applications. Notably, NPs possess unique physical (Kazemzadeh et al., 2019), structural (Ou & Wang, 2021), optical (Khan et al., 2019), thermal (Qiu et al., 2020), electrical, and magnetic (Joudeh & Linke, 2022) characteristics that distinguish them from bulk materials.

Interdisciplinary research on nanotechnology in oilfield applications, particularly within reservoir contexts, has shown significant development. Nanomaterials utilized in reservoir applications offer multiple purposes and can be tailored to meet operational, environmental, and technical prerequisites. Primary oil recovery technologies can retrieve 18%–20% of the oil reserve from the reservoir, while Enhanced Oil Recovery (EOR) extracts about 30%–60% of the crude oil from the oil reservoirs. The integration of nanotechnology into EOR has primarily focused on leveraging nanomaterials to enhance oil production. Due to their ultrafine nature, NPs can efficiently traverse reservoir pores, surpassing the limitations posed by rock pore sizes, and are required in minimal concentrations (Xi et al., 2016). To date, several reviews have been published, predominantly examining the potential of nanomaterials in EOR. However, the previous analyses predominantly focus on explaining the fundamental mechanics and behaviors of nanomaterials in the context of reservoir applications (Agi et al., 2018; Druetta et al., 2018; Eltoum et al., 2021; Foroozesh & Kumar, 2020; Kamal et al., 2017; Negin et al., 2016; Sun et al., 2017; Yakasai et al., 2022).

Nonetheless, some recent work by Gbadamosi et al. (2018) introduced the application of polymer-coated NPs in EOR, underscoring the role of polymeric coatings in reducing particle adsorption, enhancing stability, and augmenting oil recovery mechanisms. These coatings can notably enhance the viscosity of injection fluids and improve sweep efficiency. Sikiru et al. (2021) contributed insights into the application of graphene NPs in EOR, delving into synthesis methods and

DOI: 10.1201/9781032650685-9

the impact of graphene NPs on oil recovery, especially in reducing oil viscosity and altering rock wettability. Additionally, Alnarabiji and Husein (2020) presented an overview of bare silica (SiO_2) nanofluids in EOR, highlighting the advantages of SiO_2 NPs and their response to reservoir parameters such as temperature, pH, and salinity. SiO_2 NPs have gained considerable prominence in EOR due to their stability, versatile surface modification pathways, and environmentally benign nature.

Recent emphasis has shifted toward magnetic nanoparticles (MNPs), owing to their heightened magnetization, robust stability in various solutions, uniform size distribution, and superior adsorption capabilities. Among MNPs, iron oxide nanoparticles (IONPs) have garnered significant attention in multiple fields due to their distinctive attributes (Panda et al., 2024; Panda & Gangawane, 2023c, 2023a).

Notably, these eco-friendly particles find utility in various organisms, ranging from bacteria to plants and mammals (Moura & Unterlass, 2020). In an experiment done by Pereira et al. (2020), a nanofluid was created, incorporating CTAB-functionalized Fe_3O_4 NPs. This nanofluid was used to modify calcite wettability, resulting in an 80% contact angle reduction. Raman and XPS analysis showed that Fe_3O_4 NPs selectively removed less ordered asphaltene molecules, reducing their adsorption on calcite. CTAB enhanced NP mobility and stability in saline solutions while aiding wettability modification. Additionally, the nanofluid inhibited calcium carbonate ($CaCO_3$) scale formation, contributing to improved flow during nanoflooding. Nanofluid flooding, also known as nanoflooding, is an emerging chemical EOR method. It involves the injection of nanomaterial or nanocomposite fluids into oil reservoirs with the aim of displacing oil or enhancing injectivity (Peng et al., 2017). These effects boosted tertiary oil recovery, achieving a 60% recovery factor in tests within porous media (Pereira et al., 2020). Within the context of EOR, IONPs are progressively gaining traction due to their unique responses in the presence of a magnetic field. This is evidenced by the growing body of original research papers addressing MNPs in EOR (An, 2018; Izadi et al., 2019; Kazemzadeh et al., 2018; Rezvani et al., 2018).

Despite the availability of numerous laboratory research articles on IONPs in EOR, a comprehensive review addressing the thorough synthesis methods remains absent from the literature. Selecting the appropriate synthesis method for EOR is crucial because it directly impacts the effectiveness, cost efficiency, environmental considerations, reservoir compatibility, regulatory compliance, and long-term reservoir management of the oil recovery process. Making an informed choice is essential to maximize oil production while minimizing associated risks and costs. Hence, this chapter of the book aims to fill this critical gap.

9.2 IONP SYNTHESIS APPROACHES AND METHODS

There are two established approaches for producing NPs, namely the top-down and bottom-up methods. In general, top-down techniques involve reducing bulk materials into smaller particles through mechanical means such as crushing, grinding, or breaking or through processes like laser ablation, sputtering, etching, or electron

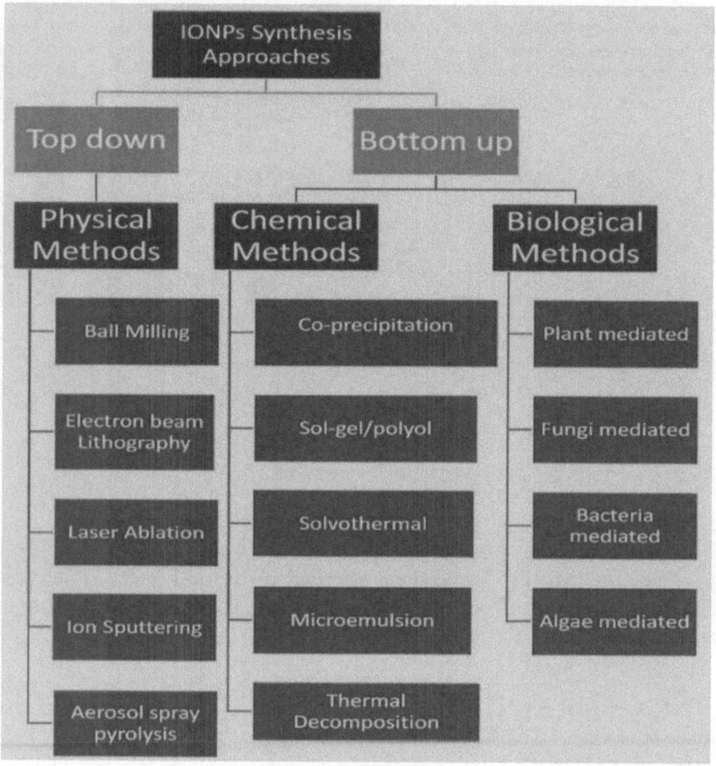

FIGURE 9.1 IONPs synthesis methods and approaches.

beam deposition (Niculescu et al., 2022; Pal Singh et al., 2020). While these methods offer an environmentally friendly option, they typically consume significant time and power (Aryal et al., 2019; Baig et al., 2021; Priyadarshana et al., 2015). On the other hand, bottom-up approaches rely on chemical reactions among specific atoms, ions, or molecules to create NPs.

These approaches can be further categorized into physical methods (associated to top-down techniques) and chemical processes (including co-precipitation, sol-gel, thermal decomposition, microemulsion, solvothermal, etc.). Additionally, there are biological methods that utilize plants or microorganisms for NP synthesis, which also fall under the bottom-up category (Mihai et al., 2020; Niculescu et al., 2022; Pal Singh et al., 2020; Rashid et al., 2020). Figure 9.1 shows the schematic representation of the synthesis methods and approaches.

Roughly 90% of techniques in the field of chemical methods are utilized for synthesizing IONPs, with physical and biological methods accounting for approximately 8% and 2%, respectively (Ali et al., 2016). Additionally, these approaches are categorized into subgroups, as illustrated in Figure 9.2, enabling us to visually gauge the distribution of methods for IONP preparation, where the numbers represent the percentage contribution of the process.

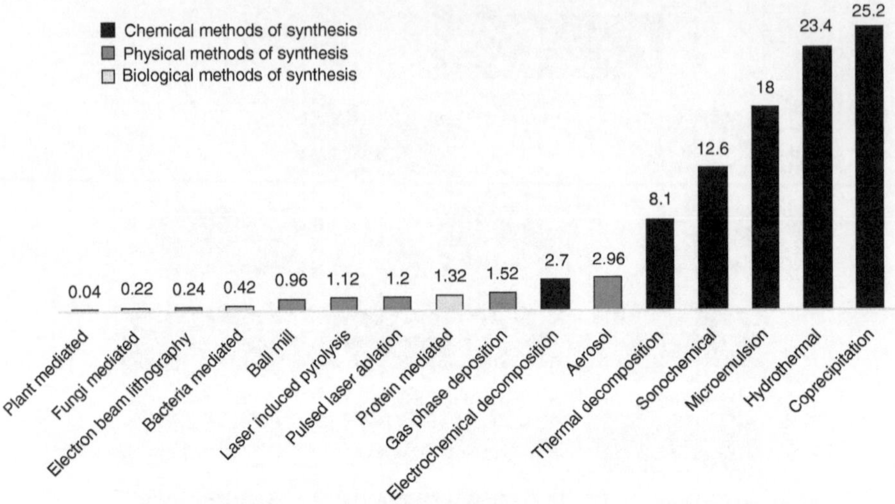

FIGURE 9.2 Basic methods of synthesizing IONPs with the detailed subcategories. Chemical are "electrochemical..." and then "thermal..." onward to the right; physical are "electron...", "ball mill" to "pulsed...", "gas...", and "aerosol"; and biological are all other methods in the graph.

9.3 METHODS OF SYNTHESIS

9.3.1 Physical Methods

Physical techniques are often favored for producing NPs on a large scale due to their straightforward processes. Nevertheless, significant drawbacks in terms of the broad distribution of particle sizes and the considerable time associated with these methods pose substantial challenges in manufacturing MNPs (Mihai et al., 2020; Sánchez-López et al., 2020).

9.3.1.1 Ball Milling

Ball milling, a widely employed industrial process known by various names such as mechanical alloying, attrition, ultrafine grinding, or nanosizing (Kumar & Alexis, 2019), stands as one of the most prevalent methods. In this solid-state mechanical size-reduction method, iron precursors transform MNPs within a stainless-steel cylindrical container filled with micron-sized grinding spheres. Solvers or excess salt species can be introduced to facilitate the milling process and prevent NP agglomeration (Marcelo et al., 2020). There are two primary categories of mechanical ball milling: the ordinary ball milling method, which involves the crushing of coarse Fe_3O_4 particles into ultrafine ones through collisions with steel balls or the tank's inner wall, and high-energy ball milling, where raw materials are mechanically alloyed to synthesize nano-spinel-type ferrite using a high-energy ball mill (Liu et al., 2020).

However, the drawbacks include the potential contamination of the powder by the steel balls. These mechanical and chemical effects can lead to the amorphization of magnetite crystals, the substantial energy input required, and the long milling duration (Jamkhande et al., 2019; Liu et al., 2020).

9.3.1.2 Laser Ablation

Laser ablation is a versatile technique that involves exposing solid target materials to laser irradiation, either by placing them under a thin layer or by utilizing a liquid fabrication approach known as "laser ablation synthesis in solution" (Amendola et al., 2011). This method employs lasers such as Nd:YAG laser (neodymium-doped yttrium aluminum garnet), Ti:Sapphire (titanium-doped sapphire), and copper vapor lasers (Jamkhande et al., 2019; Luo et al., 2018) to selectively control the phase composition, size, and shape of NPs, resulting in spherical and hexagonal NPs with relatively uniform average diameters of approximately 15 nm under various synthesis conditions (Luo et al., 2018).

On the positive side, it is a relatively straightforward and effective method for generating a substantial quantity of small particles in suspension, and the properties of these NPs can be tailored by adjusting laser parameters and the nature of the surrounding liquid, all without the need for surfactants (Simakin et al., 2004). On the downside, extended periods of laser ablation can lead to an excessive number of NPs remaining in the colloidal solution, obstructing the laser path. This results in laser energy being absorbed by the already-formed NPs rather than the target surface, causing a reduction in the ablation rate. Conversely, short exposure times to the laser can produce MNPs but result in polydisperse particles and a low production rate (Ghorbani, 2014; James et al., 2020).

9.3.1.3 Electron Beam Deposition/Electron Beam Lithography

The electron beam deposition method, often referred to as electron beam lithography (EBL), entails directing an electron beam at a high-purity bulk iron material or a thin layer of an exposure-sensitive resist material applied to a substrate. This process results in the formation of MNPs as the initial iron precursors evaporate onto the patterned resist. Subsequently, MNPs can be obtained through either a lift-off process that removes the resist or by transferring the nanopatterns to the functional substrate using an etching process (Fu et al., 2018; Marcelo et al., 2020).

EBL is capable of converting bulk high-purity iron materials or films deposited on a substrate's surface into IONPs by exposing them to an electron beam (King et al., 1996; Rishton et al., 1997). This beam induces the evaporation of the initial iron precursors, leading to the creation of one-dimensional arrays of shape-controlled superparamagnetic IONPs (SPIONs). Furthermore, it is worth noting that two-dimensional arrangements have been achieved by Fu et al. (2018) through the combination of EBL with radio-frequency sputtering.

9.3.1.4 Ion Sputtering

The ion sputtering method involves the vaporization of a solid by bombarding it with a stream of inert gas ions. Recently, this technique has been utilized to produce NPs from various metals by employing magnetron sputtering of metal targets. In this process, directed beams of NPs are generated, and these mass nanostructured films are deposited onto silicon substrates. Notably, this entire procedure takes place at relatively low pressures, around 1 mTorr (Swihart, 2003). Sputter deposition is carried out within a vacuum chamber, where sputtering gas is introduced, and a specific working pressure is maintained. A high voltage is applied to the target (cathode), causing free electrons to follow a spiral path through a magnetic system. These electrons collide with argon (Ar) gas atoms, leading to ionization and the formation of a glow discharge (plasma). Positively charged gas ions are attracted to the target, where they continually bombard it. This process is repeated, with ions approaching the target surface with energy exceeding the surface binding energy, resulting in the expulsion of atoms. Continuous collisions between metal atoms and gas molecules in the vacuum chamber lead to the scattering of atoms, forming a diffuse cloud (Vanecht, 2012).

Sputter deposition is widely used for the synthesis of IONPs where iron atoms are derived from a solid iron target material, which acts as a cathode and is subjected to bombardment by energetic particles. These iron atoms, sputtered, for example, through the use of a high-pressure magnetron-sputtering gun, are subsequently slowed down within a cooled growth chamber due to collisions with Ar and helium (He) gas atoms present in the chamber. The resulting IONPs are then expelled from the aggregation chamber via a pressure gradient and subsequently deposited onto a sample holder located in an adjacent deposition chamber. Following this, a small quantity of oxygen is introduced into the deposition chamber to facilitate the oxidation (resulting in Fe_3O_4) of the NP surface, leading to the formation of iron/iron-oxide core-shell nanoclusters (Khanal et al., 2018). By adjusting various physical parameters of the process, such as the pressure within the aggregation chamber, the ratio of He to Ar gas, and the magnetron-sputter power, the average diameter of the NPs can be finely tuned within the range of 1 to 100 nm (Xing et al., 2016, 2017). Importantly, the ferromagnetic properties of the NPs can also be altered through ion irradiation (Jiang et al., 2011; Sundararajan et al., 2014; Vernieres et al., 2017).

Advantages of this method include the preservation of the sputtered material's composition, making it suitable for refractory metals and intermetallic compounds when compared to methods like evaporation and laser ablation. It is also cost-effective as sputtering equipment is less expensive than EBL systems. Furthermore, fewer impurities are generated compared to chemical methods, and alloy NPs can be produced with better control over composition than other chemical reduction methods. The versatility of this technique allows for the synthesis of ionic NPs with various sizes and compositions that are unattainable in solution. Additionally, slow deposition of heavier or mass-selected ions offers exceptional control over parameters such as size, composition, and charge of ions deposited onto surfaces (Johnson et al., 2016). However, there are disadvantages to consider, such as the choice of sputtering gas (e.g., He, Ne, Ar, Kr, and Xe), which can impact the surface morphology, composition, texture, and optical properties of nanocrystalline metal oxide films (Chandra et al., 2006).

9.3.1.5 Aerosol/Spray Pyrolysis

The spray pyrolysis technique involves delivering NP precursors in vapor form into a heated reactor. To achieve this, a nebulizer is utilized, which dispenses the precursor directly in the form of minute tiny droplets into the hot reactor. Commonly, metal acetate, metal nitrate, or metal chloride serve as metal precursors (Okuyama & Wuled Lenggoro, 2003). The apparatus used for preparing NPs through the spray pyrolysis method consists of three primary components: (1) a fluid nebulizer for atomizing the metal precursor solution, (2) a vertically oriented tubular reactor with thermostatic control (ranging from 200°C to 1,200°C), and (3) a precipitator for collecting the NPs (Minz & Mishra, 2023; Reza et al., 2011; Santhosh & Sarkar, 2022). The apparatus can be adapted and enhanced by employing atomization techniques like two-fluid nozzles, air-assisted (pneumatic) pumps or sprayers, ultrasonic sound, vibrating orifice, and spinning disk (Okuyama & Wuled Lenggoro, 2003; Panda et al., 2023).

Ultrasonic spray pyrolysis shown in Figure 9.3 is an advanced method that utilizes ultrasound to generate atomized droplets from the precursor solution. These aerosol droplets are then transported from the atomizer to the reactor furnace by a carrier gas, leading to the formation of NPs that are subsequently collected by a collection system. The heavily diluted precursor solution is transformed into small aerosol droplets, typically ranging in size from 1 to 10 μm, depending on the ultrasound frequency and solution properties (Majerič et al., 2015).

For the synthesis of IONPs, vaporization of iron-containing materials like FeO, Fe_2O, or Fe_2O_3 with larger grain sizes produces nanosized particles through a process involving supersaturation. Various procedures describe the decomposition of these precursors in specific reactors, utilizing different energy sources, which can range from flames to lasers or plasma, leading to the development of distinct technologies (Swihart, 2003). The phase, size, shape, purity, and magnetic characteristics of the synthesized IONPs can be finely tuned by adjusting several parameters. In flame reactors, factors like O_2 content, temperature, and pressure play crucial roles (Kumar et al., 2023b; Kumfer et al., 2010; Unni et al., 2017). For laser reactors, controlling aerosol conditions and the proportion of oxidizers/stabilizers is key (Amendola et al., 2011). Meanwhile, in plasma reactors, parameters like the injection point and plasma gas composition are influential (Lee et al., 2018; Panda & Gangawane, 2023b).

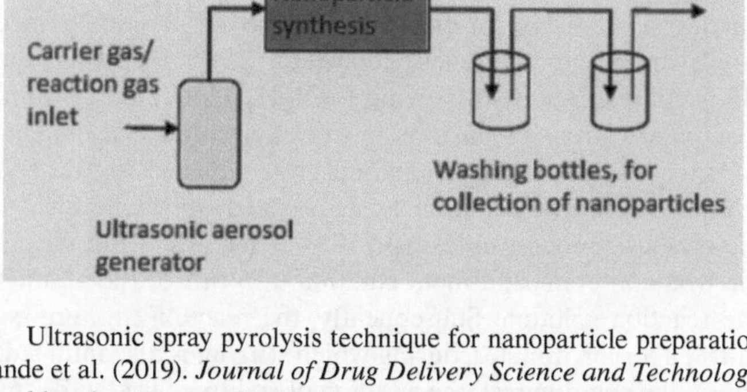

FIGURE 9.3 Ultrasonic spray pyrolysis technique for nanoparticle preparation. Reprinted from Jamkhande et al. (2019). *Journal of Drug Delivery Science and Technology*, 53, Prasad Govindrao Jamkhande, Namrata W. Ghule, Abdul Haque Bamer, Mohan G. Kalaskar, Metal nanoparticles synthesis: An overview on methods of preparation, advantages and disadvantages, and applications, 101174, 2019, with permission from Elsevier.

Within the literature, reports have documented the production of pure and mixed γFe_2O_3, α-Fe_2O_3, and Fe_3O_4 NPs using these methods. However, it is worth noting that these techniques often yield highly aggregated IONPs with typically lower saturation magnetizations, approximately around 40 emu/g, compared to those achieved through chemical approaches. Advantages of this approach include its relative simplicity, cost-effectiveness, and the ability to control and reproduce particle sizes with precision (Kumar et al., 2023a; Okuyama & Wuled Lenggoro, 2003; Reza et al., 2011).

9.3.2 Chemical Methods

Chemical approaches are commonly preferred for NP production because they offer the capability to influence size and shape by adjusting reaction conditions. Some of the common chemical methods for synthesis of IONPs are listed below.

9.3.2.1 Co-precipitation

The co-precipitation approach is widely regarded as a straightforward and very effective chemical method for the synthesis of magnetic particles. Iron oxides, namely Fe_3O_4 or γFe_2O_3, are typically synthesized through the process of ageing a stoichiometric combination of ferrous (Fe^{2+}) and ferric (Fe^{3+}) ions in an aqueous solution. The production of Fe_3O_4 can be represented as follows:

$$Fe^{2+} + 2Fe^{3+} + 8OH^{-} \rightarrow Fe_3O_4 + 4H_2O \tag{9.1}$$

The thermodynamics of this reaction predicts that complete precipitation of Fe_3O_4 will occur in a non-oxidizing oxygen atmosphere at a pH between 8 and 14, with a stoichiometric ratio of 2:1 (Fe^{3+}/Fe^{2+}). Nevertheless, magnetite (Fe_3O_4) exhibits limited stability and is susceptible to oxidation. In the presence of oxygen, magnetite transforms to maghemite (γFe_2O_3) (Laurent et al., 2008).

$$Fe_3O_4 + 2H^{+} \rightarrow \gamma Fe_2O_3 + Fe^{2+} + 2H_2O \tag{9.2}$$

The approach described below originates from the self-assembly mechanism of spiky magnetoplasmonic supraparticles (Zhou et al., 2014). It explains the formation of IONPs, which commences by dissolving $FeCl_2{\cdot}H_2O$ and $FeCl_3{\cdot}H_2O$ in 40 mL of water under continuous stirring in an oxygen-free environment (Kumar et al., 2023a). The rotational speed of the stirring apparatus was adjusted to 500 rpm to facilitate the dissolution of the iron compounds. This speed was maintained consistently throughout the synthesis process unless specified differently. After the complete dissolution of the iron compounds, a rapid addition of 5 mL of the ammonia solution is made to the reaction solution. Subsequently, the reaction mixture is allowed to remain undisturbed for ten minutes, during which stirring is discontinued. Following a period of letting the solution rest, the addition of sodium citrate takes place, subsequently raising the temperature to 90°C for 30 minutes. Continuous stirring is then recommenced. Upon completion of the heating process, the reaction is subsequently

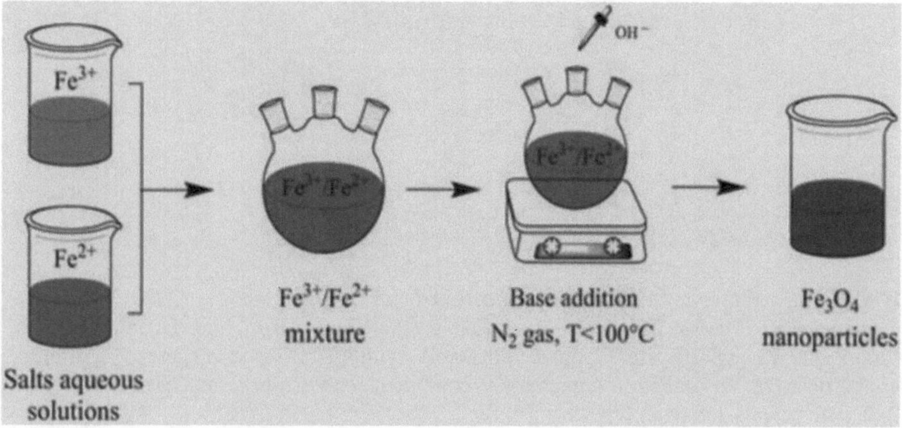

FIGURE 9.4 Co-precipitation method and reaction mechanism (Bustamante-Torres et al., 2022).

cooled to ambient temperature, resulting in the formation of a black precipitate at the base of the reaction vessel. The residue is subjected to many washes with water, employing a magnet to retain the sediment, introducing water, and then removing the uppermost layer of liquid. Following an adequate number of iterations of the washing procedure, the precipitate was subjected to the process of desiccation within an oven maintained at a temperature range of 40°C–45°C for subsequent extraction of a sample (Kumar & Gangawane, 2022a, 2022b).

One significant benefit of the co-precipitation approach is the ease to scale up for large-quantity synthesis. Nevertheless, the regulation of particle size distribution is constrained due to the exclusive influence of kinetics on crystal formation (Schwarzer & Peukert, 2004). Multiple studies have demonstrated that manipulating variables such as pH, ionic strength, speed of mixing, temperature, salt composition, and the Fe^{2+}/Fe^{3+} concentration ratio leads to modifications in both the size and structure of NPs. A general schematic of the process is depicted in Figure 9.4 (Bustamante-Torres et al., 2022). This approach enables the acquisition of particles within the size range of 2 to 17 nm (Babes et al., 1999; Bachhar & Bandyopadhyaya, 2016; Fauconnier et al., 1996, 1999; Gupta & Wells, 2004; Hasany et al., 2013; Jiang et al., 2004; Jolivet, 1992; Jolivet et al., 2002; Kumar et al., 2022; Massart, 1981; Roger et al., 1999; Tartaj et al., 2006; Vayssières et al., 1998).

9.3.2.2 Solvothermal

The solvothermal synthesis method includes the production of nanomaterials by the utilization of a solvent at high temperatures (often ranging from 100°C to 1,000°C) and pressures (ranging from 1 to 10,000 atm). The procedure is commonly conducted in an enclosed vessel, such as an autoclave, at or near the boiling point of the reaction media (Fekri Aval et al., 2016). The reaction vessels must possess chemical inertness. The solution undergoing the reaction is subjected to elevated temperatures surpassing the boiling point of the solvent contained within the sealed autoclave, resulting in the generation of significant pressure. The solvent transforms into a supercritical

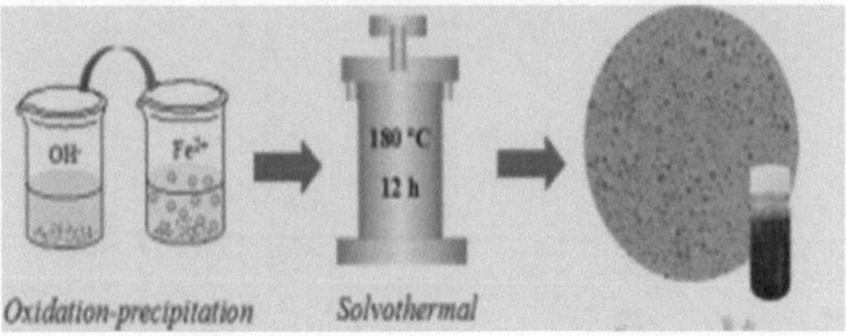

FIGURE 9.5 Solvothermal method. Reprinted from Zheng et al. (2020). *Materials Chemistry and Physics*, 252, Yuan-Yuan Zheng, Qian Sun, Yong-Hong Duan, Jing Zhai, Liang-Liang Zhang, Jie-Xin Wang, Controllable synthesis of monodispersed iron oxide nanoparticles by an oxidation-precipitation combined with solvothermal process, 123431, 2020, with permission from Elsevier.

fluid. Following the completion of the reaction, the autoclave is subjected to a cooling process until it reaches ambient temperature. Subsequently, the elimination of solvents and impurities takes place, with the ultimate objective of recovering the intended product. When water is used as the solvent, the process is referred to as hydrothermal synthesis and is typically conducted at temperatures below the supercritical point of water, which is 374°C (Gan et al., 2020). The synthesis method is schematically represented in Figure 9.5 (Zheng et al., 2020).

The procedure for IONP synthesis involves the combination of Fe^{3+} salt precursors with acetates, urea, and sodium citrate in either an aqueous solution or an organic solvent. Subsequently, the resulting mixture is transferred to an autoclave and subjected to a temperature of 200°C for a duration of 8 to 24 hours (Ali et al., 2015a; Ali et al., 2015b; Hasany et al., 2013). This method facilitates the regulation of particle size through the manipulation of various factors, including temperature, pressure, and reaction duration, so enabling the attainment of particle sizes within the range of 10 to 200 nm. The utilization of solvothermal techniques has been found to be effective in synthesizing free single-crystal particles. It has been observed that particles synthesized using solvothermal methods may exhibit higher levels of crystallinity compared to particles synthesized using alternative synthetic procedures (Adewunmi et al., 2021).

A few challenges of this method include:

1. The slow reaction kinetics at a given temperature. Microwave heating has been shown to accelerate crystallization kinetics and overcome this challenge. The microwave-hydrothermal method uses microwaves and hydrothermal conditions. Microwaves improve reaction kinetics, speeding up synthesis. Reaction kinetics can increase by two orders of magnitude with microwave heating in hydrothermal settings. This improvement is due to localized solution superheating (Faraji et al., 2010). The microwave-hydrothermal method produces γFe_2O_3 NPs and only requires one step with this method. NPs of γFe_2O_3 can be produced in 25 minutes at 150°C using this method. Its lower temperature and shorter reaction time make it better than hydrothermal methods (Sreeja & Joy, 2007).

2. Making nanocrystals with a diameter of less than 10 nm and high hydrophilicity is difficult. However, increased precursor concentration causes a rapid increase in particle size and size distribution. The mean particle size can, however, only be altered by changing the residence time (Stojanović et al., 2013).
3. Compared to co-precipitation, yields are poor in the case of hydrothermal processes (Hernández-Hernández et al., 2020).

9.3.2.3 Microemulsion Method

Microemulsions are liquid mixtures composed of oil, water, and surfactant, often with the addition of a co-surfactant. These mixtures are characterized by their clarity, thermodynamic stability, and isotropic nature. Surfactant molecules self-assemble themselves into a monolayer at the interface of oil and water. In this arrangement, the hydrophobic tails of the surfactant molecules are dissolved in the oil phase, while the hydrophilic head groups are situated in the aqueous phase. For water-soluble metallic ions, the reverse microemulsion technique is more suitable, where the soft surfactant layer separates the aqueous solution of the metal salt/ reactants, which will be insoluble in the oil phase.

These surfactant-coated emulsion droplets serve as microreactors, while the surfactants act as filters for controlling particle size (Hernández-Hernández et al., 2020). Upon the combination of two identical water-in-oil microemulsions containing the specified reactants, the microdroplets undergo a series of collision, consolidation, and subsequent breakage processes. Over time, these processes lead to the formation of a precipitate within the reverse-micelles, ultimately leading to the creation of IONPs. In general, it is possible to adjust the size and shape of IONPs by altering factors such as surfactant type, reactant concentration at the beginning, and droplet size (Amara et al., 2012; Darbandi et al., 2012; Han et al., 2011; Wongwailikhit & Horwongsakul, 2011). Several studies have also been conducted that use microemulsion method for the synthesis of IONPs used in EOR (Hu et al., 2017; Nguyen et al., 2016). Schematic representation of the process is shown in Figure 9.6 (Capek, 2004).

One notable benefit of this approach lies in its capacity to manipulate and regulate the sizes and shapes of particles effectively. Nevertheless, the co-precipitation method exhibits superior yields in comparison to alternative approaches. Moreover, the purification process of the NPs is intricate due to the inherent characteristics of the surfactants employed (Najafi & Nematipour, 2017).

9.3.2.4 Sol-Gel Method

The sol-gel synthesis technique is a well-recognized approach for producing NPs, which has also been extended to the production and coating of IONPs using metal-organic precursors (Shaker et al., 2013). This method involves the hydrolysis and condensation of metal alkoxides or alkoxide precursors, resulting in the formation of oxide particles dispersed within a "sol" that is subsequently dried or gelled by removing the solvent or through a chemical reaction (Yamashita et al., 2019). In this process, metal alkoxides are typically hydrolyzed using either an acidic or basic solution, often water. During hydrolysis, the alkoxide groups are replaced by hydroxide groups from water, leading to the formation of a three-dimensional metal oxide

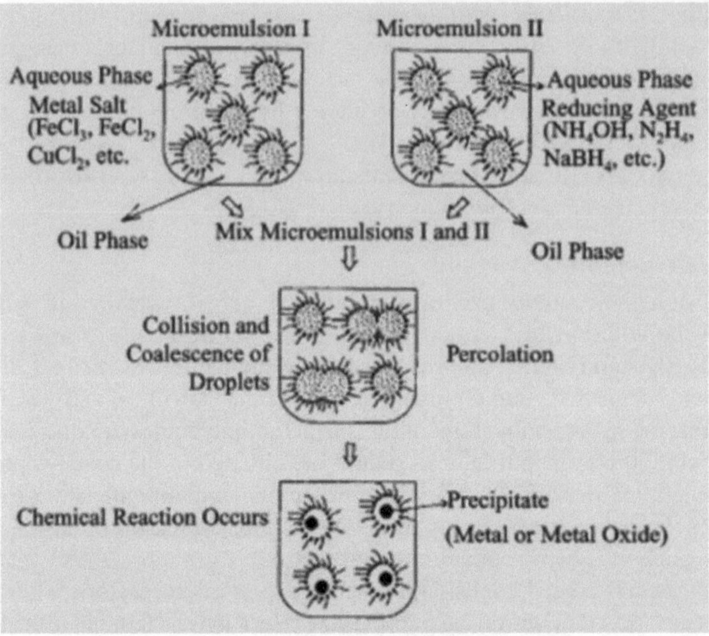

FIGURE 9.6 Mechanism for the formation of metal particles by w/o microemulsion. Reprinted from Capek (2004). *Advances in Colloid and Interface Science*, 110, I. Capek, Preparation of metal nanoparticles in water-in-oil (w/o) microemulsions, 49, 2004, with permission from Elsevier.

network through condensation and inorganic polymerization of the solvent. The NPs attain their final crystalline state after undergoing heat treatment (Laurent et al., 2008). Various factors, including reaction rates, temperature, the type of precursor, and pH, influence the size and shape of the particles. These parameters primarily affect the kinetics of hydrolysis and condensation reactions (Majidi et al., 2016). Figure 9.7 (Comite, 2017) shows the schematic representation of the process.

The sol-gel approach to synthesizing IONPs offers several advantages, including high purity due to pure amorphous phases, uniformity, control over particle size, and the potential to incorporate molecules into the sol-gel matrix, allowing them to retain their stability and characteristics (Miguel et al., 2020). However, it results in the formation of three-dimensional oxide networks, which can limit its efficiency, and it can generate pollution from reaction by-products, necessitating post-treatment of the product (Adewunmi et al., 2021; Majidi et al., 2016).

9.3.2.5 Thermal Decomposition

Thermal decomposition, also known as thermal pyrolysis, is another precise method for crafting highly crystalline and uniformly sized SPIONs with controllable dimensions and shapes. Unlike the conventional co-precipitation method, this technique involves decomposing organometallic iron precursors at elevated temperatures, approximately ranging from 200°C to 300°C, within organic solvents. Various iron

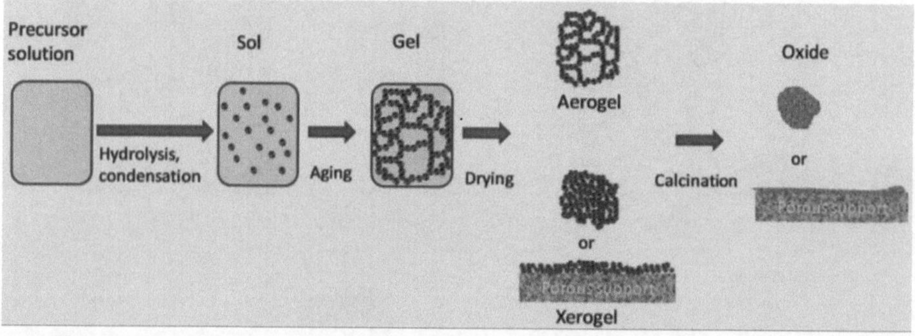

FIGURE 9.7 Sol-gel synthesis method for metal oxide NPs. Reprinted from Comite (2017). Elsevier Books, Antonio Comite, Current Trends and Future Developments on (Bio-) Membranes, 3–23, 2017, with permission from Elsevier.

precursors have been reported for this process, including $Fe(CO)_5$ (Hyeon et al., 2001; Woo et al., 2004), $Fe(acac)_3$ (acetylacetonate) (Dai et al., 2014; Sun & Zeng, 2002; Woo et al., 2004), $Fe(Cup)_3$(N-nitrophenyl hydroxylamine) (Rockenberger et al., 1999), Prussian blue ($Fe_4[Fe(CN)_6 \cdot 14H_2O]$) (Hu et al., 2010), iron-oleate (Ye et al., 2012), iron–urea complex ($[Fe(CON_2H_4)_6](NO_3)_3$) (Asuha et al., 2009), and ferrocene $[Fe(C_5H_5)_2]$ (Amara et al., 2012). The formation of SPIONs with various shapes has been found to be highly dependent on the addition of stabilizers, as well as the nature of precursors and solvents. Various studies have demonstrated the synthesis and the effect of various factors on the size of IONPs through thermal decomposition (Hyeon et al., 2001; Liu et al., 2020; Park et al., 2004; Shavel & Liz-Marzan, 2009).

Despite its merits, thermal decomposition is a resource-intensive process, demanding significant energy, materials, and time, and it suffers from a notably low production rate (James et al., 2020). Moreover, a significant drawback lies in the utilization of costly and potentially harmful substances, including high-boiling-point organic solvents (Pereira et al., 2012).

9.3.3 Biological Synthesis

In chemical methods, achieving uniform particle dispersion, preventing clumping, and controlling particle size distribution are key challenges. Moreover, chemical methods often involve the use of solvents such as sodium borohydride, hydrazine, potassium bitartrate, and sodium dodecyl sulfate, all of which are environmentally harmful due to the production of toxic by-products (Laurent et al., 2008). Additionally, physical methods, including milling, laser ablation, etc., are costly in terms of energy requirements and typically result in low yields (Ali et al., 2016).

There is a clear need to transition toward a greener, safer, biologically compatible, stable, cost-effective, rapid, and eco-friendly synthesis method (Abdul Salam et al., 2012; Adelere & Lateef, 2016; Priya et al., 2021). Biological biosynthesis of IONPs relies on the use of living organisms such as plants, bacteria, algae, actinomycetes, viruses, and fungi (Ali et al., 2015b; Latif et al., 2019; Mirza et al., 2018;

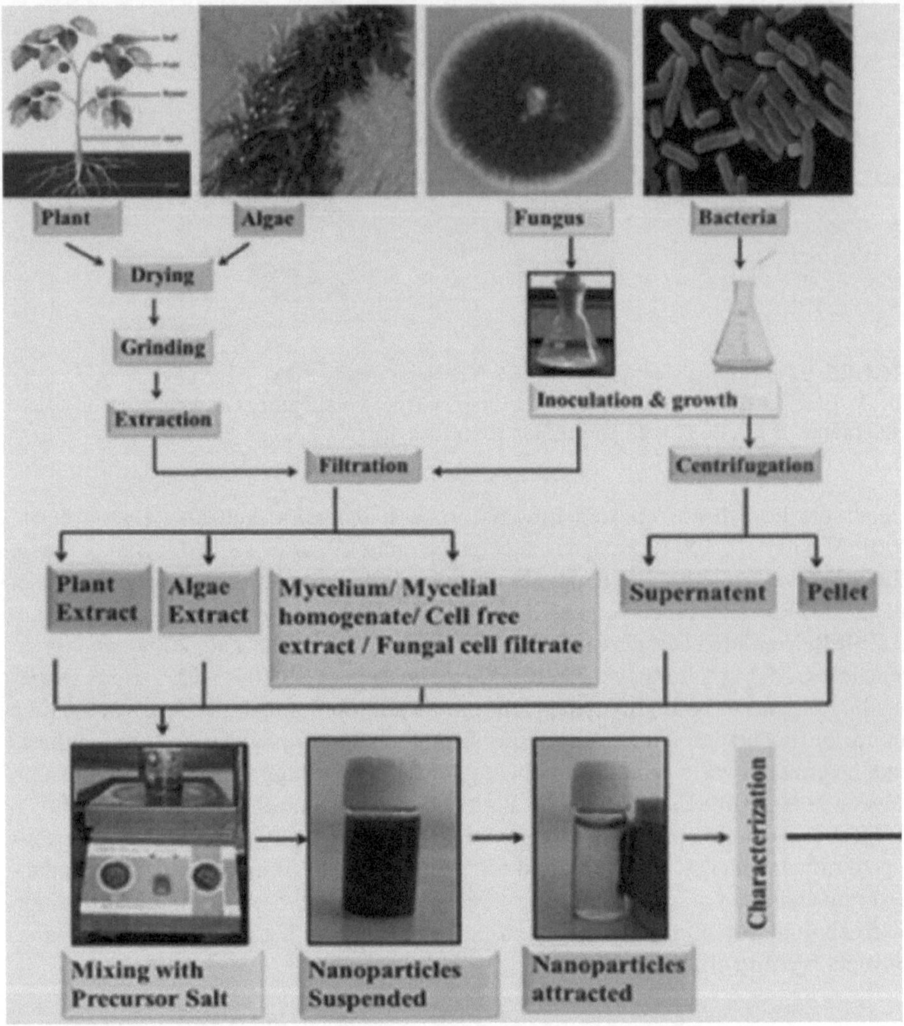

FIGURE 9.8 Biosynthesis of IONPs involving common steps for plants, algae, fungi, and bacteria meditated synthesis (Priya et al., 2021).

Schröfel et al., 2014; Sharma et al., 2019; Thakkar et al., 2010; Vasantharaj et al., 2019; Vijayaraghavan & Ashokkumar, 2017). This organic synthesis method uses the universal solvent, water, and results in NPs free from harmful chemical impurities, which will make them widely accepted in the future (Giti Emtiazi, 2015; Pantidos, 2014). A general schematic showing the biosynthesis of IONPs for plants, algae, fungi, and bacteria-meditated synthesis is shown in Figure 9.8 (Priya et al., 2021).

9.3.3.1 Biosynthesis of IONPs Using Plants

Plants are widely recognized as readily available, easily manageable, environmentally benign, and cost-effective resources for synthesizing various types of NPs (Noruzi,

2015). This biosynthesis process involves utilizing different parts of plants, such as roots, leaves, seeds, flowers, fruits, peels, petals, whole plants, and seed husks, as they are rich in diverse biomolecules like carbohydrates, amino acids, flavonoids, proteins, saponins, terpenoids, and nitrogenous compounds. These biomolecules serve as reducing agents, stabilizers, redox mediators, and capping agents in NP synthesis (Mirza et al., 2018; Mittal et al., 2013; Mondal et al., 2020; Rajeshkumar & Bharath, 2017; Roy & Das, 2015).

Currently, extensive research is dedicated to the plant-based synthesis of IONPs. One significant reason for this is the shorter incubation times required by plant-based methods compared to bacteria and fungi, attributed to the phytochemicals found in plants (Singh et al., 2018). Recent studies indicate that plant extracts from species like *Solanum trilobatum*, *Ziziphora tenuior*, *Persia americana*, *Abutilon indicum*, *Azadirachta indica*, *Camellia sinensis*, and green tea leaves have been used to synthesize IONPs of varying sizes (Arularasu et al., 2018; Khalil et al., 2017). Furthermore, these NPs, synthesized using plant extracts, are reported to be more stable than conventionally synthesized ones (Bibi et al., 2019), and they exhibit distinct shapes, including spherical, cubical, cylindrical, needle-like, stem-like, prismatic, and dendritic (Abdullah et al., 2020). The characteristics of the synthesized NPs mainly depend on factors such as the type of plant extract, the volume ratio of the extract to metallic salt solutions, and reaction conditions like pH, temperature, and incubation time (Abdullah et al., 2020; Singh et al., 2018).

A typical process for synthesizing IONPs from plant materials, such as leaves, fruits, seeds, seed coats, flowers, peels, petals, or whole plants, begins with washing the starting material, followed by extraction. The most common procedure involves collecting plant parts, either fresh – dried from the market – or from waste sources (e.g., banana peels and pruned tea), and rinsing them with tap water, followed by washing with distilled or deionized water (Kumar & Gangawane, 2023a). Subsequently, these washed plant parts are left to air dry at room temperature (Beheshtkhoo et al., 2018; Khalil et al., 2017; Lohrasbi et al., 2019) or dried in an oven at temperatures ranging from 40°C to 60°C for 1 to 4 hours (Bhuiyan et al., 2020; Izadiyan et al., 2020; Majumder et al., 2019), under shade (Razack et al., 2020), in sunlight (Arularasu et al., 2018), or simply air-dried (Kanagasubbulakshmi & Kadirvelu, 2017; Karade et al., 2019), followed by weighing. Next, the dried plant material is ground into a fine powder using various methods such as a pestle and mortar, blender, willy mill, or electric grinder (Bibi et al., 2019; Izadiyan et al., 2020; Khalil et al., 2017; Razack et al., 2020). The powdered plant material is mixed with a solvent, typically distilled water, double-distilled water, methanol, or ethanol:water (1:1), and boiled under continuous stirring using a magnetic stirrer hot plate, water bath, or heater mantle under reflux (Beheshtkhoo et al., 2018; Lohrasbi et al., 2019). The extraction mixture is then filtered using Whatman filter paper, filter membrane (0.45 μm), nylon mesh, Millipore hydrophilic filter (0.22 μm), or under vacuum. In some studies, green synthesis of NPs by plants includes an additional step of centrifugation (at 2,000 rpm for five minutes and 7,000 rpm for 15 minutes) of the filtrate after filtration (Beheshtkhoo et al., 2018; Lohrasbi et al., 2019; Razack et al., 2020). In the Carica papaya extract preparation, the filtrate obtained is concentrated using a rotary evaporator (Bhuiyan et al., 2020). The clear solution obtained, either after filtration or centrifugation, is considered the sample for the synthesis of IONPs. A portion of the extract is then mixed with precursor salt solutions (e.g., iron sulfate heptahydrate, $FeCl_3{\cdot}6H_2O$, $Fe(NO_3)_3{\cdot}9H_2O$, $FeSO_4$, $FeCl_3$, $FeCl_2{\cdot}4H_2O$, $FeCl_2$, $FeSO_4{\cdot}5H_2O$) of various molarities. The mixture is then heated (within the range of 60°C to 85°C) or vigorously stirred for a specified duration (typically 30 to 60 minutes) and under specified temperature conditions. During the incubation of plant extract with the precursor salts, the color of the solution changes, depending on the type of iron salts used, which intensifies with the formation of NPs. After synthesis, some methods proceed with continuous stirring using a magnetic stirrer (Arularasu et al., 2018; Beheshtkhoo et al., 2018; Izadiyan et al., 2020; Kanagasubbulakshmi & Kadirvelu, 2017; Kumar & Gangawane, 2023b; Lohrasbi et al., 2019; Majumder et al., 2019; Razack et al., 2020; Sharma et al., 2019), and a centrifugation step (Arularasu et al., 2018; Beheshtkhoo et al., 2018; Bibi et al., 2019; Izadiyan et al., 2020; Kanagasubbulakshmi & Kadirvelu, 2017; Khalil et al., 2017; Lohrasbi et al., 2019; Razack et al., 2020; Vasantharaj et al., 2019). The resulting pellets of IONPs are then subjected to final washing, typically with distilled water, ethanol, deionized water, methanol, or double-distilled water. The synthesis process concludes with the drying of the pellets (NPs), such as hot-air oven, vacuum drying (Kanagasubbulakshmi & Kadirvelu, 2017), air drying (Karade et al., 2019), a furnace (Bibi et al., 2019), vacuum hot-air oven (Karpagavinayagam & Vedhi, 2019), or simply in the open air (Devi et al., 2019), at specified temperature conditions and for a set duration.

9.3.3.2 Biosynthesis of IONPs Using Fungi

The extracellular synthesis of IONPs using fungal species is highly advantageous due to its scalability, cost-effective raw materials, high biomass production, simple downstream processes, low toxicity in residues, and economic feasibility (Chatterjee et al., 2020; Pal & Sen, 2024; Tarafdar & Raliya, 2013). Fungal species also possess excellent tolerance and bioaccumulation properties, facilitating the synthesis of metal NPs (Sardul et al., 2017). Fungi are known to secrete a significant amount of extracellular enzymes capable of hydrolyzing metals, and this enzymatic reduction of metal ions is believed to be responsible for the synthesis of extracellular NPs. Nitrate reductase is one such enzyme found in fungi (Abdeen et al., 2016). Although the complete mechanisms of metal nanoparticle synthesis using fungi are not fully understood, they are thought to depend on the biological material used and the applied reaction conditions. The choice of biological material includes the specific microorganism strain, growth parameters, and sample preparation methods. At the same time, reaction conditions encompass the type and concentration of metal precursor, pH, and reaction temperature (Silva et al., 2015).

In the literature, there is a wide range of variability in the starting materials used for biosynthesis. Some methods involve using fungal biomass to synthesize NPs, while others utilize cell-free filtrates (CFFs) of fungal cultures or homogenized fungal biomass. Some approaches even employ fungal cell filtrates (FCFs). The most used method involves cultivating fungi in a suitable growth medium, followed by separating fungal mycelia through centrifugation. The obtained mycelia are then mixed with a metal precursor salt and incubated at 28°C for four to five days. Filtration of this solution yields a suspension containing synthesized NPs (Bhargava et al., 2013; Kaul et al., 2012).

In an alternative method, fungi are cultured in a suitable medium; mycelia are separated by filtration, washed with sterile distilled water, and then dispensed in sterile distilled water. The mixture is incubated on a rotary shaker for one to five days. After incubation, the CFF is collected by filtering out mycelium with Whatman filter paper, and this CFF is treated with an iron precursor salt mixture. Incubation on a rotary shaker for a few hours leads to the synthesis of IONPs (Chatterjee et al., 2020).

A method combining microbial and physical processes for synthesizing IONPs was introduced. In this approach, a fine homogenate of fungal mycelium is prepared in distilled water using a cyclomixer. Metal precursor salt $FeCl_3$ or $FeSO_4$ at a concentration of 2,000 ppm is added to this homogenate. The resulting mixture is incubated for six days under static conditions. The synthesized IONPs are collected by centrifugation and thoroughly washed with ethanol. Further drying of the NPs is achieved using a supercritical pressure reactor at 300°C and 850 psi pressure (Abdeen et al., 2016).

In a simpler variation, FCF is used for IONP synthesis. In this method, a selected fungal strain is cultured in a suitable medium. Upon completion of the incubation period, the culture medium is filtered through Whatman filter paper using a vacuum pump. The filtrate obtained is centrifuged and re-filtered to get the final FCF. Equal volumes of FCF and a solution containing $FeCl_3$ and $FeCl_2$ salt solution (2:1 mM final concentration) are agitated at room temperature for just five minutes. An immediate change in color confirms the synthesis of IONPs. This method is the simplest for fungal synthesis of IONPs, eliminating the need for complex filtration procedures and expensive equipment (Mahanty et al., 2019).

In a particular investigation, the production of IONPs was carried out using the cell-free extract of *Aspergillus niger* BSC-1. The XRD examination revealed that the magnetite NPs possess a crystalline orthorhombic structure. The TEM analysis revealed that the width of IONPs ranged from 20 to 40 nm. Additionally, the average size of the IONPs was determined to be 17.29 nm using the Debye-Scherrer equation. The particles demonstrated superparamagnetic characteristics, displaying a saturation magnetization of 55.19 emu/g, a coercivity of 0.0045 Oe, and a retentivity of 1.61 emu/g (Chatterjee et al., 2020; Panda et al., 2024).

9.3.3.3 Biosynthesis of IONPs Using Bacteria

The prokaryotic system has been extensively explored as a fundamental model in nanotechnology due to its widespread presence, rapid reproduction rate, ability to thrive in challenging conditions, and the availability of cost-effective and simple cultivation media (Fariq et al., 2017; Pantidos, 2014). This system is highly regarded for its capacity to synthesize NPs with diverse shapes, size distributions, structural characteristics, and physical and chemical properties through the reduction of metal ions (Gahlawat & Choudhury, 2019). Microorganisms can accumulate and detoxify metals through the action of reductase enzymes, making this property useful for the synthesis of metallic NPs using metal salts as reaction precursors (Alam et al., 2020; Singh et al., 2016). The reduction of metal ions is a critical step in the synthesis process, and it is influenced by various factors such as functional groups on the cell wall (essential for biomineralization), the type of microbial strain, and environmental conditions including pH, temperature, growth medium, and the concentration of salts. All these factors, either directly or indirectly, impact the size, shape, composition, and yield of the NPs (Gahlawat & Choudhury, 2019; Iravani & Varma, 2020).

The synthesis of IONPs using bacteria typically begins with the selection of bacteria, either from soil or lyophilized cultures. The isolation process involves cultivating the selected sample on specific media such as BH1 agar (Sundaram et al., 2012), blood agar (Jubran et al., 2020), or nutrient agar (Fatemi et al., 2018). The incubation is carried out under controlled conditions at temperatures ranging from 35°C to 37°C for 24 to 48 hours. In the case of blood agar medium, incubation is conducted under controlled conditions with a carbon dioxide environment. Once sufficient growth is achieved, a single colony is transferred to a nutrient medium such as Luria broth, Nutrient broth, or MRS broth (Fani et al., 2018; Fatemi et al., 2018; Jubran et al., 2020; Sahoo & Mishra, 2023; Sundaram et al., 2012; Torabian et al., 2018). Subsequently, incubation and centrifugation are performed to pellet the bacterial biomass. After centrifugation, the critical step in the synthesis of IONPs begins, where either the supernatant (Fatemi et al., 2018; Jubran et al., 2020; Sundaram et al., 2012) or the cell pellet (Fani et al., 2018; Torabian et al., 2018) is used. When the pellet is utilized, additional steps are required, including centrifugation at 3,000g for ten minutes, washing, and a brief incubation in liquid nitrogen (–196°C) for five minutes. This is followed by a heat shock treatment, achieved by steam bathing the cell pellet at 37°C for 15 minutes. Subsequently, centrifugation at 12,000g for 30 minutes is performed, and the supernatant is collected as cytoplasmic extract (Fani et al., 2018; Torabian et al., 2018).

In another protocol, the pellets are dissolved in sterile distilled water. The supernatant and the pellet suspension are mixed with precursor salts (Fe_2O_3, $FeSO_4$, $FeCl_3{\cdot}6H_2O$) in the desired volume. The reaction conditions are optimized by adjusting the pH within the preferred range of 5 to 9. Finally, the mixture solution is incubated at different temperatures, depending on the bacterial strain used (Priya et al., 2021). During incubation, the initial indication of NPs' formation is observed through a color change, such as from brick red to dark brown (Sundaram et al., 2012), colorless to black (Fani et al., 2018; Torabian et al., 2018), golden yellow to turbid brown (Fatemi et al., 2018), and dark green to blackish brown (Jubran et al., 2020). The synthesized IONPs are then dried and characterized.

In the study conducted by Jubran et al. (2020), the formation of IONPs was proved by SEM analysis. The XRD pattern showed that the particles were between 12 and 32 nm in size with an average of 11.94 nm and cubic in shape.

9.3.3.4 Biosynthesis of IONPs Using Algae

Algae are defined as photosynthetic organisms lacking root and leaf structures. There are two main categories: microalgae (unicellular) and macroalgae or seaweeds (multicellular). Both microalgae and macroalgae are extensively utilized in the field of nanotechnology for the synthesis of various metallic NPs, including gold, silver, palladium, iron, and copper (Dabhade et al., 2023; Sardul et al., 2017; Subramaniyam et al., 2016). Algae, much like plants and bacteria, produce a wide range of biomolecules such as proteins, fats, carbohydrates, peptides, alkaloids, terpenes, macrolides, cell wall polysaccharides, glycoproteins (containing various functional groups like carbonyl, hydroxyl, carboxyl, sulfonate), and enzymes. These biomolecules play a crucial role in reducing, capping, fabricating, and stabilizing NPs (Siddiqi & Husen, 2016; Subramaniyam et al., 2016). Consequently, this approach is considered safe, simple, cost-effective, and environmentally friendly for NP production.

The synthesis of IONPs typically begins with the collection of algae or seaweed samples. In the case of seaweed, they are initially washed with distilled water and then dried (Salem et al., 2019). However, in the protocol by Yew et al. (2016), a bleaching step is performed immediately after washing, followed by drying the sample under sunlight for 72 hours. Next, the samples are ground, freeze-dried (Mahdavi et al., 2013), weighed, mixed with a desired volume of distilled water, and boiled (Mahdavi et al., 2013; Salem et al., 2019). In Yew et al. (2016) protocol, the boiling step is replaced by soaking the dried biomaterial in deionized water for one day. The extract obtained after boiling or soaking is subjected to a filtration step, and the resulting filtrate is labeled as the bioextract for further synthesis steps. The bioextract is then mixed with precursor salts ($FeCl_3 \cdot 6H_2O$, $FeCl_2 \cdot 4H_2O$, or $FeCl_3$) of varying molarities. The mixture is continuously stirred followed by centrifugation (Mahdavi et al., 2013; Salem et al., 2019), washing, and drying. Finally, after drying, IONPs are obtained.

Lastly, FT-IR, XRD, SEM, TEM, and VSM are used to confirm the formation of IONPs. All of these methods help to show the functional groups, crystal structure, size, form, and magnetic properties of synthesized NPs. In a study by Mahdavi et al. (2013), cubic IONPs having a size distribution of 18 ± 4 nm were synthesized using brown seaweed (*Sargassum muticum*). The NPs obtained were superparamagnetic, having a specific saturation magnetization value of 22.1 emu/g. They were studied using FT-IR, XRD, UV-visible, FESEM, EDXRF, TEM, and VSM techniques, confirming the synthesis of IONPs.

9.4 CONCLUSION

The synthesis of IONPs for EOR has been a subject of significant interest and research due to their potential to enhance oil extraction from reservoirs. IONPs have been extensively examined for a wide range of applications like reservoir sensing and imaging, magnetically driven separation, and drilling and well completion improvement. As their usage continues to grow, there is a fundamental need to enhance existing synthesis methods. To address this, new methods for creating uniform and standardized IONPs must be explored.

Various synthesis methods have been explored within the realm of EOR to produce these NPs with specific properties tailored for reservoir conditions. Some commonly employed synthesis techniques include chemical methods, physical methods, and biological methods. Chemical methods offer precise control over particle size, shape, and composition. Physical methods involve physical processes to produce IONPs and are known for their scalability. Biological methods provide a more environmentally friendly approach. In recent years, novel synthesis approaches have gained attention in the EOR community. These include microwave-assisted synthesis and biological synthesis techniques. Microwave-assisted methods offer rapid and efficient synthesis, while biological synthesis techniques prioritize environmentally friendly and sustainable methods, often using natural extracts or biocompatible materials. However, no results have been reported where IONPs synthesized by biological methods have been directly used in EOR applications (Pal & Sen, 2023a, 2023b).

The choice of synthesis method depends on various factors, including the desired properties of the IONPs, the reservoir's geological and fluid characteristics, cost considerations, and environmental concerns. Researchers are continually exploring and refining these methods to develop uniform and standardized IONPs that can be effectively applied in EOR processes. Additionally, advanced characterization techniques are being used to evaluate IONPs' size, shape, structure, crystallinity, and magnetic behavior, providing insights into their performance in oilfield applications. In conclusion, the synthesis of IONPs for EOR is a dynamic and evolving field with a wide array of methods available. These methods are essential for tailoring IONPs to specific reservoir conditions, and ongoing research in this area holds great promise for improving the efficiency and sustainability of oil recovery processes. Further, more research should be done on the effect of method of synthesis of IONPs on EOR. This should focus on the recovery rates of oil as well the NPs, reservoir sensing and imaging quality, and the economic feasibility.

Further research is imperative to comprehensively evaluate the impact of diverse synthesis methods for IONPs on EOR applications. This research should encompass several critical dimensions. Firstly, it should explore the relationship between synthesis methods and the recovery rates of both oil and NPs within reservoirs, shedding light on how these methods influence the efficiency of EOR strategies. Additionally, investigations should delve into how the unique properties of IONPs generated through different techniques affect their performance in reservoir sensing and imaging, potentially enhancing the accuracy and sensitivity of characterization processes.

Lastly, an economic and environmental assessment is vital, analyzing the cost-effectiveness and ecological implications of various synthesis approaches throughout the entire lifecycle of NP production and injection. These endeavors promise to yield comprehensive insights into the optimization and sustainability of EOR practices in the oil and gas industry.

Summary of IONPs used in EOR:

Sr. no.	Type of IONP	Method of synthesis	Applications	References
1	Hydrophobic Fe_3O_4	Thermal decomposition	1. Shields against filtrate infiltration 2. The addition of 0.5 wt% iron oxide in hexane resulted in a 70% decrease in filtrate loss and a 55% reduction in filter cake thickness 3. Brought about a 16.3% decrease in the free oil layer formed due to syneresis during aging	Alvi et al. (2021)

(Continued)

Sr. no.	Type of IONP	Method of synthesis	Applications	References
2	Fe_3O_4 coated with poly(vinyl alcohol) (PVA) or hydroxyapatite (HAp)	Co-precipitation method	1. Decreases the condensation and water imbibition rate in the vicinity of the wellbore area through wettability modification 2. HAp is more proficient at shifting wettability from being oil-wet to becoming gas-wet in comparison to PVA	Safaei et al. (2020)
3	Fe_2O_3 & KCl–Glycol–PHPA polymer	Co-precipitation method	1. Minimizes mud loss in drilling fluid and enhancing its rheological characteristics 2. Decreases filtrate volume and filter cake thickness by 13.6% and 40%, respectively	Ahmed et al. (2020)
4	Fe_3O_4@bentonite	Ultrasound-assisted co-precipitation method	1. Achieved an oil removal efficiency of 67% within 90 minutes by utilizing just 0.1 grams at an oil concentration of 100 ppm	Ewis et al. (2020)
5	Fe_3O_4 hydrophobic EC and hydrophilic CMC	Sequential adsorption of cellulosic material	1. Destabilization of water in crude oil emulsion 2. Improves demulsification speed 3. Reusability and stability of Janus Fe_3O_4	He et al. (2019)
6	Fe_3O_4@3 (aminopropyl) triethoxysilane	Co-precipitation method	1. Reduced crude oil content in the separated water by 99.9% within five minutes, employing a minimal MNP concentration of just 0.04 wt%	Ko et al. (2016)
7	Fe_3O_4@poly-methylmethacrylate-acrylicacid-divinylbenzene P(MMA-AA-DVB)	Solvothermal method	1. 98% separation efficiency was attained after 1 h by demulsification of water and crude oil	Ali et al. (2015a)
8	Fe_3O_4@polyethylene glycol (PEG-400)	Solvothermal method	1. Fe_3O_4@PEG relaxtivity value is 2.07 times greater than that of bare Fe_3O_4	Ali et al. (2018)
9	Fe_3O_4@citric acid and Fe_3O_4@PEG-grafted	Thermal decomposition	1. Reduces NMR T2 relaxation time for reservoir characterization	Park et al. (2014)

(Continued)

Sr. no.	Type of IONP	Method of synthesis	Applications	References
10	Fe_2O_3@citric acid	Co-precipitation method	1. Enhancement in oil recovery in the presence of a magnetic field when compared to bare Fe_2O_3 and polymer solutions	Divandari et al. (2019)
11	Bare Fe_3O_4	Modified co-precipitation method	1. Improved oil recovery owing to improved sweeping area and wettability alteration in the presence of magnetic field	Esmaeilnezhad et al. (2018)
12	Fe_3O_4@AEAPFS	Solvothermal method	Separation of emulsified oil-water Fe_3O_4@AEAPFS MNPs can be reused for up to five cycles without experiencing a notable decrease in their separation efficiency	Lü et al. (2018)

REFERENCES

Abdeen, M., Sabry, S., Ghozlan, H., El-Gendy, A. A., & Carpenter, E. E. (2016). Microbial-physical synthesis of Fe and Fe3O4 magnetic nanoparticles using *Aspergillus niger* YESM1 and supercritical condition of ethanol. *Journal of Nanomaterials, 2016*, 1–7. https://doi.org/10.1155/2016/9174891.

Abdul Salam, H., Gunalan, S., & Sivaraj, R. (2012). Plants: Green route for nanoparticle synthesis. *International Research Journal of Biological Sciences*, *1*(5), 85–90.

Abdullah, J. A. A., Salah Eddine, L., Abderrhmane, B., Alonso-González, M., Guerrero, A., & Romero, A. (2020). Green synthesis and characterization of iron oxide nanoparticles by pheonix dactylifera leaf extract and evaluation of their antioxidant activity. *Sustainable Chemistry and Pharmacy*, *17*, 100280. https://doi.org/10.1016/j.scp.2020.100280.

Adelere, I. A., & Lateef, A. (2016). A novel approach to the green synthesis of metallic nanoparticles: The use of agro-wastes, enzymes, and pigments. *Nanotechnology Reviews*, *5*(6), 567–587. https://doi.org/10.1515/ntrev-2016-0024.

Adewunmi, A. A., Kamal, M. S., & Solling, T. I. (2021). Application of magnetic nanoparticles in demulsification: A review on synthesis, performance, recyclability, and challenges. *Journal of Petroleum Science and Engineering*, *196*, 107680. https://doi.org/10.1016/j.petrol.2020.107680.

Agi, A., Junin, R., & Gbadamosi, A. (2018). Mechanism governing nanoparticle flow behaviour in porous media: Insight for enhanced oil recovery applications. *International Nano Letters*, *8*(2), 49–77. https://doi.org/10.1007/s40089-018-0237-3.

Ahmed, N., Alam, M. S., & Salam, M. A. (2020). Experimental analysis of drilling fluid prepared by mixing iron (III) oxide nanoparticles with a KCl–Glycol–PHPA polymer-based mud used in drilling operation. *Journal of Petroleum Exploration and Production Technology*, *10*(8), 3389–3397. https://doi.org/10.1007/s13202-020-00933-1.

Alam, H., Khatoon, N., Khan, M. A., Husain, S. A., Saravanan, M., & Sardar, M. (2020). Synthesis of selenium nanoparticles using probiotic bacteria Lactobacillus acidophilus and their enhanced antimicrobial activity against resistant bacteria. *Journal of Cluster Science*, *31*(5), 1003–1011. https://doi.org/10.1007/s10876-019-01705-6.

Ali, A., Zafar, H., Zia, M., ul Haq, I., Phull, A. R., Ali, J. S., & Hussain, A. (2016). Synthesis, characterization, applications, and challenges of iron oxide nanoparticles. *Nanotechnology, Science and Applications*, *9*, 49–67. https://doi.org/10.2147/NSA.S99986.

Ali, N., Zhang, B., Zhang, H., Li, W., Zaman, W., Tian, L., & Zhang, Q. (2015a). Novel Janus magnetic micro particle synthesis and its applications as a demulsifier for breaking heavy crude oil and water emulsion. *Fuel*, *141*, 258–267. https://doi.org/10.1016/j.fuel.2014.10.026.

Ali, N., Zhang, B., Zhang, H., Zaman, W., Li, X., Li, W., & Zhang, Q. (2015b). Interfacially active and magnetically responsive composite nanoparticles with raspberry like structure; synthesis and its applications for heavy crude oil/water separation. *Colloids and Surfaces A: Physicochemical and Engineering Aspects*, *472*, 38–49. https://doi.org/10.1016/j.colsurfa.2015.01.087.

Ali, S., Khan, S. A., Eastoe, J., Hussaini, S. R., Morsy, M. A., & Yamani, Z. H. (2018). Synthesis, characterization, and relaxometry studies of hydrophilic and hydrophobic superparamagnetic Fe3O4 nanoparticles for oil reservoir applications. *Colloids and Surfaces A: Physicochemical and Engineering Aspects*, *543*(February), 133–143. https://doi.org/10.1016/j.colsurfa.2018.02.002.

Alnarabiji, M. S., & Husein, M. M. (2020). Application of bare nanoparticle-based nanofluids in enhanced oil recovery. *Fuel*, *267*, 117262. https://doi.org/10.1016/j.fuel.2020.117262.

Alvi, M. A. A., Belayneh, M., Fjelde, K. K., Saasen, A., & Bandyopadhyay, S. (2021). Effect of hydrophobic iron oxide nanoparticles on the properties of oil-based drilling fluid. *Journal of Energy Resources Technology*, *143*(4), 1–10. https://doi.org/10.1115/1.4048231.

Amara, D., Grinblat, J., & Margel, S. (2012). Solventless thermal decomposition of ferrocene as a new approach for one-step synthesis of magnetite nanocubes and nanospheres. *Journal of Material Chemistry*, *22*(5), 2188–2195. https://doi.org/10.1039/C1JM13942H.

Amendola, V., Riello, P., & Meneghetti, M. (2011). Magnetic nanoparticles of iron carbide, iron oxide, iron@iron oxide, and metal iron synthesized by laser ablation in organic solvents. *The Journal of Physical Chemistry C*, *115*(12), 5140–5146. https://doi.org/10.1021/jp109371m.

An, E. (2018). Investigating the effect of different nanoparticles on the interfacial tension reduction. *Petroleum & Petrochemical Engineering Journal*, *2*(4). https://doi.org/10.23880/PPEJ-16000176.

Arularasu, M. V., Devakumar, J., & Rajendran, T. V. (2018). An innovative approach for green synthesis of iron oxide nanoparticles: Characterization and its photocatalytic activity. *Polyhedron*, *156*, 279–290. https://doi.org/10.1016/j.poly.2018.09.036.

Aryal, S., Park, H., Leary, J. F., & Key, J. (2019). Top-down fabrication-based nano/microparticles for molecular imaging and drug delivery. *International Journal of Nanomedicine*, *14*, 6631–6644. https://doi.org/10.2147/IJN.S212037.

Asuha, S., Zhao, S., Wu, H. Y., Song, L., & Tegus, O. (2009). One step synthesis of maghemite nanoparticles by direct thermal decomposition of Fe–urea complex and their properties. *Journal of Alloys and Compounds*, *472*(1–2), L23–L25. https://doi.org/10.1016/j.jallcom.2008.05.028.

Babes, L., Denizot, B., Tanguy, G., Le Jeune, J. J., & Jallet, P. (1999). Synthesis of iron oxide nanoparticles used as MRI contrast agents: A parametric study. *Journal of Colloid and Interface Science*, *212*(2), 474–482. https://doi.org/10.1006/jcis.1998.6053.

Bachhar, N., & Bandyopadhyaya, R. (2016). Role of coating agent in iron oxide nanoparticle formation in an aqueous dispersion: Experiments and simulation. *Journal of Colloid and Interface Science*, *464*, 254–263. https://doi.org/10.1016/j.jcis.2015.11.006.

Baig, N., Kammakakam, I., & Falath, W. (2021). Nanomaterials: A review of synthesis methods, properties, recent progress, and challenges. *Materials Advances*, *2*(6), 1821–1871. https://doi.org/10.1039/D0MA00807A.

Beheshtkhoo, N., Kouhbanani, M. A. J., Savardashtaki, A., Amani, A. M., & Taghizadeh, S. (2018). Green synthesis of iron oxide nanoparticles by aqueous leaf extract of Daphne mezereum as a novel dye removing material. *Applied Physics A*, *124*(5), 363. https://doi.org/10.1007/s00339-018-1782-3.

Bhargava, A., Jain, N., Barathi L., M., Akhtar, M. S., Yun, Y.-S., & Panwar, J. (2013). Synthesis, characterization and mechanistic insights of mycogenic iron oxide nanoparticles. *Journal of Nanoparticle Research*, *15*(11), 2031. https://doi.org/10.1007/s11051-013-2031-5.

Bhuiyan, M. S. H., Miah, M. Y., Paul, S. C., Aka, T. Das, Saha, O., Rahaman, M. M., Sharif, M. J. I., Habiba, O., & Ashaduzzaman, M. (2020). Green synthesis of iron oxide nanoparticle using Carica papaya leaf extract: Application for photocatalytic degradation of remazol yellow RR dye and antibacterial activity. *Heliyon*, *6*(8), e04603. https://doi.org/10.1016/j.heliyon.2020.e04603.

Bibi, I., Nazar, N., Ata, S., Sultan, M., Ali, A., Abbas, A., Jilani, K., Kamal, S., Sarim, F. M., Khan, M. I., Jalal, F., & Iqbal, M. (2019). Green synthesis of iron oxide nanoparticles using pomegranate seeds extract and photocatalytic activity evaluation for the degradation of textile dye. *Journal of Materials Research and Technology*, *8*(6), 6115–6124. https://doi.org/10.1016/j.jmrt.2019.10.006.

Bustamante-Torres, M., Romero-Fierro, D., Estrella-Nuñez, J., Arcentales-Vera, B., Chichande-Proaño, E., & Bucio, E. (2022). Polymeric composite of magnetite iron oxide nanoparticles and their application in biomedicine: A review. *Polymers*, *14*(4). https://doi.org/10.3390/polym14040752.

Capek, I. (2004). Preparation of metal nanoparticles in water-in-oil (w/o) microemulsions. *Advances in Colloid and Interface Science*, *110*(1–2), 49–74. https://doi.org/10.1016/j.cis.2004.02.003.

Chandra, R., Chawla, A. K., & Ayyub, P. (2006). Optical and structural properties of sputter-deposited nanocrystalline Cu2O films: Effect of sputtering gas. *Journal of Nanoscience and Nanotechnology*, *6*(4), 1119–1123. https://doi.org/10.1166/jnn.2006.176.

Chatterjee, S., Mahanty, S., Das, P., Chaudhuri, P., & Das, S. (2020). Biofabrication of iron oxide nanoparticles using manglicolous fungus Aspergillus niger BSC-1 and removal of Cr(VI) from aqueous solution. *Chemical Engineering Journal*, *385*, 123790. https://doi.org/10.1016/j.cej.2019.123790.

Comite, A. (2017). Preparation of silica membranes by sol-gel method. In *Current Trends and Future Developments on (Bio-) Membranes: Silica Membranes: Preparation, Modelling, Application, and Commercialization*. Elsevier B.V. https://doi.org/10.1016/B978-0-444-63866-3.00001-7.

Dabhade, A. H., Verma, R. P., Paramasivan, B., Kumawat, A., & Saha, B. (2023). Development of silver nanoparticles and aptamer conjugated biosensor for rapid detection of E. coli in a water sample. *3 Biotech*, *13*(7), 244. https://doi.org/10.1007/s13205-023-03663-3.

Dai, L., Liu, Y., Wang, Z., Guo, F., Shi, D., & Zhang, B. (2014). One-pot facile synthesis of PEGylated superparamagnetic iron oxide nanoparticles for MRI contrast enhancement. *Materials Science and Engineering: C*, *41*, 161–167. https://doi.org/10.1016/j.msec.2014.04.041.

Darbandi, M., Stromberg, F., Landers, J., Reckers, N., Sanyal, B., Keune, W., & Wende, H. (2012). Nanoscale size effect on surface spin canting in iron oxide nanoparticles synthesized by the microemulsion method. *Journal of Physics D: Applied Physics*, *45*(19), 195001. https://doi.org/10.1088/0022-3727/45/19/195001.

Devi, H. S., Boda, M. A., Shah, M. A., Parveen, S., & Wani, A. H. (2019). Green synthesis of iron oxide nanoparticles using Platanus orientalis leaf extract for antifungal activity. *Green Processing and Synthesis*, *8*(1), 38–45. https://doi.org/10.1515/gps-2017-0145.

Divandari, H., Hemmati-Sarapardeh, A., Schaffie, M., & Ranjbar, M. (2019). Integrating synthesized citric acid-coated magnetite nanoparticles with magnetic fields for enhanced oil recovery: Experimental study and mechanistic understanding. *Journal of Petroleum Science and Engineering*, *174*, 425–436. https://doi.org/10.1016/j.petrol.2018.11.037.

Druetta, P., Raffa, P., & Picchioni, F. (2018). Plenty of room at the bottom: Nanotechnology as solution to an old issue in enhanced oil recovery. *Applied Sciences*, *8*(12), 2596. https://doi.org/10.3390/app8122596.

Eltoum, H., Yang, Y.-L., & Hou, J.-R. (2021). The effect of nanoparticles on reservoir wettability alteration: A critical review. *Petroleum Science*, *18*(1), 136–153. https://doi.org/10.1007/s12182-020-00496-0.

Esmaeilnezhad, E., Van, S. Le, Chon, B. H., Choi, H. J., Schaffie, M., Gholizadeh, M., & Ranjbar, M. (2018). An experimental study on enhanced oil recovery utilizing nanoparticle ferrofluid through the application of a magnetic field. *Journal of Industrial and Engineering Chemistry*, *58*, 319–327. https://doi.org/10.1016/j.jiec.2017.09.044.

Ewis, D., Benamor, A., Ba-Abbad, M. M., Nasser, M., El-Naas, M., & Qiblawey, H. (2020). Removal of oil content from oil-water emulsions using iron oxide/bentonite nano adsorbents. *Journal of Water Process Engineering*, *38*, 1–11. https://doi.org/10.1016/j.jwpe.2020.101583.

Fani, M., Ghandehari, F., & Rezaee, M. (2018). biosynthesis of iron oxide nanoparticles by cytoplasmic extract of bacteria Lactobacillus fermentum. *Journal of Medicinal and Chemical Sciences*, *1*(1), 28–30. https://doi.org/https://doi.org/10.26655/jmchemsci.2018.9.2.

Faraji, M., Yamini, Y., & Rezaee, M. (2010). Magnetic nanoparticles: Synthesis, stabilization, functionalization, characterization, and applications. *Journal of the Iranian Chemical Society*, *7*(1), 1–37.

Fariq, A., Khan, T., & Yasmin, A. (2017). Microbial synthesis of nanoparticles and their potential applications in biomedicine. *Journal of Applied Biomedicine*, *15*(4), 241–248. https://doi.org/10.1016/j.jab.2017.03.004.

Fatemi, M., Mollania, N., Momeni-Moghaddam, M., & Sadeghifar, F. (2018). Extracellular biosynthesis of magnetic iron oxide nanoparticles by Bacillus cereus strain HMH1: Characterization and in vitro cytotoxicity analysis on MCF-7 and 3T3 cell lines. *Journal of Biotechnology*, *270*, 1–11. https://doi.org/10.1016/j.jbiotec.2018.01.021.

Fauconnier, N., Bee, A., Roger, J., & Pons, J. N. (1996). Adsorption of gluconic and citric acids on maghemite particles in aqueous medium. In *Trends in Colloid and Interface Science X* (Vol. 100, pp. 212–216). Steinkopff. https://doi.org/10.1007/BFb0115782.

Fauconnier, N., Bée, A., Roger, J., & Pons, J. N. (1999). Synthesis of aqueous magnetic liquids by surface complexation of maghemite nanoparticles. *Journal of Molecular Liquids*, *83*(1–3), 233–242. https://doi.org/10.1016/S0167-7322(99)00088-4.

Fekri Aval, S., Akbarzadeh, A., Yamchi, M. R., Zarghami, F., Nejati-Koshki, K., & Zarghami, N. (2016). Gene silencing effect of SiRNA-magnetic modified with biodegradable copolymer nanoparticles on hTERT gene expression in lung cancer cell line. *Artificial Cells, Nanomedicine, and Biotechnology*, *44*(1), 188–193. https://doi.org/10.3109/21691401.2014.934456.

Foroozesh, J., & Kumar, S. (2020). Nanoparticles behaviors in porous media: Application to enhanced oil recovery. *Journal of Molecular Liquids*, *316*, 113876. https://doi.org/10.1016/j.molliq.2020.113876.

Fu, X., Cai, J., Zhang, X., Li, W.-D., Ge, H., & Hu, Y. (2018). Top-down fabrication of shape-controlled, monodisperse nanoparticles for biomedical applications. *Advanced Drug Delivery Reviews*, *132*, 169–187. https://doi.org/10.1016/j.addr.2018.07.006.

Gahlawat, G., & Choudhury, A. R. (2019). A review on the biosynthesis of metal and metal salt nanoparticles by microbes. *RSC Advances*, *9*(23), 12944–12967. https://doi.org/10.1039/C8RA10483B.

Gan, Y. X., Jayatissa, A. H., Yu, Z., Chen, X., & Li, M. (2020). hydrothermal synthesis of nanomaterials. *Journal of Nanomaterials*, *2020*, 1–3. https://doi.org/10.1155/2020/8917013.

Gbadamosi, A. O., Junin, R., Manan, M. A., Yekeen, N., Agi, A., & Oseh, J. O. (2018). Recent advances and prospects in polymeric nanofluids application for enhanced oil recovery. *Journal of Industrial and Engineering Chemistry*, *66*, 1–19. https://doi.org/10.1016/j.jiec.2018.05.020.

Ghorbani, H. (2014). A review of methods for synthesis of Al nanoparticles. *Oriental Journal of Chemistry*, *30*(4), 1941–1949. https://doi.org/10.13005/ojc/300456.

Giti Emtiazi, N. G. (2015). The influence of Microbacterium hominis and Bacillus licheniformis extracellular polymers on silver and iron oxide nanoparticles production; green biosynthesis and mechanism of bacterial nano production. *Journal of Nanomaterials and Molecular Nanotechnology*, *04*(02). https://doi.org/10.4172/2324-8777.1000160.

Gupta, A. K., & Wells, S. (2004). Surface-modified superparamagnetic nanoparticles for drug delivery: Preparation, characterization, and cytotoxicity studies. *IEEE Transactions on Nanobioscience*, *3*(1), 66–73. https://doi.org/10.1109/TNB.2003.820277.

Han, L. H., Liu, H., & Wei, Y. (2011). In situ synthesis of hematite nanoparticles using a low-temperature microemulsion method. *Powder Technology*, *207*(1–3), 42–46. https://doi.org/10.1016/j.powtec.2010.10.008.

Hasany, S. F., Ahmed, I., Rajan, J., & Rehman, A. (2013). Systematic review of the preparation techniques of iron oxide magnetic nanoparticles. *Nanoscience and Nanotechnology*, *2*(6), 148–158. https://doi.org/10.5923/J.NN.20120206.01.

He, X., Liang, C., Liu, Q., & Xu, Z. (2019). Magnetically responsive Janus nanoparticles synthesized using cellulosic materials for enhanced phase separation in oily wastewaters and water-in-crude oil emulsions. *Chemical Engineering Journal*, *378*(March), 122045. https://doi.org/10.1016/j.cej.2019.122045.

Hernández-Hernández, A. A., Aguirre-Álvarez, G., Cariño-Cortés, R., Mendoza-Huizar, L. H., & Jiménez-Alvarado, R. (2020). Iron oxide nanoparticles: synthesis, functionalization, and applications in diagnosis and treatment of cancer. *Chemical Papers*, *74*(11), 3809–3824. https://doi.org/10.1007/s11696-020-01229-8.

Hu, M., Jiang, J.-S., & Zeng, Y. (2010). Prussian blue microcrystals prepared by selective etching and their conversion to mesoporous magnetic iron(III) oxides. *Chem. Commun.*, *46*(7), 1133–1135. https://doi.org/10.1039/B917702G.

Hu, Z., Nourafkan, E., Gao, H., & Wen, D. (2017). Microemulsions stabilized by in-situ synthesized nanoparticles for enhanced oil recovery. *Fuel*, *210*, 272–281. https://doi.org/10.1016/j.fuel.2017.08.004.

Hyeon, T., Lee, S. S., Park, J., Chung, Y., & Na, H. Bin. (2001). Synthesis of highly crystalline and monodisperse maghemite nanocrystallites without a size-selection process. *Journal of the American Chemical Society*, *123*(51), 12798–12801. https://doi.org/10.1021/ja016812s.

Iravani, S., & Varma, R. S. (2020). Bacteria in heavy metal remediation and nanoparticle biosynthesis. *ACS Sustainable Chemistry & Engineering*, *8*(14), 5395–5409. https://doi.org/10.1021/acssuschemeng.0c00292.

Izadi, N., Koochi, M. M., Amrollahi, A., & Pourkhalil, M. (2019). Investigation of functionalized polyelectrolyte polymer-coated Fe3O4 nanoparticles stabilized in high salinity brine at high temperatures as an EOR agent. *Journal of Petroleum Science and Engineering*, *178*, 1079–1091. https://doi.org/10.1016/j.petrol.2019.01.074.

Izadiyan, Z., Shameli, K., Miyake, M., Hara, H., Mohamad, S. E. B., Kalantari, K., Taib, S. H. M., & Rasouli, E. (2020). Cytotoxicity assay of plant-mediated synthesized iron oxide nanoparticles using Juglans regia green husk extract. *Arabian Journal of Chemistry*, *13*(1), 2011–2023. https://doi.org/10.1016/j.arabjc.2018.02.019.

James, M., Revia, R. A., Stephen, Z., & Zhang, M. (2020). Microfluidic synthesis of iron oxide nanoparticles. *Nanomaterials*, *10*(11), 1–19. https://doi.org/10.3390/nano10112113.

Jamkhande, P. G., Ghule, N. W., Bamer, A. H., & Kalaskar, M. G. (2019). Metal nanoparticles synthesis: An overview on methods of preparation, advantages and disadvantages, and applications. *Journal of Drug Delivery Science and Technology*, *53*, 1–11. https://doi.org/10.1016/j.jddst.2019.101174.

Jiang, W., McCloy, J. S., Lea, A. S., Sundararajan, J. A., Yao, Q., & Qiang, Y. (2011). Magnetization and susceptibility of ion-irradiated granular magnetite films. *Physical Review B*, *83*(13), 1–7. https://doi.org/10.1103/PhysRevB.83.134435.

Jiang, W., Yang, H. C., Yang, S. Y., Horng, H. E., Hung, J. C., Chen, Y. C., & Hong, C.-Y. (2004). Preparation and properties of superparamagnetic nanoparticles with narrow size distribution and biocompatible. *Journal of Magnetism and Magnetic Materials*, *283*(2–3), 210–214. https://doi.org/10.1016/j.jmmm.2004.05.022.

Johnson, G. E., Moser, T., Engelhard, M., Browning, N. D., & Laskin, J. (2016). Fabrication of electrocatalytic Ta nanoparticles by reactive sputtering and ion soft landing. *Journal of Chemical Physics*, *145*(17), 1–12. https://doi.org/10.1063/1.4966199.

Jolivet, J. P. (1992). Influence of Fe(II) on the formation of the spinel iron oxide in alkaline medium. *Clays and Clay Minerals*, *40*(5), 531–539. https://doi.org/10.1346/CCMN.1992.0400506.

Jolivet, J.-P., Tronc, É., & Chanéac, C. (2002). Synthesis of iron oxide-based magnetic nanomaterials and composites. *Comptes Rendus Chimie*, *5*(10), 659–664. https://doi.org/10.1016/S1631-0748(02)01422-4.

Joudeh, N., & Linke, D. (2022). Nanoparticle classification, physicochemical properties, characterization, and applications: A comprehensive review for biologists. *Journal of Nanobiotechnology*, *20*(1), 1–29. https://doi.org/10.1186/s12951-022-01477-8.

Jubran, A. S., Al-Zamely, O. M., & Al-Ammar, M. H. (2020). A study of iron oxide nanoparticles synthesis by using bacteria. *International Journal of Pharmaceutical Quality Assurance*, *11*(01), 88–92. https://doi.org/10.25258/ijpqa.11.1.13.

Kamal, M. S., Adewunmi, A. A., Sultan, A. S., Al-Hamad, M. F., & Mehmood, U. (2017). Recent advances in nanoparticles enhanced oil recovery: Rheology, interfacial tension, oil recovery, and wettability alteration. *Journal of Nanomaterials*, *2017*, 1–15. https://doi.org/10.1155/2017/2473175.

Kanagasubbulakshmi, S., & Kadirvelu, K. (2017). Green synthesis of Iron oxide nanoparticles using Lagenaria siceraria and evaluation of its antimicrobial activity. *Defence Life Science Journal*, *2*(4), 422.

Karade, V. C., Parit, S. B., Dawkar, V. V., Devan, R. S., Choudhary, R. J., Kedge, V. V., Pawar, N. V., Kim, J. H., & Chougale, A. D. (2019). A green approach for the synthesis of α-Fe2O3 nanoparticles from Gardenia resinifera plant and it's In vitro hyperthermia application. *Heliyon*, *5*(7), e02044. https://doi.org/10.1016/j.heliyon.2019.e02044.

Karpagavinayagam, P., & Vedhi, C. (2019). Green synthesis of iron oxide nanoparticles using Avicennia marina flower extract. *Vacuum*, *160*, 286–292. https://doi.org/10.1016/j.vacuum.2018.11.043.

Kaul, R. K., Kumar, P., Burman, U., Joshi, P., Agrawal, A., Raliya, R., & Tarafdar, J. C. (2012). Magnesium and iron nanoparticles production using microorganisms and various salts. *Materials Science-Poland*, *30*(3), 254–258. https://doi.org/10.2478/s13536-012-0028-x.

Kazemzadeh, Y., Sharifi, M., Riazi, M., Rezvani, H., & Tabaei, M. (2018). Potential effects of metal oxide/SiO2 nanocomposites in EOR processes at different pressures. *Colloids and Surfaces A: Physicochemical and Engineering Aspects*, *559*, 372–384. https://doi.org/10.1016/j.colsurfa.2018.09.068.

Kazemzadeh, Y., Shojaei, S., Riazi, M., & Sharifi, M. (2019). Review on application of nanoparticles for EOR purposes: A critical review of the opportunities and challenges. *Chinese Journal of Chemical Engineering*, *27*(2), 237–246. https://doi.org/10.1016/j.cjche.2018.05.022.

Khalil, A. T., Ovais, M., Ullah, I., Ali, M., Shinwari, Z. K., & Maaza, M. (2017). Biosynthesis of iron oxide (Fe2O3) nanoparticles via aqueous extracts of *Sageretia thea* (Osbeck.) and their pharmacognostic properties. *Green Chemistry Letters and Reviews*, *10*(4), 186–201. https://doi.org/10.1080/17518253.2017.1339831.

Khan, I., Saeed, K., & Khan, I. (2019). Nanoparticles: Properties, applications and toxicities. *Arabian Journal of Chemistry*, *12*(7), 908–931. https://doi.org/10.1016/j.arabjc.2017.05.011.

Khanal, L. R., Williams, T., & Qiang, Y. (2018). High-temperature investigation on morphology, phase and size of iron/iron-oxide core–shell nanoclusters for radiation nanodetector. *Journal of Physics D: Applied Physics*, *51*(25), 1–8. https://doi.org/10.1088/1361-6463/aac47e.

King, J. G., Williams, W., Wilkinson, C. D. W., McVitie, S., & Chapman, J. N. (1996). Magnetic properties of magnetite arrays produced by the method of electron beam lithography. *Geophysical Research Letters*, *23*(20), 2847–2850. https://doi.org/10.1029/96GL01371.

Ko, S., Kim, E. S., Park, S., Daigle, H., Milner, T. E., Huh, C., Bennetzen, M. V., & Geremia, G. A. (2016). Oil droplet removal from produced water using nanoparticles and their magnetic separation. *SPE Annual Technical Conference and Exhibition*, *2*, 1–18. https://doi.org/10.2118/181893-MS.

Kumar, A., & Gangawane, K. M. (2022a). Effect of precipitating agents on the magnetic and structural properties of the synthesized ferrimagnetic nanoparticles by co-precipitation method. *Powder Technology*, *401*, 117298. https://doi.org/10.1016/j.powtec.2022.117298.

Kumar, A., & Gangawane, K. M. (2022b). Synthesis and effect on the surface morphology & magnetic properties of ferrimagnetic nanoparticles by different wet chemical synthesis methods. *Powder Technology*, *410*(June), 117867. https://doi.org/10.1016/j.powtec.2022.117867.

Kumar, A., & Gangawane, K. M. (2023a). Advanced computational approaches for water treatment. In *Advanced Computational Approaches for Water Treatment* (pp. 1–12). CRC Press. https://doi.org/10.1201/9781003325147-1.

Kumar, A., & Gangawane, K. M. (2023b). Molecular dynamics for the membrane process. In *Advanced Computational Approaches for Water Treatment* (pp. 13–38). CRC Press. https://doi.org/10.1201/9781003325147-2.

Kumar, A., Gangawane, K. M., & Ramanjaneyulu, B. (2022). Ferrofluids for waste-water treatment. In *International Conference on Chemical, Bio and Environmental Engineering* (pp. 723–744). https://link.springer.com/chapter/10.1007/978-3-030-96554-9_48.

Kumar, A., Jena, H. M., & Gangawane, K. M. (2023a). Effect of ferrimagnetic nanoparticles as an oxygen absorber on the shelf life of tomatoes. In *9th International Food Convention (IFCoN)*, "TRIMSAFE – Technology Re-engineering for Innovation and Mitigating the risk for a Safe, sustainable Affordable & secure Food Eco-system", CSIR-CFTRI campus, Mysuru, 7–10 December 2023. http://hdl.handle.net/2080/4172.

Kumar, A., Jena, H. M., & Gangawane, K. M. (2023b). Variations in magnetic behavior of ferrimagnetic nanoparticles with temperature. In *76th Indian Chemical Engineering Congress (CHEMCON)*, Kolkata, India, 27–30 December 2023. http://hdl.handle.net/2080/4265.

Kumar, P. S. S. R., & Alexis, S. J. (2019). Synthesized carbon nanotubes and their applications. In *Carbon-Based Nanofillers and Their Rubber Nanocomposites* (pp. 109–122). Elsevier. https://doi.org/10.1016/B978-0-12-813248-7.00004-3.

Kumfer, B. M., Shinoda, K., Jeyadevan, B., & Kennedy, I. M. (2010). Gas-phase flame synthesis and properties of magnetic iron oxide nanoparticles with reduced oxidation state. *Journal of Aerosol Science*, *41*(3), 257–265. https://doi.org/10.1016/j.jaerosci.2010.01.003.

Latif, M. S., Abbas, S., Kormin, F., & Mustafa, M. K. (2019). Green synthesis of plant-mediated metal nanoparticles: The role of polyphenols. *Asian Journal of Pharmaceutical and Clinical Research*, 75–84. https://doi.org/10.22159/ajpcr.2019.v12i7.33211.

Laurent, S., Forge, D., Port, M., Roch, A., Robic, C., Vander Elst, L., & Muller, R. N. (2008). Magnetic iron oxide nanoparticles: Synthesis, stabilization, vectorization, physicochemical characterizations, and biological applications. *Chemical Reviews*, *108*(6), 2064–2110.

Lee, H., Lee, W.-J., Park, Y.-K., Ki, S., Kim, B.-J., & Jung, S.-C. (2018). Liquid phase plasma synthesis of iron oxide nanoparticles on nitrogen-doped activated carbon resulting in nanocomposite for supercapacitor applications. *Nanomaterials*, *8*(4), 1–13. https://doi.org/10.3390/nano8040190.

Liu, S., Yu, B., Wang, S., Shen, Y., & Cong, H. (2020). Preparation, surface functionalization and application of Fe3O4 magnetic nanoparticles. *Advances in Colloid and Interface Science*, *281*, 102165. https://doi.org/10.1016/j.cis.2020.102165.

Lohrasbi, S., Kouhbanani, M. A. J., Beheshtkhoo, N., Ghasemi, Y., Amani, A. M., & Taghizadeh, S. (2019). Green synthesis of iron nanoparticles using plantago major leaf extract and their application as a catalyst for the decolorization of azo dye. *BioNanoScience*, *9*(2), 317–322. https://doi.org/10.1007/s12668-019-0596-x.

Lü, T., Qi, D., Zhang, D., Lin, S., Mao, Y., & Zhao, H. (2018). Facile synthesis of N-(aminoethyl)-aminopropyl functionalized core-shell magnetic nanoparticles for emulsified oil-water separation. *Journal of Alloys and Compounds*, *769*, 858–865. https://doi.org/10.1016/j.jallcom.2018.08.071.

Luo, X., Al-Antaki, A. H. M., Alharbi, T. M. D., Hutchison, W. D., Zou, Y., Zou, J., Sheehan, A., Zhang, W., & Raston, C. L. (2018). Laser-ablated vortex fluidic-mediated synthesis of superparamagnetic magnetite nanoparticles in water under flow. *ACS Omega*, *3*(9), 11172–11178. https://doi.org/10.1021/acsomega.8b01606.

Mahanty, S., Bakshi, M., Ghosh, S., Chatterjee, S., Bhattacharyya, S., Das, P., Das, S., & Chaudhuri, P. (2019). Green synthesis of iron oxide nanoparticles mediated by filamentous fungi isolated from sundarban mangrove ecosystem, India. *BioNanoScience*, *9*(3), 637–651. https://doi.org/10.1007/s12668-019-00644-w.

Mahdavi, M., Namvar, F., Ahmad, M., & Mohamad, R. (2013). Green biosynthesis and characterization of magnetic iron oxide (Fe3O4) nanoparticles using seaweed (Sargassum muticum) Aqueous Extract. *Molecules*, *18*(5), 5954–5964. https://doi.org/10.3390/molecules18055954.

Majerič, P., Friedrich, B., & Rudolf, R. (2015). Au-nanoparticle synthesis via ultrasonic spray pyrolysis with a separate evaporation zone. *Materiali in Tehnologije*, *49*(5), 791–796. https://doi.org/10.17222/mit.2014.264.

Majidi, S., Sehrig, F. Z., Farkhani, S. M., Goloujeh, M. S., & Akbarzadeh, A. (2016). Current methods for synthesis of magnetic nanoparticles. *Artificial Cells, Nanomedicine and Biotechnology*, *44*(2), 722–734. https://doi.org/10.3109/21691401.2014.982802.

Majumder, A., Ramrakhiani, L., Mukherjee, D., Mishra, U., Halder, A., Mandal, A. K., & Ghosh, S. (2019). Green synthesis of iron oxide nanoparticles for arsenic remediation in water and sludge utilization. *Clean Technologies and Environmental Policy*, *21*(4), 795–813. https://doi.org/10.1007/s10098-019-01669-1.

Marcelo, G. A., Lodeiro, C., Capelo, J. L., Lorenzo, J., & Oliveira, E. (2020). Magnetic, fluorescent and hybrid nanoparticles: From synthesis to application in biosystems. *Materials Science and Engineering C*, *106*, 1–31. https://doi.org/10.1016/j.msec.2019.110104.

Massart, R. (1981). Preparation of aqueous magnetic liquids in alkaline and acidic media. *IEEE Transactions on Magnetics*, *17*(2), 1247–1248. https://doi.org/10.1109/TMAG.1981.1061188.

Miguel, M. G., Lourenço, J. P., & Faleiro, M. L. (2020). Superparamagnetic iron oxide nanoparticles and essential oils: A new tool for biological applications. *International Journal of Molecular Sciences*, *21*(18), 6633. https://doi.org/10.3390/ijms21186633.

Mihai, A. D., Chircov, C., Grumezescu, A. M., & Holban, A. M. (2020). Magnetite nanoparticles and essential oils systems for advanced antibacterial therapies. *International Journal of Molecular Sciences*, *21*(19), 7355. https://doi.org/10.3390/ijms21197355.

Minz, P., & Mishra, S. (2023). Jackfruit peel derived ZnCl2-impregnated activated carbon: Optimization, characterization, and application in dye removal. *Biomass Conversion and Biorefinery*, 0123456789. https://doi.org/10.1007/s13399-023-04592-0.

Mirza, A. U., Kareem, A., Nami, S. A. A., Khan, M. S., Rehman, S., Bhat, S. A., Mohammad, A., & Nishat, N. (2018). Biogenic synthesis of iron oxide nanoparticles using Agrewia optiva and Prunus persica phyto species: Characterization, antibacterial and antioxidant activity. *Journal of Photochemistry and Photobiology B: Biology*, *185*, 262–274. https://doi.org/10.1016/j.jphotobiol.2018.06.009.

Mittal, A. K., Chisti, Y., & Banerjee, U. C. (2013). Synthesis of metallic nanoparticles using plant extracts. *Biotechnology Advances*, *31*(2), 346–356. https://doi.org/10.1016/j.biotechadv.2013.01.003.

Mondal, P., Anweshan, A., & Purkait, M. K. (2020). Green synthesis and environmental application of iron-based nanomaterials and nanocomposite: A review. *Chemosphere*, *259*, 127509. https://doi.org/10.1016/j.chemosphere.2020.127509.

Moura, H. M., & Unterlass, M. M. (2020). Biogenic metal oxides. *Biomimetics*, *5*(2), 29. https://doi.org/10.3390/biomimetics5020029.

Najafi, A., & Nematipour, K. (2017). Synthesis and magnetic properties evaluation of monosized FeCo alloy nanoparticles through microemulsion method. *Journal of Superconductivity and Novel Magnetism*, *30*(9), 2647–2653. https://doi.org/10.1007/s10948-017-4052-2.

Negin, C., Ali, S., & Xie, Q. (2016). Application of nanotechnology for enhancing oil recovery – A review. *Petroleum*, *2*(4), 324–333. https://doi.org/10.1016/j.petlm.2016.10.002.

Nguyen, T. P., Le, U. T. P., Ngo, K. T., Pham, K. D., & Dinh, L. X. (2016). Synthesis of polymer-coated magnetic nanoparticles from red mud waste for enhanced oil recovery in offshore reservoirs. *Journal of Electronic Materials*, *45*(7), 3801–3808. https://doi.org/10.1007/s11664-016-4513-6.

Niculescu, A.-G., Chircov, C., & Grumezescu, A. M. (2022). Magnetite nanoparticles: Synthesis methods – A comparative review. *Methods*, *199*, 16–27. https://doi.org/10.1016/j.ymeth.2021.04.018.

Noruzi, M. (2015). Biosynthesis of gold nanoparticles using plant extracts. *Bioprocess and Biosystems Engineering*, *38*(1), 1–14. https://doi.org/10.1007/s00449-014-1251-0.

Okuyama, K., & Wuled Lenggoro, I. (2003). Preparation of nanoparticles via spray route. *Chemical Engineering Science*, *58*(3–6), 537–547. https://doi.org/10.1016/S0009-2509(02)00578-X.

Ou, C., & Wang, D. (2021). Structural performance characteristics of nanomaterials and its application in traditional architectural cultural design and landscape planning. *Advances in Civil Engineering*, *2021*. https://doi.org/10.1155/2021/5531679.

Pal, D., & Sen, S. (2023a). Emerging petroleum pollutants and their adverse effects on the environment. In I. D. Behera, & A. P. Das (Eds.), *Impact of Petroleum Waste on Environmental Pollution and Its Sustainable Management through Circular Economy. Environmental Science and Engineering* (pp. 103–137). Springer Nature Switzerland. https://doi.org/10.1007/978-3-031-48220-5_5.

Pal, D., & Sen, S. (2023b). In-depth coverage of petroleum waste sources, characteristics, environmental impact, and sustainable remediation process. In I. D. Behera, & A. P. Das (Eds.), *Impact of Petroleum Waste on Environmental Pollution and Its Sustainable Management through Circular Economy* (pp. 1–38). Springer Nature Switzerland. https://doi.org/10.1007/978-3-031-48220-5_1.

Pal, D., & Sen, S. (2024). Optimal synthesis of dolochar derived faujasite zeolite X for highly effective Cd(II) removal. *Environmental Research*, *240*, 117494. https://doi.org/10.1016/j.envres.2023.117494.

Pal Singh, J., Kumar, M., Sharma, A., Pandey, G., Chae, K. H., & Lee, S. (2020). Bottom-up and top-down approaches for MgO. In *Sonochemical Reactions*. IntechOpen. https://doi.org/10.5772/intechopen.91182.

Panda, D., & Gangawane, K. M. (2023a). Development of superhydrophobic hybrid silica-cellulose aerogel as promising thermal insulation and sound absorption. *International Journal of Materials Research*, *114*(4–5), 405–417. https://doi.org/10.1515/ijmr-2022-0066.

Panda, D., & Gangawane, K. M. (2023b). Hybrid NiFe2O4-NiCO4O4 nanoparticles-based eutectic phase change materials for enhancement of thermal efficiency of pin-fin heat sink arrangement. *Journal of Energy Storage*, *60*(January), 102417. https://doi.org/10.1016/j.est.2023.106644.

Panda, D., & Gangawane, K. M. (2023c). Recycled cellulose–silica hybrid aerogel for effective oil adsorption: Optimization and kinetics study. *Sadhana - Academy Proceedings in Engineering Sciences*, 48(3). https://doi.org/10.1007/s12046-023-02161-9.

Panda, D., Sahu, A. K., & Gangawane, K. M. (2024). Effect of molar concentration and drying methodologies on monodispersed silica sol for synthesis of silica aerogels with temperature-resistant characteristics. *Materials Science and Engineering: B*, *299*, 117062. https://doi.org/10.1016/j.mseb.2023.117062.

Panda, D., Sharma, S., & Gangawane, K. M. (2023). Synthesis of novel biocatalyst from organic waste protonated with acid treatment for hydrogen production. *International Journal of Hydrogen Energy*, *48*(63), 24242–24254. https://doi.org/10.1016/j.ijhydene.2023.03.117.

Pantidos, N. (2014). Biological synthesis of metallic nanoparticles by bacteria, fungi and plants. *Journal of Nanomedicine & Nanotechnology*, *05*(05). https://doi.org/10.4172/2157-7439.1000233.

Park, J., An, K., Hwang, Y., Park, J.-G., Noh, H.-J., Kim, J.-Y., Park, J.-H., Hwang, N.-M., & Hyeon, T. (2004). Ultra-large-scale syntheses of monodisperse nanocrystals. *Nature Materials*, *3*(12), 891–895. https://doi.org/10.1038/nmat1251.

Park, Y. C., Paulsen, J., Nap, R. J., Whitaker, R. D., Mathiyazhagan, V., Song, Y. Q., Hürlimann, M., Szleifer, I., & Wong, J. Y. (2014). Adsorption of superparamagnetic iron oxide nanoparticles on silica and calcium carbonate sand. *Langmuir*, *30*(3), 784–792. https://doi.org/10.1021/la404387t.

Peng, B., Zhang, L., Luo, J., Wang, P., Ding, B., Zeng, M., & Cheng, Z. (2017). A review of nanomaterials for nanofluid enhanced oil recovery. *RSC Advances*, *7*(51), 32246–32254. https://doi.org/10.1039/c7ra05592g.

Pereira, C., Pereira, A. M., Fernandes, C., Rocha, M., Mendes, R., Fernández-García, M. P., Guedes, A., Tavares, P. B., Grenèche, J.-M., Araújo, J. P., & Freire, C. (2012). Superparamagnetic MFe2O4 (M = Fe, Co, Mn) nanoparticles: Tuning the particle size and magnetic properties through a novel one-step coprecipitation route. *Chemistry of Materials*, *24*(8), 1496–1504. https://doi.org/10.1021/cm300301c.

Pereira, M. L. d. O., Maia, K. C. B., Silva, W. C., Leite, A. C., Francisco, A. D. d. S., Vasconcelos, T. L., Nascimento, R. S. V., & Grasseschi, D. (2020). Fe3O4 nanoparticles as surfactant carriers for enhanced oil recovery and scale prevention. *ACS Applied Nano Materials*, *3*(6), 5762–5772. https://doi.org/10.1021/acsanm.0c00939.

Priya, N., Kaur, K., & Sidhu, A. K. (2021). Green synthesis: An eco-friendly route for the synthesis of iron oxide nanoparticles. *Frontiers in Nanotechnology*, *3*. https://doi.org/10.3389/fnano.2021.655062.

Priyadarshana, G., Kottegoda, N., Senaratne, A., de Alwis, A., & Karunaratne, V. (2015). Synthesis of magnetite nanoparticles by top-down approach from a high purity ore. *Journal of Nanomaterials*, 2015, 1–8. https://doi.org/10.1155/2015/317312.

Qiu, L., Zhu, N., Feng, Y., Michaelides, E. E., Żyła, G., Jing, D., Zhang, X., Norris, P. M., Markides, C. N., & Mahian, O. (2020). A review of recent advances in thermophysical properties at the nanoscale: From solid state to colloids. *Physics Reports*, *843*, 1–81. https://doi.org/10.1016/j.physrep.2019.12.001.

Rajeshkumar, S., & Bharath, L. V. (2017). Mechanism of plant-mediated synthesis of silver nanoparticles – A review on biomolecules involved, characterisation and antibacterial activity. *Chemico-Biological Interactions*, *273*, 219–227. https://doi.org/10.1016/j.cbi.2017.06.019.

Rashid, H., Mansoor, M. A., Haider, B., Nasir, R., Abd Hamid, S. B., & Abdulrahman, A. (2020). Synthesis and characterization of magnetite nano particles with high selectivity using in-situ precipitation method. *Separation Science and Technology*, *55*(6), 1207–1215. https://doi.org/10.1080/01496395.2019.1585876.

Razack, S. A., Suresh, A., Sriram, S., Ramakrishnan, G., Sadanandham, S., Veerasamy, M., Nagalamadaka, R. B., & Sahadevan, R. (2020). Green synthesis of iron oxide nanoparticles using Hibiscus rosa-sinensis for fortifying wheat biscuits. *SN Applied Sciences*, *2*(5), 898. https://doi.org/10.1007/s42452-020-2477-x.

Reza, H., Ali, M., & Morad, A. (2011). Synthesis of ZnO nanoparticles by spray pyrolysis method. *Iranian Journal of Chemistry and Chemical Engineering*, *30*(1), 1–6.

Rezvani, H., Riazi, M., Tabaei, M., Kazemzadeh, Y., & Sharifi, M. (2018). Experimental investigation of interfacial properties in the EOR mechanisms by the novel synthesized Fe3O4@ Chitosan nanocomposites. *Colloids and Surfaces A: Physicochemical and Engineering Aspects*, *544*(February), 15–27. https://doi.org/10.1016/j.colsurfa.2018.02.012.

Rishton, S. A., Lu, Y., Altman, R. A., Marley, A. C., Bian, X. P., Jahnes, C., Viswanathan, R., Xiao, G., Gallagher, W. J., & Parkin, S. S. P. (1997). Magnetic tunnel junctions fabricated at tenth-micron dimensions by electron beam lithography. *Microelectronic Engineering*, *35*(1–4), 249–252. https://doi.org/10.1016/S0167-9317(96)00107-4.

Rockenberger, J., Scher, E. C., & Alivisatos, A. P. (1999). A new nonhydrolytic single-precursor approach to surfactant-capped nanocrystals of transition metal oxides. *Journal of the American Chemical Society*, *121*(49), 11595–11596. https://doi.org/10.1021/ja993280v.

Roger, J., Pons, J. N., Massart, R., Halbreich, A., & Bacri, J. C. (1999). Some biomedical applications of ferrofluids. *The European Physical Journal Applied Physics*, *5*(3), 321–325. https://doi.org/10.1051/epjap:1999144.

Roy, S., & Das, T. K. (2015). Plant mediated green synthesis of silver nanoparticles-A review. *International Journal of Plant Biology and Research*, *3*(3), 1044.

Safaei, A., Esmaeilzadeh, F., Sardarian, A., Mousavi, S. M., & Wang, X. (2020). Experimental investigation of wettability alteration of carbonate gas-condensate reservoirs from oil-wetting to gas-wetting using Fe3O4 nanoparticles coated with poly(vinyl alcohol), (PVA) or hydroxyapatite (HAp). *Journal of Petroleum Science and Engineering*, *184*, 106530. https://doi.org/10.1016/j.petrol.2019.106530.

Sahoo, A., & Mishra, S. (2023). Synthesis of microporous-activated carbon via direct H3PO4 and KOH activation of Pterospermum acerifolium fruits: process optimization and characterization. *Biomass Conversion and Biorefinery*, 0123456789. https://doi.org/10.1007/s13399-023-05072-1.

Salem, D. M. S. A., Ismail, M. M., & Aly-Eldeen, M. A. (2019). Biogenic synthesis and antimicrobial potency of iron oxide (Fe3O4) nanoparticles using algae harvested from the Mediterranean Sea, Egypt. *The Egyptian Journal of Aquatic Research*, *45*(3), 197–204. https://doi.org/10.1016/j.ejar.2019.07.002.

Sánchez-López, E., Gomes, D., Esteruelas, G., Bonilla, L., Lopez-Machado, A. L., Galindo, R., Cano, A., Espina, M., Ettcheto, M., Camins, A., Silva, A. M., Durazzo, A., Santini, A., Garcia, M. L., & Souto, E. B. (2020). Metal-based nanoparticles as antimicrobial agents: An overview. *Nanomaterials*, *10*(2), 1–39. https://doi.org/10.3390/nano10020292.

Santhosh, R., & Sarkar, P. (2022). Jackfruit seed starch/tamarind kernel xyloglucan/zinc oxide nanoparticles-based composite films: Preparation, characterization, and application on tomato (Solanum lycopersicum) fruits. *Food Hydrocolloids*, *133*(April), 107917. https://doi.org/10.1016/j.foodhyd.2022.107917.

Sardul, S. S., Harshita, S., & Shyamji, S. (2017). Biosynthesis of silver nanoparticles by endophytic fungi: Its mechanism, characterization techniques and antimicrobial potential. *African Journal of Biotechnology*, *16*(14), 683–698. https://doi.org/10.5897/AJB2017.15873.

Schröfel, A., Kratošová, G., Šafařík, I., Šafaříková, M., Raška, I., & Shor, L. M. (2014). Applications of biosynthesized metallic nanoparticles – A review. *Acta Biomaterialia*, *10*(10), 4023–4042. https://doi.org/10.1016/j.actbio.2014.05.022.

Schwarzer, H.-C., & Peukert, W. (2004). Tailoring particle size through nanoparticle precipitation. *Chemical Engineering Communications*, *191*(4), 580–606. https://doi.org/10.1080/00986440490270106.

Shaker, S., Zafarian, S., Chidurala, S., & Rao, K. V. (2013). Preparation and characterization of magnetite nanoparticles by sol-gel method for water treatment. *International Journal of Innovative Research in Science, Engineering and Technology*, *2*(7), 2969–2973.

Sharma, D., Kanchi, S., & Bisetty, K. (2019). Biogenic synthesis of nanoparticles: A review. *Arabian Journal of Chemistry*, *12*(8), 3576–3600. https://doi.org/10.1016/j.arabjc.2015.11.002.

Shavel, A., & Liz-Marzan, L. (2009). Chemistry and physics of metal oxide nanostructures. *Physical Chemistry Chemical Physics*, *11*(19), 3607. https://doi.org/10.1039/b905768d.

Siddiqi, K. S., & Husen, A. (2016). Fabrication of metal nanoparticles from fungi and metal salts: Scope and application. *Nanoscale Research Letters*, *11*(1), 98. https://doi.org/10.1186/s11671-016-1311-2.

Sikiru, S., Rostami, A., Soleimani, H., Yahya, N., Afeez, Y., Aliu, O., Yusuf, J. Y., & Oladosu, T. L. (2021). Graphene: Outlook in the enhance oil recovery (EOR). *Journal of Molecular Liquids*, *321*, 114519. https://doi.org/10.1016/j.molliq.2020.114519.

Silva, L. P., Reis, I. G., & Bonatto, C. C. (2015). Green synthesis of metal nanoparticles by plants: Current trends and challenges. In *Green Processes for Nanotechnology* (pp. 259–275). Springer International Publishing. https://doi.org/10.1007/978-3-319-15461-9_9.

Simakin, A. V., Voronov, V. V., Kirichenko, N. A., & Shafeev, G. A. (2004). Nanoparticles produced by laser ablation of solids in liquid environment. *Applied Physics A*, *79*(4–6), 1127–1132. https://doi.org/10.1007/s00339-004-2660-8.

Singh, J., Dutta, T., Kim, K.-H., Rawat, M., Samddar, P., & Kumar, P. (2018). 'Green' synthesis of metals and their oxide nanoparticles: Applications for environmental remediation. *Journal of Nanobiotechnology*, *16*(1), 84. https://doi.org/10.1186/s12951-018-0408-4.

Singh, P., Kim, Y.-J., Zhang, D., & Yang, D.-C. (2016). Biological synthesis of nanoparticles from plants and microorganisms. *Trends in Biotechnology*, *34*(7), 588–599. https://doi.org/10.1016/j.tibtech.2016.02.006.

Sreeja, V., & Joy, P. A. (2007). Microwave–hydrothermal synthesis of γ-Fe2O3 nanoparticles and their magnetic properties. *Materials Research Bulletin*, *42*(8), 1570–1576. https://doi.org/10.1016/j.materresbull.2006.11.014.

Stojanović, Z., Otoničar, M., Lee, J., Stevanović, M. M., Hwang, M. P., Lee, K. H., Choi, J., & Uskoković, D. (2013). The solvothermal synthesis of magnetic iron oxide nanocrystals and the preparation of hybrid poly(l-lactide)–polyethyleneimine magnetic particles. *Colloids and Surfaces B: Biointerfaces*, *109*, 236–243. https://doi.org/10.1016/j.colsurfb.2013.03.053.

Subramaniyam, V., Subashchandrabose, S. R., Ganeshkumar, V., Thavamani, P., Chen, Z., Naidu, R., & Megharaj, M. (2016). Cultivation of Chlorella on brewery wastewater and nano-particle biosynthesis by its biomass. *Bioresource Technology*, *211*, 698–703. https://doi.org/10.1016/j.biortech.2016.03.154.

Sun, S., & Zeng, H. (2002). Size-controlled synthesis of magnetite nanoparticles. *Journal of the American Chemical Society*, *124*(28), 8204–8205. https://doi.org/10.1021/ja026501x.

Sun, X., Zhang, Y., Chen, G., & Gai, Z. (2017). Application of nanoparticles in enhanced oil recovery: A critical review of recent progress. *Energies*, *10*(3). https://doi.org/10.3390/en10030345.

Sundaram, P. A., Augustine, R., & Kannan, M. (2012). Extracellular biosynthesis of iron oxide nanoparticles by Bacillus subtilis strains isolated from rhizosphere soil. *Biotechnology and Bioprocess Engineering*, *17*(4), 835–840. https://doi.org/10.1007/s12257-011-0582-9.

Sundararajan, J. A., Kaur, M., Jiang, W., McCloy, J. S., & Qiang, Y. (2014). Oxide shell reduction and magnetic property changes in core-shell Fe nanoclusters under ion irradiation. *Journal of Applied Physics*, *115*(17), 1–3. https://doi.org/10.1063/1.4862520.

Swihart, M. T. (2003). Vapor-phase synthesis of nanoparticles. *Current Opinion in Colloid & Interface Science*, *8*(1), 127–133. https://doi.org/10.1016/S1359-0294(03)00007-4.

Tarafdar, J. C., & Raliya, R. (2013). Rapid, low-cost, and ecofriendly approach for iron nanoparticle synthesis using *Aspergillus oryzae* TFR9. *Journal of Nanoparticles*, *2013*, 1–4. https://doi.org/10.1155/2013/141274.

Tartaj, P., Morales, M. P., Veintemillas-Verdaguer, S., Gonzalez-Carreño, T., & Serna, C. J. (2006). Synthesis, properties and biomedical applications of magnetic nanoparticles. In *Handbook of Magnetic Materials* (Vol. 16, pp. 403–482). https://doi.org/10.1016/S1567-2719(05)16005-3.

Thakkar, K. N., Mhatre, S. S., & Parikh, R. Y. (2010). Biological synthesis of metallic nanoparticles. *Nanomedicine: Nanotechnology, Biology and Medicine*, *6*(2), 257–262. https://doi.org/10.1016/j.nano.2009.07.002.

Torabian, P., Ghandehari, F., & Fatemi, M. (2018). Biosynthesis of iron oxide nanoparticles by cytoplasmic extracts of bacteria lactobacillus casei. *Asian Journal of Green Chemistry*, *2*, 181–188. https://doi.org/https://dx. doi.org/10.22034/ajgc.2018.57914.

Unni, M., Uhl, A. M., Savliwala, S., Savitzky, B. H., Dhavalikar, R., Garraud, N., Arnold, D. P., Kourkoutis, L. F., Andrew, J. S., & Rinaldi, C. (2017). Thermal decomposition synthesis of iron oxide nanoparticles with diminished magnetic dead layer by controlled addition of oxygen. *ACS Nano*, *11*(2), 2284–2303. https://doi.org/10.1021/acsnano.7b00609.

Vanecht, E. (2012). *Gold Nanoparticles in Ionic Liquids Prepared by Sputter Deposition*. Katholieke Universiteit Leuven.

Vasantharaj, S., Sathiyavimal, S., Senthilkumar, P., LewisOscar, F., & Pugazhendhi, A. (2019). Biosynthesis of iron oxide nanoparticles using leaf extract of Ruellia tuberosa: Antimicrobial properties and their applications in photocatalytic degradation. *Journal of Photochemistry and Photobiology B: Biology*, *192*, 74–82. https://doi.org/10.1016/j.jphotobiol.2018.12.025.

Vayssières, L., Chanéac, C., Tronc, E., & Jolivet, J. P. (1998). Size tailoring of magnetite particles formed by aqueous precipitation: An example of thermodynamic stability of nanometric oxide particles. *Journal of Colloid and Interface Science*, *205*(2), 205–212. https://doi.org/10.1006/jcis.1998.5614.

Vernieres, J., Steinhauer, S., Zhao, J., Chapelle, A., Menini, P., Dufour, N., Diaz, R. E., Nordlund, K., Djurabekova, F., Grammatikopoulos, P., & Sowwan, M. (2017). Gas phase synthesis of multifunctional Fe-based nanocubes. *Advanced Functional Materials*, *27*(11), 1–12. https://doi.org/10.1002/adfm.201605328.

Vijayaraghavan, K., & Ashokkumar, T. (2017). Plant-mediated biosynthesis of metallic nanoparticles: A review of literature, factors affecting synthesis, characterization techniques and applications. *Journal of Environmental Chemical Engineering*, *5*(5), 4866–4883. https://doi.org/10.1016/j.jece.2017.09.026.

Wongwailikhit, K., & Horwongsakul, S. (2011). The preparation of iron(III) oxide nanoparticles using W/O microemulsion. *Materials Letters*, *65*(17–18), 2820–2822. https://doi.org/10.1016/j.matlet.2011.05.063.

Woo, K., Hong, J., Choi, S., Lee, H.-W., Ahn, J.-P., Kim, C. S., & Lee, S. W. (2004). Easy synthesis and magnetic properties of iron oxide nanoparticles. *Chemistry of Materials*, *16*(14), 2814–2818. https://doi.org/10.1021/cm049552x.

Xi, K., Cao, Y., Haile, B. G., Zhu, R., Jahren, J., Bjørlykke, K., Zhang, X., & Hellevang, H. (2016). How does the pore-throat size control the reservoir quality and oiliness of tight sandstones? The case of the Lower Cretaceous Quantou Formation in the southern Songliao Basin, China. *Marine and Petroleum Geology*, *76*, 1–15. https://doi.org/10.1016/j.marpetgeo.2016.05.001.

Xing, L., ten Brink, G. H., Chen, B., Schmidt, F. P., Haberfehlner, G., Hofer, F., Kooi, B. J., & Palasantzas, G. (2016). Synthesis and morphology of iron–iron oxide core–shell nanoparticles produced by high pressure gas condensation. *Nanotechnology*, *27*(215703), 1–10. https://doi.org/10.1088/0957-4484/27/21/215703.

Xing, L., ten Brink, G. H., Kooi, B. J., & Palasantzas, G. (2017). Preparation of tunable-sized iron nanoparticles based on magnetic manipulation in inert gas condensation (IGC). *Journal of Applied Physics*, *121*(2), 1–5. https://doi.org/10.1063/1.4974052.

Yakasai, F., Jaafar, M. Z., Bandyopadhyay, S., Agi, A., & Sidek, M. A. (2022). Application of iron oxide nanoparticles in oil recovery – A critical review of the properties, formulation, recent advances and prospects. *Journal of Petroleum Science and Engineering*, *208*(PC), 109438. https://doi.org/10.1016/j.petrol.2021.109438.

Yamashita, K., Okada, S., & Sawada, H. (2019). Preparation of fluoroalkyl end-capped vinyltrimethoxysilane oligomeric silica/magnetite composites – Application to separation of oil and water. *Colloids and Surfaces A: Physicochemical and Engineering Aspects*, *581*, 1–11. https://doi.org/10.1016/j.colsurfa.2019.123668.

Ye, F., Laurent, S., Fornara, A., Astolfi, L., Qin, J., Roch, A., Martini, A., Toprak, M. S., Muller, R. N., & Muhammed, M. (2012). Uniform mesoporous silica coated iron oxide nanoparticles as a highly efficient, nontoxic MRI T_2 contrast agent with tunable proton relaxivities. *Contrast Media & Molecular Imaging*, *7*(5), 460–468. https://doi.org/10.1002/cmmi.1473.

Yew, Y. P., Shameli, K., Miyake, M., Kuwano, N., Bt Ahmad Khairudin, N. B., Bt Mohamad, S. E., & Lee, K. X. (2016). Green synthesis of magnetite (Fe3O4) nanoparticles using seaweed (Kappaphycus alvarezii) Extract. *Nanoscale Research Letters*, *11*(1), 276. https://doi.org/10.1186/s11671-016-1498-2.

Zheng, Y.-Y., Sun, Q., Duan, Y.-H., Zhai, J., Zhang, L.-L., & Wang, J.-X. (2020). Controllable synthesis of monodispersed iron oxide nanoparticles by an oxidation-precipitation combined with solvothermal process. *Materials Chemistry and Physics*, *252*, 123431. https://doi.org/10.1016/j.matchemphys.2020.123431.

Zhou, H., Kim, J.-P., Bahng, J. H., Kotov, N. A., & Lee, J. (2014). Self-assembly mechanism of spiky magnetoplasmonic supraparticles. *Advanced Functional Materials*, *24*(10), 1439–1448. https://doi.org/10.1002/adfm.201302405.

10 CFD of Fluid Flow for EOR Using Nanoparticles

Krishna Raghav Chaturvedi, Abhishek Kumar and Anand Raghav Chaubey

10.1 INTRODUCTION

Computational fluid dynamics (CFD) is a specialised field of numerical analysis that employs mathematical models and algorithms to simulate and analyse fluid flow phenomena. It involves the division of a fluid domain into discrete elements, or cells, allowing for the mathematical representation of fluid behaviour (Vaziri et al., 2020). By solving the governing equations for fluid motion, such as the Navier-Stokes equations, CFD provides insights into complex fluid dynamics, including velocity distributions, pressure variations, and temperature gradients. Its computational approach enables researchers and engineers to study fluid flow in diverse scenarios, ranging from aerodynamics in aviation to the behaviour of fluids in industrial processes (Kumar et al., 2023b).

In the hydrocarbon industry, CFD finds crucial applications across various domains like reservoir simulation, where CFD aids in simulating fluid flow within reservoirs, offering insights into the displacement of hydrocarbons and the effectiveness of recovery methods, and in wellbore dynamics, where it is used to understand the fluid behaviour in wellbores that is vital for optimising drilling operations, ensuring well integrity, and enhancing production efficiency (Dewangan & Sinha, 2016; Medhi et al., 2020; Alade et al., 2023). Furthermore, it is also utilised for flow assurance, which can be used to predict and mitigate challenges related to flow assurance, such as hydrate formation and pipeline blockages, ensuring uninterrupted hydrocarbon transport. The midstream engineers used CFD for the design of gas-liquid separation, wherein it can be used to aid the design of efficient separators for separating gas and liquid phases, a critical process in hydrocarbon processing facilities, and heat exchanger design, wherein the CFD software is routinely used to optimise the design of heat exchangers, ensuring efficient heat transfer in processes like refining and petrochemical production. Separately, pipeline design engineers use CFD to help in designing pipelines that minimise pressure drop and optimise fluid flow, contributing to efficient and cost-effective transportation of hydrocarbons (Hu & Cheng, 2016; Tang et al., 2021; Yeo et al., 2021). Drilling engineers have used CFD processes and techniques to model the behaviour of drilling fluids to explore hole cleaning and potential additive effects (Dewangan & Sinha, 2016; Medhi et al., 2020). In previous work by Medhi et al. (2020), the hydraulic performance of

 DOI: 10.1201/9781032650685-10

zirconium dioxide (ZrO_2) nanoparticle (NP)-based non-damaging drilling fluid was evaluated using numerical formulation and CFD simulations, with the results showing a positive impact of NPs on NDDF presented in terms of cutting volume fraction contours, velocity profiles, and pressure drops. In Enhanced Oil Recovery (EOR), CFD plays a pivotal role in optimising EOR techniques by simulating the injection of various substances, including nanomaterials, to enhance fluid displacement within reservoirs (Panda et al., 2024; Panda & Gangawane, 2023c).

By leveraging CFD in these applications, the hydrocarbon industry benefits from enhanced efficiency, reduced costs, and improved understanding of fluid dynamics, ultimately contributing to the sustainable and effective extraction and processing of hydrocarbon resources. The focus of this chapter is to explore, in detail, the various processes involved in the simulation of fluid flow in porous media where the application of nanotechnology has been investigated.

10.2 FLOW SIMULATION IN POROUS MEDIA

Porous media simulation in CFD involves modelling and simulating the behaviour of fluids as they flow through a porous structure (Figure 10.1). Porous media refer to materials or substances containing void spaces (pores) that allow the passage of fluids.

This simulation technique is particularly relevant in various engineering applications, including oil and gas reservoirs, groundwater aquifers, filters, and biological tissues.

In a porous media simulation, the key aspects include:

- *Geometry Representation*: Defining the geometry of the porous media, including the solid matrix and void spaces.
- *Mesh Generation*: Discretising the geometry into a computational mesh, dividing the porous structure into discrete elements to facilitate numerical calculations.

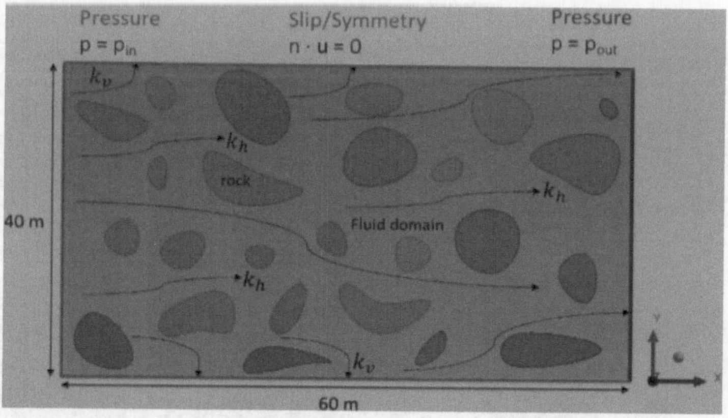

FIGURE 10.1 Schematic of an infinite model geometry showing flow and no-flow regions. Adapted with permission from Ejeh et al. (2020).

- *Fluid Flow Modelling*: Applying fluid flow equations, typically the Navier-Stokes equations, to describe the movement of fluids within the porous structure.
- *Porous Media Properties*: Specifying properties such as porosity and permeability. Porosity represents the fraction of void space in the porous media, while permeability characterises its ability to allow fluid flow.
- *Boundary Conditions*: Implementing appropriate boundary conditions to represent the interaction between the porous media and external fluid sources or sinks (Figure 10.2).
- *Fluid–Structure Interaction*: Modelling the interaction between the fluid and the solid matrix, considering how the presence of the fluid affects the movement and deformation of the solid structure.
- *Transport Phenomena*: Incorporating additional transport phenomena, such as heat transfer or species transport, if relevant to the specific application.
- *Solver Selection*: Choosing a solver capable of handling the complexities associated with fluid flow through porous media. It may involve specialised solvers or numerical techniques.
- *Post-Processing*: Analysing simulation results and extracting relevant data such as velocity profiles, pressure distributions, and concentration gradients.

CFD plays a crucial role in simulating particle movement in porous media by solving governing equations that describe fluid flow and particle transport within the porous structure. CFD provides a powerful tool for simulating particle movement in porous media by integrating fluid dynamics and particle transport models. This approach allows for a comprehensive understanding of the complex interactions between fluids and particles within porous structures, aiding in the design and optimisation of processes such as filtration, contaminant remediation, and enhanced resource recovery in various industries, including oil and gas, environmental engineering, and materials science. The simulation involves discretising the porous media into computational

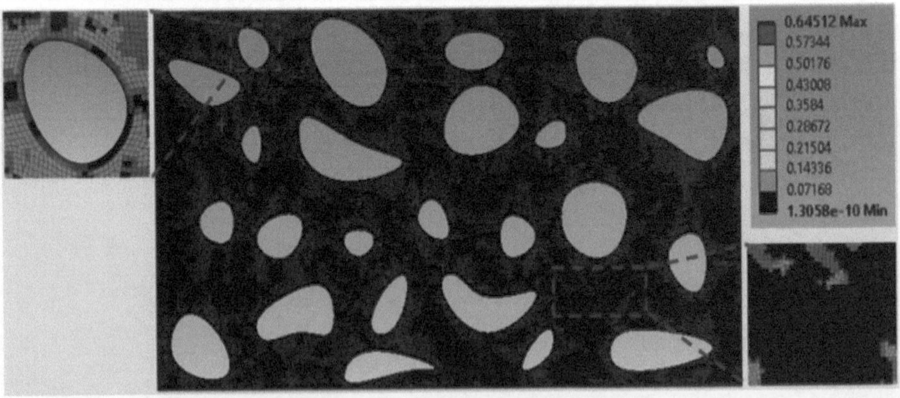

FIGURE 10.2 Visual of a discretisation model detailing the treatment of boundary layers and flow media. Adapted with permission from Ejeh et al. (2020).

cells and applying mathematical models to represent fluid flow and particle interactions (Kumar et al., 2022; Kumar & Gangawane, 2023a, 2023b). Here's a breakdown of how CFD facilitates the simulation of particle movement in porous media:

- *Fluid Flow Modelling*: CFD employs fluid dynamics equations, such as the Navier-Stokes equations, to simulate the movement of fluids within the porous structure. This includes considering the porosity and permeability of the media, which influence how fluids flow through the interconnected void spaces.
- *Particle Transport*: CFD incorporates Lagrangian particle tracking methods to model the movement of individual particles within the fluid. This involves accounting for the forces acting on particles, such as drag forces, gravity, and Brownian motion, to predict their trajectories.
- *Interactions with Porous Media*: CFD allows for the inclusion of interactions between particles and the porous media. This includes considerations for particle deposition on the porous surfaces, filtration effects, and changes in particle size distribution as particles move through the media.
- *Pore-Scale Simulation*: CFD can operate at the pore scale, providing detailed insights into the complex interactions occurring at the microscopic level. This allows for a more accurate representation of particle-fluid interactions within the intricate pore geometry of porous media.
- *Parameter Sensitivity Analysis*: CFD simulations enable sensitivity analysis of various parameters, such as porosity, permeability, and particle characteristics, to understand their impact on particle transport within the porous structure.
- *Optimisation of Particle Injection*: By using CFD, researchers and engineers can optimise the injection of particles for specific applications, such as EOR or groundwater remediation. It involves assessing the distribution of particles and their effectiveness in targeted areas of the porous media.

These processes can be understood in more detail in the Ansys Fluent Theory Guide. Porous media simulations in CFD find applications in a range of fields, which have been listed as under:

- *Oil and Gas Industry*: Understanding fluid flow in reservoir rocks and optimising EOR processes.
- *Environmental Engineering*: Modelling groundwater flow and contaminant transport in porous aquifers.
- *Biomedical Engineering*: Simulating blood flow through tissues and porous implants.
- *Filtration and Separation*: Designing and optimising filters for various applications.
- *Geotechnical Engineering*: Analysing soil permeability and fluid movement in porous soils.

By accurately representing the interaction between fluid and porous structures, these simulations provide valuable insights into the behaviour of fluids in porous media, aiding in the design and optimisation of processes across diverse industries.

10.3 FLOW SIMULATION OF NPS IN POROUS MEDIA

Modelling NP behaviour in porous media during EOR applications involves leveraging CFD to simulate the intricate interactions between fluids, NPs, and the porous structure (Kumar et al., 2023a; Panda et al., 2023; Panda & Gangawane, 2023a). In a lucid study on the use of CFD for flow simulation of NPs in porous media, Ejeh et al. (2020) visually reported the use of numerical models to properly show the resolved interfacial properties of the interacting phases and their transport (Figure 10.3).

Here are the critical steps involved in this process and how CFD facilitates the modelling:

- *Geometry Representation*: Define the geometry of the porous reservoir, including the solid matrix and void spaces. Specify the injection points for NPs within the reservoir.
- *Mesh Generation*: Generate a computational mesh that discretises the porous media, allowing for an accurate representation of fluid flow and NP movement.
- *Fluid Flow Modelling*: Apply fluid dynamics equations, such as the Navier-Stokes equations, to describe fluid flow within the porous media. Incorporate the effects of NP injection on fluid behaviour.
- *Porous Media Properties*: Specify properties such as porosity, permeability, and initial fluid saturation. Introduce parameters related to NP transport, such as deposition rates and attachment efficiency.

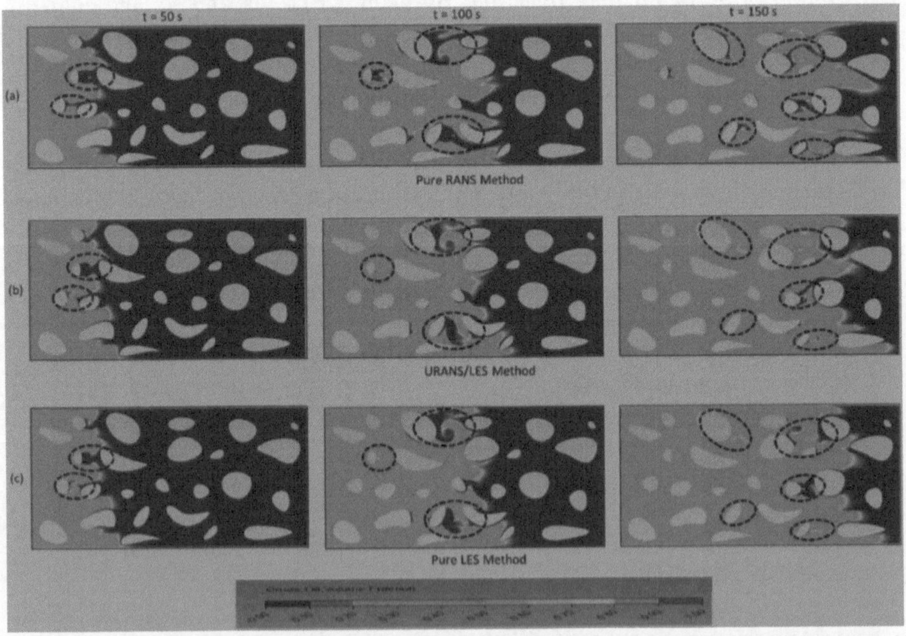

FIGURE 10.3 Flow visualisation of nanofluid injection in porous media. Adapted with permission from Ejeh et al. (2020).

- *NP Transport Modelling*: Implement Lagrangian particle tracking methods within the CFD framework to simulate the movement of individual NPs. Consider forces like Brownian motion, fluid drag, and gravitational effects.
- *Fluid–NP Interactions*: Model interactions between NPs and fluid, including aggregation, adsorption, and detachment processes. Account for the impact of NP size, surface charge, and surface chemistry on their behaviour.
- *Transport and Retention Mechanisms*: Incorporate mechanisms that influence NP transport and retention, such as straining, electrostatic interactions, and pore-scale heterogeneity.
- *Chemical Reactions*: Integrate any relevant chemical reactions involving NPs and the reservoir fluids. This could include reactions affecting NP stability or interactions with the oil.
- *Solver Selection*: Choose a CFD solver capable of handling multiphase flow, particle tracking, and complex fluid-NP interactions. Specialised solvers or modules may be employed for accurate representation.
- *Validation and Calibration*: Validate the model by comparing simulation results with experimental data or analytical solutions. Calibrate model parameters to improve agreement between simulations and real-world observations.

The use of CFD streamlines this modelling process by providing a computational framework that can handle the complex physics involved. CFD allows for the simultaneous simulation of fluid flow, NP transport, and interactions within the porous media. The advantages of using CFD in this context include:

- *Accuracy*: CFD enables a detailed representation of fluid dynamics and NP behaviour at the pore scale, capturing the physics of transport and retention mechanisms.
- *Flexibility*: The modular nature of CFD allows researchers to incorporate various physical phenomena and adapt the model to specific NP and porous media characteristics.
- *Efficiency*: CFD simulations provide insights into the performance of NP-EOR strategies without the need for extensive and costly experimental campaigns.

By utilising CFD, researchers can gain a comprehensive understanding of how NPs interact with fluids and porous media in the context of EOR, facilitating the optimisation of NP injection strategies for improved oil displacement and recovery efficiency (Kumar & Gangawane, 2022a, 2022b; Panda & Gangawane, 2023b).

10.4 STEPS TO SIMULATE POROUS MEDIA IN CFD

Simulating porous media in CFD involves modelling the flow of fluids through a porous structure, accounting for the complex interactions between the fluid and the solid matrix. Here's an overview of the critical steps and processes involved in simulating porous media in CFD that can be followed when performing the porous media simulation in CFD:

- *Porous Structure Definition*: Create a computational domain that represents the porous media geometry. This involves defining the solid matrix and void spaces that make up the porous structure.
- *Mesh Generation and Discretisation*: Generate a computational mesh that divides the porous media geometry into discrete elements (cells or elements). This mesh defines the spatial resolution of the simulation and is crucial for accurate representation of fluid flow and particle movement.
- *Fluid Flow Modelling*: Apply the Navier-Stokes equations to model fluid flow within the porous media. These equations describe the conservation of mass and momentum for the fluid phase.
- *Porosity and Permeability*: Define the porosity of the porous media, representing the fraction of void space relative to the total volume. Next, permeability is specified, which characterises the ability of the porous media to allow fluid flow. This property is crucial for determining the resistance to flow within the porous structure.
- *Fluid-Structure Interaction*: Implement fluid-structure interaction models to account for the dynamic interaction between the fluid and solid matrix. This involves considering how the presence of the fluid affects the motion and deformation of the solid matrix.
- *Particle Transport Modelling*: Incorporate Lagrangian particle tracking methods to simulate the movement of particles within the fluid. Account for forces acting on particles, such as fluid drag, gravity, and any additional relevant forces.
- *Boundary Conditions*: Apply appropriate boundary conditions at the inlet and outlet of the porous media to represent fluid injection and extraction points.
- *No-Slip Condition*: Implement no-slip conditions at solid surfaces to account for the interaction between the fluid and the solid matrix.
- *Solver Selection*: Choose a solver that can handle the specific challenges of simulating fluid flow through porous media. Iterative solvers and specialised algorithms may be necessary for efficient convergence.
- *Post-Processing*: Perform post-processing to extract relevant information, such as velocity profiles, pressure distributions, and particle trajectories.
- *Visualisation*: Use visualisation tools to interpret and communicate simulation results effectively.
- *Validation and Calibration*: Validate the simulation results by comparing them with experimental data or analytical solutions. If needed, calibrate model parameters to improve the agreement between simulation and experimental observations.

However, it is essential to note that simulating porous media in CFD requires a comprehensive understanding of fluid dynamics, porous structure characteristics, and the interplay between fluid and solid phases. The process involves a combination of mathematical modelling, numerical methods, and computational techniques to capture the complex behaviours occurring within the porous medium.

10.5 IMPORTANT CONSIDERATIONS

When conducting CFD simulations for porous media, several important considerations should be made to ensure accurate and meaningful results. Here are some of the key considerations to keep in mind while CFD simulation to ensure that accurate results are obtained:

- *Porous Media Representation*: Accurately represent the porous media geometry, including the complex structure of the solid matrix and void spaces. It is critical for capturing realistic flow patterns.
- *Mesh Quality*: Generate a high-quality computational mesh that adequately captures the porous structure. Ensure appropriate resolution, especially near regions of interest, such as injection or extraction point.
- *Relevance of Equations*: Choose or develop fluid flow equations (e.g., Darcy's law or the Navier-Stokes equations) that are appropriate for the flow regime and characteristics of the porous media under consideration.
- *Porous Media Properties*: Accurately specify porosity and permeability values, as these properties have a significant impact on fluid flow behaviour within the porous structure.
- *Boundary Conditions*: Define realistic boundary conditions, taking into account factors such as injection/extraction rates, fluid properties, and any external influences on the porous media system.
- *Fluid-Structure Interaction*: If applicable, model fluid-structure interaction accurately, considering how the presence of fluid affects the deformation or movement of the solid matrix.
- *Numerical Methods*: Choose a numerical solver suitable for porous media simulations. Implicit solvers may be preferred for stability and convergence, especially when dealing with multiphase flow.
- *Pore-Scale Modelling*: If aiming for high-resolution simulations, consider pore-scale modelling to capture detailed interactions within the porous structure. This may involve resolving individual pores and particles.
- *Particle Transport*: Implement Lagrangian particle tracking methods for accurate simulation of particle transport within the porous media. Consider relevant forces, such as fluid drag and particle-particle interactions.
- *Transport Phenomena*: Include additional transport phenomena, such as heat transfer or species transport, if relevant to the specific application of porous media.
- *Model Validation*: Validate the CFD model by comparing simulation results with experimental data or analytical solutions. This helps ensure the model's accuracy and reliability.
- *Sensitivity Analysis*: Conduct sensitivity analyses to understand how variations in model parameters, such as porosity, permeability, or particle properties, impact simulation results.

- *Computational Resources*: Consider the computational resources required for the simulations. High-resolution simulations may demand substantial computing power and storage.

By carefully addressing these considerations, researchers and engineers can enhance the accuracy and reliability of CFD simulations for porous media, leading to valuable insights into fluid flow, particle transport, and other phenomena within these complex structures.

10.6 ADVANTAGES OF USING CFD FOR POROUS MEDIA SIMULATIONS

The use of CFD for reservoir simulation, particularly in scenarios involving expensive chemicals like nanomaterials for field applications, offers several significant benefits:

- *Reduced Experimental Costs*: CFD allows for virtual testing and optimisation of reservoir processes, minimising the need for expensive experimental setups and reducing overall research and development costs.
- *Process Optimisation*: CFD enables the exploration of a wide range of parameters, such as injection rates, NP concentrations, and reservoir conditions. It facilitates the optimisation of EOR processes without the need for numerous costly field trials.
- *Risk Assessment*: CFD provides a platform to assess the effectiveness of nanomaterial applications and other chemical processes in a controlled virtual environment. It helps identify potential risks and optimise strategies before deploying them in the field.
- *Understanding Complex Interactions*: CFD offers detailed insights into fluid flow patterns, interactions between injected nanomaterials and reservoir fluids, and the impact of various parameters on the success of EOR. This understanding is crucial for making informed decisions in the design and execution of field applications.
- *Time Efficiency*: CFD simulations provide relatively quick results, allowing for rapid evaluation and iteration of different scenarios. This efficiency accelerates the decision-making process and reduces the time required for the development and implementation of field applications.
- *Optimised Nanomaterial Design*: CFD facilitates the study of how different nanomaterial properties (size, shape, surface chemistry) influence their behaviour in the reservoir. This information aids in the design and optimisation of nanomaterials for EOR.
- *Environmental Impact Assessment*: CFD allows for the prediction of nanomaterial dispersion in the reservoir and the assessment of potential environmental impacts. This proactive approach helps in designing applications that minimise unintended consequences.
- *Data-Driven Decision-Making*: CFD simulations provide detailed and quantitative data, empowering decision-makers with a robust foundation for assessing the feasibility and efficacy of using nanomaterials in reservoirs.

- *Reduced Field Trials*: By leveraging CFD to evaluate different scenarios extensively, the need for extensive and costly field trials can be minimised. It is especially advantageous when dealing with expensive chemicals like nanomaterials.
- *Flexibility in Scenario Testing*: CFD allows for sensitivity analysis, enabling the examination of how changes in various parameters affect outcomes. This flexibility is invaluable for identifying optimal conditions for nanomaterial application.

In essence, the integration of CFD into reservoir simulation not only streamlines the decision-making process but also enhances the overall efficiency, cost-effectiveness, and environmental sustainability of deploying expensive chemicals like nanomaterials in field applications. It offers a valuable tool set for optimising strategies and mitigating risks associated with EOR processes (Robbana et al., 2012; Pal & Sen, 2023a, 2023b, 2024).

10.7 CONCLUSION

In conclusion, the application of CFD for simulating nanomaterial use in EOR represents a pivotal advancement in the quest for efficient and sustainable oil extraction. CFD serves as a powerful tool, allowing researchers and engineers to delve into the intricate dynamics of fluid flow and NP behaviour within porous media. The ability to model the transport, interactions, and deposition of nanomaterials at a microscopic scale provides unprecedented insights into the mechanisms influencing EOR processes. By leveraging CFD, the oil and gas industry can optimise the design and implementation of nanotechnology-enhanced strategies, leading to increased oil production and reduced environmental impact. As we navigate the evolving landscape of energy exploration and production, CFD emerges as a crucial ally, enabling informed decision-making and fostering innovations that propel the industry towards more sustainable and technologically advanced practices.

REFERENCES

Alade, O., & Mahmoud, M., & Al-Nakhli, A. (2023). Rheological studies and numerical investigation of barite sag potential of drilling fluids with thermochemical fluid additive using computational fluid dynamics (CFD). *Journal of Petroleum Science and Engineering*, 220, 111179. https://doi.org/10.1016/j.petrol.2022.111179.

Ansys Fluent Theory Guide, ANSYS Inc., USA, 2013.

Dewangan, S. K., & Sinha, S. L. (2016). Exploring the hole cleaning parameters of horizontal wellbore using two-phase Eulerian CFD approach. *The Journal of Computational Multiphase Flows*, 8(1), 15–39. https://doi.org/10.1177/1757482X16634218.

Ejeh, C., Afgan, I., AlMansob, H., Brantson, E., Fekala, J., Odiator, M., Stanley, P., Anumah, P., Onyekperem, C., & Boah, E. (2020). Computational fluid dynamics for ameliorating oil recovery using silicon-based nanofluids and ethanol in oil-wet reservoirs. *Energy Reports*, 6, 3023–3035. https://doi.org/10.1016/j.egyr.2020.10.028.

Hu, H., & Cheng, Y. F. (2016). Modeling by computational fluid dynamics simulation of pipeline corrosion in CO2-containing oil-water two phase flow. *Journal of Petroleum Science and Engineering*, 146, 134–141. https://doi.org/10.1016/j.petrol.2016.04.030.

Kumar, A., & Gangawane, K. M. (2022a). Effect of precipitating agents on the magnetic and structural properties of the synthesised ferrimagnetic nanoparticles by co-precipitation method. *Powder Technology*, 401, 117298. https://doi.org/10.1016/j.powtec.2022.117298.

Kumar, A., & Gangawane, K. M. (2022b). Synthesis and effect on the surface morphology & magnetic properties of ferrimagnetic nanoparticles by different wet chemical synthesis methods. *Powder Technology*, 410(June), 117867. https://doi.org/10.1016/j.powtec.2022.117867.

Kumar, A., & Gangawane, K. M. (2023a). Advanced computational approaches for water treatment. In *Advanced Computational Approaches for Water Treatment* (pp. 1–12). CRC Press. https://doi.org/10.1201/9781003325147-1.

Kumar, A., & Gangawane, K. M. (2023b). Molecular dynamics for the membrane process. In *Advanced Computational Approaches for Water Treatment* (pp. 13–38). CRC Press. https://doi.org/10.1201/9781003325147-2.

Kumar, A., Gangawane, K. M., & Ramanjaneyulu, B. (2022). Ferrofluids for waste-water treatment. In *International Conference on Chemical, Bio and Environmental Engineering* (pp. 723–744). https://link.springer.com/chapter/10.1007/978-3-030-96554-9_48.

Kumar, A., Jena, H. M., & Gangawane, K. M. (2023a). Effect of ferrimagnetic nanoparticles as an oxygen absorber on the shelf life of tomatoes. Poster presented at the *9th International Food Convention (IFCoN)*, CSIR-CFTRI, Mysuru, India, 9 December 2023.

Kumar, A., Jena, H. M., & Gangawane, K. M. (2023b). Variations in magnetic behavior of ferrimagnetic nanoparticles with temperature. Poster presented at the *75th Indian Chemical Engineering Congress (CHEMCON 2023)*, Dr. H L Roy Building, Jadavpur University Campus, Kolkata, India, 28 December 2023.

Medhi, S., Chowdhury, S., Kumar, A., Gupta, D. K., Aswal, Z., & Sangwai, J. S. (2020). Zirconium oxide nanoparticle as an effective additive for non-damaging drilling fluid: A study through rheology and computational fluid dynamics investigation. *Journal of Petroleum Science and Engineering*, 187, 106826. https://doi.org/10.1016/j.petrol.2019.106826.

Pal, D., & Sen, S. (2023a). Emerging petroleum pollutants and their adverse effects on the environment. In I. D. Behera, & A. P. Das (Eds.), *Impact of Petroleum Waste on Environmental Pollution and Its Sustainable Management through Circular Economy. Environmental Science and Engineering* (pp. 103–137). Springer Nature Switzerland. https://doi.org/10.1007/978-3-031-48220-5_5.

Pal, D., & Sen, S. (2023b). In-depth coverage of petroleum waste sources, characteristics, environmental impact, and sustainable remediation process. In I. D. Behera, & A. P. Das (Eds.), *Impact of Petroleum Waste on Environmental Pollution and Its Sustainable Management through Circular Economy* (pp. 1–38). Springer Nature Switzerland. https://doi.org/10.1007/978-3-031-48220-5_1.

Pal, D., & Sen, S. (2024). Optimal synthesis of dolochar derived faujasite zeolite X for highly effective Cd(II) removal. *Environmental Research*, 240, 117494. https://doi.org/10.1016/j.envres.2023.117494.

Panda, D., & Gangawane, K. M. (2023a). Development of superhydrophobic hybrid silica-cellulose aerogel as promising thermal insulation and sound absorption. *International Journal of Materials Research*, 114(4–5), 405–417. https://doi.org/10.1515/ijmr-2022-0066.

Panda, D., & Gangawane, K. M. (2023b). Hybrid NiFe2O4-NiCO4O4 nanoparticles-based eutectic phase change materials for enhancement of thermal efficiency of pin-fin heat sink arrangement. *Journal of Energy Storage*, 60(January), 102417. https://doi.org/10.1016/j.est.2023.106644.

Panda, D., & Gangawane, K. M. (2023c). Recycled cellulose–silica hybrid aerogel for effective oil adsorption: optimisation and kinetics study. *Sadhana - Academy Proceedings in Engineering Sciences*, 48(3). https://doi.org/10.1007/s12046-023-02161-9.

Panda, D., Sahu, A. K., & Gangawane, K. M. (2024). Effect of molar concentration and drying methodologies on monodispersed silica sol for synthesis of silica aerogels with temperature-resistant characteristics. *Materials Science and Engineering: B*, 299, 117062. https://doi.org/10.1016/j.mseb.2023.117062.

Panda, D., Sharma, S., & Gangawane, K. M. (2023). Synthesis of novel biocatalyst from organic waste protonated with acid treatment for hydrogen production. *International Journal of Hydrogen Energy*, 48(63), 24242–24254. https://doi.org/10.1016/j.ijhydene.2023.03.117.

Robbana, E., Buikema, T., Mair, C., Williams, D., Mercer, D., Webb, K., Hewson, A., & Reddick, C. (2012). Low salinity enhanced oil recovery - Laboratory to day one field implementation - LoSal EOR into the Clair Ridge Project. *All Days*, 4, 2582–2596. https://doi.org/10.2118/161750-MS.

Tang, H. W., Foo, H. C. Y., Tan, I. S., & Lam, M. K. (2021). Photogrammetry based computational fluid dynamics of erosion of sand particles in water pipeline: Dynamic shape factors of 3D particles and minimisation of erosion activity. *Journal of Petroleum Science and Engineering*, 108794. https://doi.org/10.1016/j.petrol.2021.108794.

Vaziri, E., Simjoo, M., & Chahardowli, M. (2020). Application of foam as drilling fluid for cuttings transport in horizontal and inclined wells: A numerical study using computational fluid dynamics. *Journal of Petroleum Science and Engineering*, 194, 107325. https://doi.org/10.1016/j.petrol.2020.107325.

Yeo, L., Feng, Y., Seibi, A., Temani, A., & Liu, N. (2021). Optimization of hole cleaning in horizontal and inclined wellbores: A study with computational fluid dynamics. *Journal of Petroleum Science and Engineering*, 205, 108993. https://doi.org/10.1016/j.petrol.2021.108993.

11 Simulation of Nanoparticle-Assisted Enhanced Oil Recovery

A Coupled Multiphase Fluid Flow and Nanoparticle Transport Model

P. Sivasankar

11.1 INTRODUCTION

Nanoparticle-assisted Enhanced Oil Recovery (EOR) is a technique in which nanoparticles are used to enhance the residual oil from hydrocarbon reservoirs either by altering the properties of reservoir fluids or by altering the rock-fluid interaction properties or by a combination of both. Depending on the type of nanoparticles used, the reservoir fluid properties are changed to enhance the oil recovery. The nanoparticles can enhance the oil recovery by any of the following mechanisms: (a) enhancement in viscosity of injected water, which effectively displaces the crude oil; (b) lowering of crude oil viscosity within the reservoir, which enhances the oil mobility; and (c) reduction in interfacial tension (IFT) between the oil and water, which improves the oil flow by weakening the capillary force acting on the oil. The nanoparticles also enhance the oil recovery by altering the reservoir rock-fluid property, such as rock wettability from unfavourable oil-wet to favourable water-wet condition, which allows effective displacement of oil by injected water (Kumar & Gangawane, 2022b).

This nanoparticle-assisted EOR is a complex technique that involves multiple processes co-occurring within the reservoir. The major processes that are involved during NP EOR are as follows: (a) multiphase flow of fluids (water, oil, gas); (b) transport of nanoparticles, which involves advection, dispersion, sorption, and reactive processes; and (c) oil recovery process due to wettability alteration, alteration in reservoir fluid properties, and IFT reduction depending on the type of nanoparticles used (Sepehri et al., 2019). These processes are, in turn, governed by different parameters, such as type of nanoparticle, size of nanoparticles, the salinity of injection and formation water, oil viscosity, rock wettability, reservoir temperature, rate of injection, rock type

 DOI: 10.1201/9781032650685-11

(i.e., sandstone, carbonate), rock pore size, and concentration of nanoparticle injected. To efficiently implement the nanoparticle EOR technique in the field, it is necessary to quantify and analyse the effects of these parameters on the nanoparticle EOR process through experimental and simulation studies. Core-flooding experimental studies are widely carried out at laboratories to evaluate the performance of nanoparticle EOR on enhancing oil recovery through experiments. Core-flooding experiments are conducted to mimic the oil recovery in a reservoir at a lab scale. In core-flooding studies, the nanoparticle solution is injected/flooded through an inlet into a rock core sample, which is saturated with water and residual oil, and subsequently, the oil that is produced through the outlet of the core sample will be quantified and analysed. Hence, core-flooding experimental studies help quantify oil recovery and understand the underlying mechanisms/physics that cause it. Though core-flooding experimental studies possess several advantages, they have the limitation of being expensive and time-consuming. For instance, to quantify and analyse the influence of different governing parameters on final oil recovery, sensitivity analysis is performed, for which numerous experiments need to be completed. In such cases, experimental core-flooding studies are practically infeasible to complete owing to increased cost and time. In these circumstances, the constraint in time and cost can be overcome by conducting mathematical modelling and simulation studies on EOR (Pal & Sen, 2023a, 2023b).

Modelling and simulation studies can act as a proxy to the experimental and field studies on EOR by mimicking the actual oil recovery processes and predicting the resultant oil recovery at a relatively lesser cost and time compared to experimental and field studies. Furthermore, modelling and simulation studies help quantify the interdependence of different parameters influencing the EOR process and its associated oil recovery, which can provide more insights into the underlying mechanisms that influence oil recovery. In addition, modelling and simulation studies help to predict the oil under different combinations of operational conditions and, thereby, assist in identifying a suitable operation strategy that would yield higher oil recovery at a relatively lower cost during its field implementation. Moreover, modelling and simulation studies at the field scale help the operating companies to make an economic feasibility analysis before implementing an EOR project in the field, based on which the operating companies decide whether to implement the EOR project or not. Suppose the decision was to implement the EOR project. In that case, subsequently, the results from modelling and simulation studies help the operating companies plan the resources (i.e., financial, material, and workforce) in a timely manner. Holistically, modelling and simulation studies serve as an effective reservoir management tool to design, plan, and implement EOR techniques efficiently in the field. Hence, the modelling and simulation study is a critical step in the successful implementation of an EOR technique, which requires special attention. So, in this chapter, modelling and simulation of nanoparticles, EOR is discussed owing to its importance.

11.2 OIL RECOVERY MECHANISMS OF NANOPARTICLE EOR TECHNIQUE

In this section, the different mechanisms by which the nanoparticles enhance the oil recovery from the reservoir have been discussed. For the implementation of the nanoparticle EOR technique, the nanoparticles are widely employed as nanocatalysts, nanoemulsions, and nanofluids to enhance the oil recovery, and each of these plays a distinctive role in improving the oil recovery. The role of nanocatalysts, nanoemulsions, and nanofluids in enhancing oil recovery has been discussed here as follows (El-Diasty & Aly, 2015; Kumar et al., 2022; Panda et al., 2024).

11.2.1 Role of Nanocatalysts

Nanocatalysts are metal particles in nanosize that are used along with the steam injection EOR process as catalysts to reduce the viscosity of heavy oil. These nanocatalysts catalyse the aquathermolysis reaction between the steam and oil, which lowers the oil viscosity. As a result, oil mobility and the consequent oil recovery are enhanced in heavy oil reservoirs using nanocatalysts. Examples of nanocatalysts are iron and nickel (Kumar & Gangawane, 2022a; Panda & Gangawane, 2023b).

11.2.2 Role of Nanoemulsions

Nanoparticles are used to synthesise nanoemulsions, the role of which is to stabilise the emulsion instead of using surfactants under harsher reservoir temperatures and salinity conditions. The viscosity of this synthesised nanoemulsion is relatively higher than that of the oil, which, in turn, helps to displace the oil towards the production well effectively. This nanoemulsion flooding can serve as an alternative to the polymer-flooding EOR technique, whereby it lowers the mobility ratio value to less than 1 (favourable for oil recovery), and it enhances the macroscopic displacement, resulting in the enhancement of volumetric sweep efficiency and oil recovery. Some studies also reported that nanoemulsions could also improve oil recovery by lowering the IFT between oil and water (i.e., enhancing microscopic displacement). However, the role of nanoemulsion in improving the oil recovery by viscosity enhancement (i.e., macroscopic displacement) is found to be more significant than the IFT reduction (i.e., microscopic displacement) (Kumar & Gangawane, 2023a, 2023b).

11.2.3 Role of Nanofluids

Nanofluids are synthesised by mixing nanoparticles in a base fluid (generally, water is considered), in which the nanoparticles are suspended as suspension colloids due to their lightweight and small size. Nanoparticles help to recover the additional oil from the reservoir by altering the wettability of rock from an unfavourable oil-wet state to a favourable water-wet state. During the nanofluid-flooding EOR process, hydrophilic nanoparticles at a particular concentration and size are mixed with water (i.e., nanofluid) and, subsequently, it is injected into the oil-wet reservoir at a specific rate. Within the pore space of the reservoir, the injected nanofluid creates structural disjoining pressure at the interface of oil and rock surfaces. This causes the nanofluid to wet the rock surface by displacing the oil from the rock surface, which will eventually be recovered (Kumar et al., 2023a, 2023b; Panda & Gangawane, 2023c, 2023a; Zhang et al., 2016). Hence, the nanofluid helps to heal the oil from an oil-wet reservoir by wettability alteration with the aid of structural disjoining pressure. The effectiveness of nanofluid in enhancing oil recovery depends on different parameters, such as concentration and size of nanoparticle used, wettability of nanoparticles used (i.e., hydrophilic or hydrophobic), the salinity of base fluid and formation water, the temperature of the reservoir, rock grain size, and injection rate of the nanofluid during flooding. Nanofluid-flooding EOR is widely used for oil recovery purposes. Hence, in the present chapter, modelling and simulation of the nanofluid-flooding EOR technique has been discussed.

11.3 MATHEMATICAL MODELLING AND NUMERICAL SIMULATION OF NANOFLUID-FLOODING EOR PROCESS

In this chapter, we shall consider modelling the injection/flooding of nanofluid into the oil-wet reservoir to enhance the recovery of crude oil by altering the wettability of rock from oil-wet to water-wet. We shall also consider that the hydrophilic nanoparticles are used in the nanofluids so that the injected nanoparticles are always suspended in the water phase, which helps to alter the rock wettability towards more water-wet.

The assumptions that are considered in the present modelling and simulation study are as follows:

a. Darcy's law governs the flow of fluids in porous media.
b. The effect of temperature, pH, and salinity variation on nanoparticle transport and its associated oil recovery process is considered insignificant. Hence, an isothermal condition at constant salinity and pH is considered.
c. The fluid flow is considered horizontal, so the gravity effects are neglected.
d. The rock is considered incompressible, while the fluid is considered slightly compressible.
e. Oil is recovered only by wettability alteration mechanism.
f. Nanoparticles are always present in the water phase due to their hydrophilic nature, and they are always considered to be in the suspended state without undergoing any gravitational settling.

11.3.1 Physical System and Governing Equations

Consider a three-dimensional oil-wet reservoir that is saturated with oil and water. Hence, in this study, only two fluid phases (i.e., oil and water) are considered. To implement the nanofluid-flooding EOR technique in the reservoir, the nanofluid is flooded into the reservoir at a flow rate through an injection well. Here, the nanofluid is prepared by mixing the hydrophilic nanoparticles with the water. Upon the injection of nanofluid into the reservoir, the pressure gradient will be developed within the reservoir, which causes the injected water to flow from the injection well towards the production well. As the fluid flows within the reservoir, the saturation of fluid phases and its corresponding relative permeability change within the reservoir. This change in fluid saturation in porous media is simulated using saturation equations, which are mathematically represented by Eqs. (11.1 & 11.2) (Irfan et al.,

2019; Panda et al., 2023; Sivasankar & Suresh Kumar, 2018). Eq. (11.1) is called the mass balance equation (or) mass conservation equation (or) continuity equation for two-phase fluid flow in porous media. While Eq. (11.2) is the momentum equation, and in the present modelling, Darcy's equation is used as the momentum equation to model the multiphase fluid flow in porous media under a pressure gradient.

Saturation equations:

$$\phi\left(\frac{\partial S_w}{\partial t}\right) = -\nabla\cdot(u_w); \phi\left(\frac{\partial S_o}{\partial t}\right) = -\nabla\cdot(u_o); S_w + S_o = 1 \tag{11.1}$$

$$u_w = -\frac{kk_{rw}}{\mu_w}\nabla P_w; u_o = -\frac{kk_{ro}}{\mu_o}\nabla P_o \tag{11.2}$$

In Eqs. (11.1 & 11.2), terms ϕ and k represent rock's porosity and absolute permeability, respectively; S_w and S_o represent water saturation and oil saturation, respectively; u_w and u_o represent Darcy velocity of water phase and oil phase, respectively; k_{rw} and k_{ro} represent relative permeability of water and relative permeability of oil, respectively; P_w and P_o represent pressure of water phase and oil phase, respectively.

The spatiotemporal distribution of water pressure within the reservoir that occurs due to the injection of nanofluid is modelled using the pressure equation for the water phase, which is presented through Eqs. (11.3 & 11.4) (Sepehri et al., 2019). This pressure equation is deduced from the saturation equations (i.e., Eqs. 11.1 & 11.2).

Pressure equation:

$$\phi c_t\left(\frac{\partial P_w}{\partial t}\right) = \nabla\cdot(\lambda_t\nabla P_w) + \nabla\cdot(\lambda_o\nabla P_c) \tag{11.3}$$

In Eq. (11.3), the terms λ_w, λ_o, and λ_t represent the water mobility, oil mobility and total mobility, respectively; and c_t represents the total compressibility. These mobility terms are dependent on the relative permeability of water (k_{rw}) and relative permeability of oil (k_{ro}), and it is mathematically expressed through Eq. (11.4).

$$\lambda_w = \frac{kk_{rw}}{\mu_w}; \lambda_o = \frac{kk_{ro}}{\mu_o}; \lambda_t = \lambda_w + \lambda_o \tag{11.4}$$

The relative permeability of water (k_{rw}) and relative permeability of oil (k_{ro}) within the reservoir are strongly influenced by the saturation of water. Hence, the variation of k_{rw} and k_{ro} with respect to water saturation is mathematically modelled using the Brooks & Correy model as given Eqs. (11.5–11.7) (Ahmed, 2018).

$$k_{rw} = k_{rw,max}\left(S_w^*\right)^{nw} \tag{11.5}$$

$$k_{ro} = k_{ro,max}\left(1 - S_w^*\right)^{no} \tag{11.6}$$

$$S_w^* = \left[(S_w - S_{wir})/(1 - S_{or} - S_{wir})\right] \tag{11.7}$$

In Eqs. (11.5–11.7), the terms $k_{rw,max}$ and $k_{ro,max}$ correspondingly represent the maximum relative permeability values of water phase and oil phase; *nw* and *no* are the exponential constants of k_{rw} and k_{ro} curves, respectively; S_{wir} represents the irreducible water saturation; and S_{or} represents the residual oil saturation.

Like the relative permeability values, the capillary pressure (P_c) also varies with respect to water saturation and wettability of rock. Since nanofluid alters the wettability of rock surface, the capillary pressure term is calculated here by considering the water saturation and wettability using the Eq. (11.8) (El-Amin et al., 2013).

$$P_c = Y_w\left(\frac{S_w - S_{wir}}{1 - S_{wir}}\right)^{-a} + Y_o\left(\frac{S_o - S_{or}}{1 - S_{or}}\right)^{-b} \tag{11.8}$$

In Eq. (11.8), the terms Y_o and Y_w represent the entry pressure for drainage and imbibition, respectively; $1/b$ and $1/a$ are the constants for drainage and imbibition, respectively.

The initial condition and boundary conditions for the pressure equation (11.3) are presented in Eq. (11.9) and Eq. (11.10), respectively.

Initial condition for pressure equation:

$$P_w\left(x, y, z, t = 0\right) = P_{initial} \tag{11.9}$$

Boundary conditions for pressure equation:

$$\left(\frac{\partial P_w}{\partial x}\right)_{x=0,y,z,t} = -\left(\frac{Q_t}{A_r}\right)\left(\frac{\mu_w}{kk_{rw}}\right) \tag{11.10}$$

$$\nabla P_w\left(x = l_x, y, z, t\right) = 0 \tag{11.11}$$

In Eqs. (11.9–11.11), the term $P_{initial}$ represents the initial pressure within the reservoir; x, y, z represents the direction of the reservoir in x, y, z axis; Q_t represents the total injection rate of the injected fluid; A_r represents the cross-sectional area of the reservoir across which the fluid is injected; and l_x is the length of the reservoir along x -direction.

The saturation and pressure equations that are represented through Eqs. (11.1–11.11) are combinedly referred to as multiphase fluid flow equations, which help to determine the spatiotemporal distribution of each fluid phase saturation and pressure within the reservoir due to fluid flow.

11.4 TRANSPORT OF NANOPARTICLES IN MULTIPHASE-SATURATED POROUS MEDIA

Due to flow of water within the reservoir, the nanoparticles that are suspended in the water phase get transported within the reservoir. As the nanoparticles used are hydrophilic in nature, they will only be present in the water phase during their transport. During the transport of nanoparticles in porous media, the nanoparticles undergo diffusion process, advection process, sorption process, and entrapment process. This transport process of nanoparticles that occurs within the water phase in reservoir is mathematically represented through Eqs. (11.12–11.19) (Irfan et al., 2019; Ju & Fan, 2009; Sepehri et al., 2019).

Equation (11.12) models the spatiotemporal distribution of nanoparticle concentration (C) within the reservoir due to diffusion, advection, sorption, and entrapment processes. In Eq. (11.12), the term in LHS represents the accumulation term, which simulates the time-dependent variation of nanoparticle concentration within the reservoir. Furthermore, in RHS of Eq. (11.12),

$$\text{The first term} \left\{ \text{i.e., } \phi S_w \frac{\partial C}{\partial t} = \nabla \cdot (\phi S_w D_m \ \nabla C) \right\} \text{ and}$$

$$\text{The second term} \left\{ \text{i.e., } \quad \phi S_w \frac{\partial C}{\partial t} = -\nabla \cdot (u_w C) \right\} \quad \text{and}$$

correspondingly represent the diffusion and advection processes. Here, the term D_m represents the diffusion coefficient. The last term in RHS of Eq. (11.12) (i.e., R) represents the loss in nanoparticle concentration during its transport.

$$\phi S_w \frac{\partial C}{\partial t} = \nabla \cdot (\phi S_w D_m \ \nabla C) - \nabla \cdot (u_w C) - R \tag{11.12}$$

The initial and boundary conditions for the transport of nanoparticle concentration are given in Eqs. (11.13 & 11.14), respectively. In Eq. (11.14), the C_{inj} term represents the concentration of nanoparticles injected into the reservoir.

Initial condition for nanoparticle transport in reservoir:

$$C(x, y, z, t = 0) = 0; \tag{11.13}$$

Boundary conditions for nanoparticle transport in reservoir:

$$C(x = 0, y, z, t) = C_{inj}; \nabla C(x = l_x, y, z, t) = 0 \tag{11.14}$$

The loss in nanoparticle concentration (i.e., R) happens due to (a) sorption of nanoparticles to the pore surface; and (b) entrapment of nanoparticles in the pore throat, which is given through Eqs. (11.15–11.19) (Binshan et al., 2002; Ju et al., 2012; Sepehri et al., 2019).

$$R = C\left(\frac{\partial v_w}{\partial t} + \frac{\partial v_w^*}{\partial t} \right) \tag{11.15}$$

In Eq. (11.15), the term v_w represents the volume of nanoparticles per unit bulk volume of rock that are sorbed onto the pore surface; v_w^* represents the volume of nanoparticles per unit bulk volume of rock that are entrapped in the pore throats because of plugging.

The time-dependent variation of v_w and v_w^* within the reservoir during nanoparticle transport has been mathematically expressed through the Eqs. (11.16–11.19) (Binshan et al., 2002).

$$C\frac{\partial v_w}{\partial t} = \begin{cases} \alpha_d u_w C, \ u_w < u_{lw} \\ \alpha_d u_w C - \alpha_e v_w (u_w - u_{lw}) C, u_w > u_{lw} \end{cases} \tag{11.16}$$

$$v_w(x, y, z, t = 0) = 0 \tag{11.17}$$

The magnitude of loss in nanoparticles mainly depends on the fluid velocity (u_w) along which it gets transported. At flow velocity lesser than critical velocity (u_{lw}), only sorption of nanoparticles occurs, while at velocity higher than the critical velocity, both sorption and entrapment happen simultaneously, which is given through Eqs. (11.16 & 11.17), respectively. The time-dependent nanoparticle entrapment process that occurs due to blocking of pore throats is modelled through Eqs. (11.18 & 11.19).

$$\frac{\partial v_w^*}{\partial t} = \alpha_p u_w \tag{11.18}$$

$$v_w^*(x, y, z, t = 0) = 0 \tag{11.19}$$

In Eqs. (11.16–11.19), the terms α_d, α_e, and α_p represent the rate coefficient for nanoparticle sorption, entrainment, and pore throat blocking process, respectively.

11.4.1 Change in Rock's Porosity and Permeability

Due to the deposition of nanoparticles on the rock surface (i.e., sorption) and entrapment in the pore throats, the rock's porosity and permeability will get reduced, which is calculated by using the Eqs. (11.20–11.22) (Binshan et al., 2002; Irfan et al., 2019). The reduction in rock's porosity at any given location within the reservoir due to sorption and entrapment is determined by using the Eqs. (11.20 & 11.21).

$$\phi = \phi_{initial} - \sum \Delta\phi \tag{11.20}$$

$$\sum \Delta\phi = v_w + v_w^* \tag{11.21}$$

In Eqs. (11.20 & 11.21), the term $\phi_{initial}$ represents the initial porosity before the start of nanofluid-flooding EOR process; $\sum \Delta\phi$ represents the reduction in porosity due to sorption and entrapment.

Similar to the porosity, the rock's permeability is also reduced, and the updated permeability at any given location within the reservoir is quantified using the Eq. (11.22).

$$k = k_{initial}\left[(1 - f)k_f + f\phi/\phi_{initial}\right]^n \tag{11.22}$$

In Eq. (11.22), the term $k_{initial}$ represents the initial permeability of the reservoir before the start of nanofluid-flooding EOR process; k_f represents the seepage constant; f is the fraction of cross-sectional area in the pore that is available for the fluid to flow; and n is the exponential constant.

11.4.2 Effect of Nanoparticle on Relative Permeability of Oil and Water

As discussed earlier, the nanofluid enhances the oil recovery, thereby improving the relative permeability of oil (i.e., oil flow) within the reservoir by altering the wettability of rock from oil-wet to water-wet. This alteration of wettability happens due to structural disjoining pressure, which causes the nanoparticle in the water phase to sorb onto the rock surface by displacing or segregating the oil phase from the rock surface. The oil that is displaced from the rock surface will be subsequently recovered and produced, which is denoted by the increase in relative permeability of oil (i.e., oil flow) in the reservoir. While the sorption of nanoparticles onto the rock surface will make the rock surface as water-wet, and it will be denoted by the decrease in relative permeability of oil (i.e., oil flow) in the reservoir. This mechanism of change in relative permeability of oil and water due to nanoparticle sorption over the rock surface is evaluated through Eqs. (11.23–11.28) (Binshan et al., 2002; Ju et al., 2012; Sepehri et al., 2019).

$$S_i = (v_w + v_w^*)\frac{6}{d} \tag{11.23}$$

$$S = \beta \sum_{i=1}^{m} S_i \tag{11.24}$$

In Eqs. (11.23 & 11.24), the term S_i represents the surface area of nanoparticles that are sorbed onto the rock's surface as a single layer at any given location point '*i*' within the reservoir; *d* represents the diameter of nanoparticle used in nanofluid flooding; *S* represents the total surface area of nanoparticles in water phase which is in contact with the rock's surface; β represents the coefficient for the surface area; and *m* represents the total number of location points considered in the reservoir.

The total specific area of the porous media that is available for the fluid to wet its surface is measured as S_v, and it is calculated by using the Eq. (11.25).

$$S_v = 7{,}000\phi\sqrt{\frac{\phi}{k}} \tag{11.25}$$

Now, the ratio between the parameters S and S_v is denoted by the term R_s (Eq. 11.26), which measures the fraction of rock surface in the reservoir that is sorbed with the nanoparticles.

$$R_s = \frac{S}{S_v} \tag{11.26}$$

The R_s value varies from 0 to 1. Here, the R_s value of 0 represents that there is no sorption of nanoparticles onto the rock surface. This illustrates that the rock surface remains as oil-wet reservoir and no oil is recovered by nanoparticles. While the R_s value of 1 represents that the entire surface area within the reservoir is sorbed by the nanoparticles, which means that the wettability of rock is altered to water-wet and favours the oil flow.

Now, based on the R_s value due to nanoparticle sorption, the change in relative permeability of water and oil phases are updated using the Eqs. (11.27 & 11.28).

$$k'_{rwjp} = k_{rwj} + \left(k'_{rwj} - k_{rwj}\right) R_s \tag{11.27}$$

$$k'_{rwjp} = k_{roj} + \left(k'_{roj} - k_{roj}\right) R_s \tag{11.28}$$

In Eqs. (11.27 & 11.28), the terms k'_{rwj} and k'_{roj} correspondingly represent the relative permeability of the water phase and oil phase at a given saturation. After the nanoparticle (i.e., nanofluid) injection; k_{rwj} and k_{roj} correspondingly, it represents the relative permeability of the water phase and oil phase at a given water saturation. 'S_{wj}' before the nanoparticle (i.e., nanofluid) injection; k'_{rwjp} and k'_{rwjp} correspondingly denote the updated relative permeability of the water phase and oil phase at a given saturation. 'S_{wj}'. Here, the values of k'_{rwjp} and k'_{rwjp} are calculated by considering a linear function with R_s value.

11.5 SOLUTION PROCEDURE TO SOLVE THE MATHEMATICAL MODEL

The mathematical modelling of the nanofluid-flooding EOR process has several processes (i.e., multiphase fluid flow process, transport process, sorption and entrapment process, and wettability alteration process), which act simultaneously within the reservoir and are non-linear.

The processes that are mathematically represented through non-linear coupled partial differential equations (PDEs) can be solved numerically by adopting numerical discretisation techniques such as finite difference method, finite volume method, and finite element method. These discretisation techniques help convert the governing PDEs into algebraic equations, which shall be subsequently solved using efficient numerical methods. In the case of using the finite difference method for discretising the PDEs, unconditionally numerically stable discretisation schemes must be used to attain numerically stable results. To attain such numerically stable results, first-order PDE terms that appear in the transport equation (i.e., advection process in Eq. 11.12) shall be discretised spatially by an upwind scheme and temporally by a fully implicit scheme. In case of second-order PDE, terms that exist in saturation equation (i.e., Eq. 11.11), pressure equation (i.e., Eq. 11.13), and transport equation (i.e., Eq. 11.12) shall be spatially discretised by the central differencing scheme and temporally by Crank–Nicholson scheme. In addition to numerical stability, it is also necessary to ensure that the obtained numerical results are numerically consistent and achieve numerical convergence.

To numerically solve the mathematical model that is presented through Eqs. (11.1–11.28), following are the sequence of steps to be adopted (Badwal et al., 2014; Binshan et al., 2002; Ju et al., 2012).

Initially, fluid flow equations (i.e., both saturation and pressure equations) that are presented through Eqs. (11.1–11.11) are solved. As the saturation equation and pressure equation are non-linearly coupled equations, the Implicit Pressure and Explicit Saturation (IMPES) method shall be adopted to solve the saturation and pressure equations. By solving these equations, the spatiotemporal distribution of saturation (S_w, S_o) and pressure (P_w, P_o) of both the fluid phases (i.e., oil and water) is determined.

a. Upon determination of P_w and P_o, the Darcy velocity for the water phase (u_w) is determined.
b. Now, the nanoparticle transport equations (Eqs. 11.12–11.19) are solved. Solving these equations helps to determine the spatiotemporal distribution of nanoparticle concentration (C) within the reservoir; v_w and v_w^*.
c. Based on v_w and v_w^* values, the change in rock's porosity and absolute permeability are determined using Eqs. (11.20–11.22).
d. Finally, the wettability alteration mechanism is simulated by computing the change in relative permeability of water and oil phases due to nanofluid flooding using Eqs. (11.23–11.28).
e. The above steps from 11.1 to 11.5 are repeated till the final time step in the simulation is reached.

11.6 VALIDATION AND VERIFICATION OF MATHEMATICAL MODELLING AND NUMERICAL STUDY

The results obtained by numerical simulation of the mathematical model have to be verified for its prediction accuracy. The accuracy of the numerical simulation results is lost due to the prevalence of numerical errors and the lack of development of an appropriate mathematical model mimicking the underlying mechanism in a process. To ensure that the obtained results from the simulation studies are free of errors, the obtained numerical results must be validated against the analytical solution and verified with the experimental/field observations. Validating the numerical results with the analytical solution ensures that the obtained results are numerically stable and free of any numerical errors that arise due to rounding off the values, truncating of higher order terms during discretisation, and choosing inappropriate sizes of time step and spatial step during discretisation. In this case of nanofluid-flooding EOR, the numerical simulation of fluid flow equations can be validated with Buckley and Leverett's semi-analytic solution by comparing the spatial distribution of water saturation at different time levels.

Further, to ensure that the obtained numerical results are closer to reality and to verify the validity of the adopted mathematical model, the obtained oil recovery results from the simulation can be verified with the oil recovery results obtained from core-flooding experiments. Also, to verify the prediction accuracy of the simulated results and its ability to mimic the reality at the field scale, the simulated results can be history-matched with the field results by matching the oil recovery profile, pressure profile, and water-cut profile data with the simulation results. The history-matching process is a tedious task, and it is carried out by fine-tuning several parameters that govern the flow process, transport process, and wettability alteration. Once the numerical results are validated and verified with the benchmark results, further simulations are carried out with confidence to predict the oil recovery under different scenarios for better decision-making.

11.7 APPLICATION OF MODELLING AND SIMULATION STUDIES ON NANOFLUID-FLOODING EOR

The numerical modelling and simulation of the nanofluid-flooding EOR process help to investigate and quantify the influence of different parameters on the spatiotemporal distribution of pressure, saturation, nanoparticle concentration, and final oil recovery. This helps the operators to optimise their operations during implementation. Moreover, this simulation study also assists in evaluating different operational scenarios on oil recovery, such as adopting different injection durations, different arrangements of injection and production wells, cyclic or continuous injection, etc. These results from the simulation studies aid the operators in identifying the efficient method to maximise the oil recovery from the reservoir. Furthermore, based on the oil recovery profile obtained from the simulation results, an economic analysis can be made, which will help the operators plan the material, human resources, and financial resources well ahead of time for an identified production scenario. Hence, simulation studies help improve asset value and create value for the operating company.

11.8 CURRENT CHALLENGES AND WAY AHEAD

Currently, modelling and simulation studies on nanofluid-flooding EOR are carried out with a limited understanding of the underlying mechanism that causes oil recovery from the reservoir (Irfan et al., 2019). Moreover, the mathematical models that are developed to simulate the nanofluid-flooding EOR process have limited ability to replicate/mimic reality. Also, the coupled effect of temperature, salinity, pH, and pressure on nanoparticle sorption and its associated influence on final oil recovery have not yet been mathematically modelled and quantified. Furthermore, the uncertainty involved in predicting the oil recovery by nanofluid-flooding EOR has not been quantified yet. Hence, to overcome these challenges, the understanding of the nanofluid EOR process must be improved through pore scale, micro-scale, and mesoscale studies. Accordingly, better mathematical models are to be developed by linking the oil recovery mechanism at the pore scale with the macroscopic scale, which will allow the model to be simulated at the field scale. Probabilistic prediction studies need to be performed under uncertainty to evaluate the uncertainty in the nanofluid-flooding EOR process using Monte Carlo simulations to evaluate the uncertainty in the nanofluid-flooding EOR process. Since performing probabilistic prediction studies using the mechanistic modelling approach is time-consuming, the Physics Informed Machine Learning (PIML) approach can be deployed to get accurate, physically reliable, and faster results (Pavan et al., 2022). These uncertainty quantification studies will immensely help the operating companies plan their resources under uncertainty.

11.9 CONCLUSIONS

Nanoparticles are used as nanocatalysts, nanoemulsions, and nanofluids for the EOR process. Among all, nanofluids are widely studied for their ability to recover additional oil by altering the wettability of rock from unfavourable oil-wet conditions to favourable water-wet conditions.

For EOR purposes, hydrophilic nanoparticles are used in the nanofluids, which are permanently suspended in the water phase. The nanoparticles present in the nanofluid create structural disjoining pressure, causing the nanofluid to wet the rock surface by displacing the oil from the rock surface. This sorption of nanoparticles on the rock surface alters the wettability from oil to water.

In this chapter, the mathematical modelling and numerical simulation of the nanofluid-flooding EOR process have been discussed. The simulation of the mathematical model presented in this chapter helps to predict and investigate the spatiotemporal distribution of fluid phase pressure, fluid saturation, and nanoparticle concentration in the reservoir. Additionally, during the transport of nanoparticles, the loss of surfactant due to sorption and entrapment process has been considered, and its subsequent effect on reduction in porosity and absolute permeability of the rock has also been modelled. Ultimately, the discussed model can simulate the final oil recovery from wettability alteration by evaluating the change in relative permeability of fluids (i.e., oil and water). The wettability alteration and the change in relative permeability of fluids are modelled based on the sorption of nanoparticles on the pore surface.

The discussed mathematical model for nanofluid-flooding EOR is non-linear, having coupled PDEs. The numerically stable scheme to discretise the first-order PDE term is a fully implicit upwind scheme, and the second-order PDE term is a central differencing Crank–Nicholson scheme. To simulate the multiphase fluid flow in reservoirs, the IMPES scheme is considered to solve the non-linearly coupled pressure and saturation equations.

The accuracy of the simulated results from the mathematical model is to be validated with the analytic solution to ensure the numerical stability of the simulated results. Validation of simulation results with Buckley and Leverett's semi-analytic solution can ensure the numerical stability of fluid flow results obtained from numerical simulation. Furthermore, the prediction accuracy of the simulated results and its ability to mimic the reality at the field scale can be verified with the core-flooding experimental results and by history matching with the past production data from the field.

The results from this simulation study assist the operators in efficiently designing, planning, and implementing the nanofluid-flooding EOR technique in the field to maximise oil recovery and minimise costs. This could help the operators identify ways to improve the asset value.

The current challenges in modelling and simulation of nanofluid-flooding EOR technique are as follows: lack of understanding of underlying mechanisms that recover the oil; lack of appropriate mathematical models to simulate the effect of temperature, pH, salinity, and pressure on nanoparticle sorption and its associated oil recovery due to wettability alteration; and lack in quantifying the uncertainty involved during nanofluid-flooding EOR process.

NOMENCLATURE

A_r	Cross-sectional area of the reservoir across which the fluid is injected, m^2
$1/a$	Constant for imbibition

$1/b$	Constant for drainage
C	Concentration of nanoparticles, g/l
C_{inj}	Concentration of nanoparticles injected into the reservoir, g/l
c_t	Total compressibility, m^2/N
d	Diameter of nanoparticle, m
f	Fraction of cross-sectional area in the pore that is available for the fluid to flow
k	Rock's absolute permeability, m^2
k_f	Seepage constant, N/m^2
$k_{initial}$	Initial permeability of the reservoir before the start of nanofluid-flooding EOR process
k_{ro}	Relative permeability of oil, *fraction*
$k_{ro,max}$	Maximum relative permeability values of oil phase, *fraction*
k_{rw}	Relative permeability of water, *fraction*
$k_{rw,max}$	Maximum relative permeability values of water phase, *fraction*
l_x	Length of the reservoir along x – direction
m	Total number of location points considered in the reservoir
n	Exponential constant
no	Exponential constant of k_{ro} curves
nw	Exponential constant of k_{rw} curves
$P_{initial}$	Initial pressure within the reservoir, N/m^2
P_o	Pressure of oil phase, N/m^2
P_w	Pressure of water phase, N/m^2
Q_t	Total injection rate of the injected fluid, m^3/s
R	Loss in nanoparticle concentration during its transport
S	Total surface area of nanoparticles in contact with the rock's surface, m^2/m^3
S_i	Surface area of nanoparticles that are sorbed onto the rock's surface as a single layer at any given location point 'i' within the reservoir, m^2/m^3
S_o	Oil saturation, *fraction*
S_{or}	Residual oil saturation, *fraction*
S_w	Water saturation, *fraction*
S_{wir}	Irreducible water saturation, *fraction*
u_{lw}	Critical velocity of the fluid, m/s
u_o	Darcy velocity of oil phase, m/s
u_w	Darcy velocity of water phase, m/s
v_w	Volume of nanoparticles per unit bulk volume of rock that are sorbed onto the pore surface, m^3/m^3
v_w^*	Volume of nanoparticles per unit bulk volume of rock that are entrapped in the pore throats because of plugging, m^3/m^3
x, y, z	Direction of the reservoir in x, y, z axis
Y_o	Entry pressure for drainage, N/m^2
Y_w	Entry pressure for imbibition, N/m^2
α	Rate coefficient for nanoparticle, m^{-1}
β	Coefficient of surface area
λ_o	Oil mobility
λ_t	Total mobility
λ_w	Water mobility

ϕ Rock's porosity, *fraction*
$\phi_{initial}$ Initial porosity before the start of nanofluid-flooding EOR process

REFERENCES

Ahmed, T. (2018). Reservoir engineering handbook. In *Reservoir Engineering Handbook*. Gulf Professional Publishing. https://doi.org/10.1016/C2016-0-04718-6.

Badwal, S. P. S., Giddey, S. S., Munnings, C., Bhatt, A. I., & Hollenkamp, A. F. (2014). Emerging electrochemical energy conversion and storage technologies. *Frontiers in Chemistry*, *2*(September), 1–28. https://doi.org/10.3389/fchem.2014.00079.

Binshan, J., Shugao, D., Zhian, L., Tiangao, Z., Xiantao, S., & Xiaofeng, Q. (2002). A study of wettability and permeability change caused by adsorption of nanometer structured polysilicon on the surface of porous media. In *SPE - Asia Pacific Oil and Gas Conference* (pp. 915–926). https://doi.org/10.2118/77938-ms.

El-Amin, M. F., Sun, S., & Salama, A. (2013). Enhanced oil recovery by nanoparticles injection: Modeling and simulation. *SPE Middle East Oil and Gas Show and Conference, MEOS, Proceedings*, *2*, 1428–1437. https://doi.org/10.2118/164333-ms.

El-Diasty, A. I., & Aly, A. M. (2015). Understanding the mechanism of nanoparticles applications in enhanced oil recovery. In *Society of Petroleum Engineers - SPE North Africa Technical Conference and Exhibition 2015, NATC 2015* (pp. 944–962). https://doi.org/10.2118/175806-ms.

Irfan, S. A., Shafie, A., Yahya, N., & Zainuddin, N. (2019). Mathematical modeling and simulation of nanoparticle-assisted enhanced oil recovery - A review. *Energies*, *12*(8), 1575. https://doi.org/10.3390/en12081575.

Ju, B., & Fan, T. (2009). Experimental study and mathematical model of nanoparticle transport in porous media. *Powder Technology*, *192*(2), 195–202. https://doi.org/10.1016/j.powtec.2008.12.017.

Ju, B., Fan, T., & Li, Z. (2012). Improving water injectivity and enhancing oil recovery by wettability control using nanopowders. *Journal of Petroleum Science and Engineering*, *86–87*, 206–216. https://doi.org/10.1016/j.petrol.2012.03.022.

Kumar, A., & Gangawane, K. M. (2022a). Effect of precipitating agents on the magnetic and structural properties of the synthesised ferrimagnetic nanoparticles by co-precipitation method. *Powder Technology*, *401*, 117298. https://doi.org/10.1016/j.powtec.2022.117298.

Kumar, A., & Gangawane, K. M. (2022b). Synthesis and effect on the surface morphology & magnetic properties of ferrimagnetic nanoparticles by different wet chemical synthesis methods. *Powder Technology*, *410*(June), 117867. https://doi.org/10.1016/j.powtec.2022.117867.

Kumar, A., & Gangawane, K. M. (2023a). Advanced computational approaches for water treatment. In *Advanced Computational Approaches for Water Treatment* (pp. 1–12). CRC Press. https://doi.org/10.1201/9781003325147-1.

Kumar, A., & Gangawane, K. M. (2023b). Molecular dynamics for the membrane process. In *Advanced Computational Approaches for Water Treatment* (pp. 13–38). CRC Press. https://doi.org/10.1201/9781003325147-2.

Kumar, A., Gangawane, K. M., & Ramanjaneyulu, B. (2022). Ferrofluids for waste-water treatment. In *International Conference on Chemical, Bio and Environmental Engineering* (pp. 723–744). https://link.springer.com/chapter/10.1007/978-3-030-96554-9_48.

Kumar, A., Jena, H. M., & Gangawane, K. M. (2023a). Effect of ferrimagnetic nanoparticles as an oxygen absorber on the shelf life of tomatoes. Poster presented at the *9th International Food Convention (IFCoN)*, CSIR-CFTRI, Mysuru, India, 9 December 2023.

Kumar, A., Jena, H. M., & Gangawane, K. M. (2023b). Variations in magnetic behavior of ferrimagnetic nanoparticles with temperature. Poster presented at the *75th Indian Chemical Engineering Congress (CHEMCON 2023)*, Dr. H L Roy Building, Jadavpur University Campus, Kolkata, India, 28 December 2023.

Pal, D., & Sen, S. (2023a). Emerging petroleum pollutants and their adverse effects on the environment. In I. D. Behera, & A. P. Das (Eds.), *Impact of Petroleum Waste on Environmental Pollution and Its Sustainable Management through Circular Economy. Environmental Science and Engineering* (pp. 103–137). Springer Nature Switzerland. https://doi.org/10.1007/978-3-031-48220-5_5.

Pal, D., & Sen, S. (2023b). In-depth coverage of petroleum waste sources, characteristics, environmental impact, and sustainable remediation process. In I. D. Behera, & A. P. Das (Eds.), *Impact of Petroleum Waste on Environmental Pollution and Its Sustainable Management through Circular Economy* (pp. 1–38). Springer Nature Switzerland. https://doi.org/10.1007/978-3-031-48220-5_1.

Panda, D., & Gangawane, K. M. (2023a). Development of superhydrophobic hybrid silica-cellulose aerogel as promising thermal insulation and sound absorption. *International Journal of Materials Research, 114*(4–5), 405–417. https://doi.org/10.1515/ijmr-2022-0066.

Panda, D., & Gangawane, K. M. (2023b). Hybrid NiFe2O4-NiCO4O4 nanoparticles-based eutectic phase change materials for enhancement of thermal efficiency of pin-fin heat sink arrangement. *Journal of Energy Storage*, *60*(January), 102417. https://doi.org/10.1016/j.est.2023.106644.

Panda, D., & Gangawane, K. M. (2023c). Recycled cellulose–silica hybrid aerogel for effective oil adsorption: optimisation and kinetics study. *Sadhana - Academy Proceedings in Engineering Sciences,* 48(3). https://doi.org/10.1007/s12046-023-02161-9.

Panda, D., Sahu, A. K., & Gangawane, K. M. (2024). Effect of molar concentration and drying methodologies on monodispersed silica sol for synthesis of silica aerogels with temperature-resistant characteristics. *Materials Science and Engineering: B*, *299*, 117062. https://doi.org/10.1016/j.mseb.2023.117062.

Panda, D., Sharma, S., & Gangawane, K. M. (2023). Synthesis of novel biocatalyst from organic waste protonated with acid treatment for hydrogen production. *International Journal of Hydrogen Energy*, *48*(63), 24242–24254. https://doi.org/10.1016/j.ijhydene.2023.03.117.

Pavan, P. S., Arvind, K., Nikhil, B., & Sivasankar, P. (2022). Predicting performance of in-situ microbial enhanced oil recovery process and screening of suitable microbe-nutrient combination from limited experimental data using physics informed machine learning approach. *Bioresource Technology*, *351*, 127023. https://doi.org/10.1016/j.biortech.2022.127023.

Sepehri, M., Moradi, B., Emamzadeh, A., & Mohammadi, A. H. (2019). Experimental study and numerical modeling for enhancing oil recovery from carbonate reservoirs by nanoparticle flooding. *Oil and Gas Science and Technology*, *74*, 5. https://doi.org/10.2516/ogst/2018080.

Sivasankar, P., & Suresh Kumar, G. (2018). Modelling the influence of interaction between injection and formation brine salinities on in-situ microbial enhanced oil recovery processes by coupling of multiple-ion exchange transport model with multiphase fluid flow and multi-species reactive transpo. *Journal of Petroleum Science and Engineering*, *163*, 435–452. https://doi.org/10.1016/j.petrol.2018.01.004.

Zhang, H., Ramakrishnan, T. S., Nikolov, A., & Wasan, D. (2016). Enhanced oil recovery driven by nanofilm structural disjoining pressure: Flooding experiments and microvisualization. *Energy and Fuels*, *30*(4), 2771–2779. https://doi.org/10.1021/acs.energyfuels.6b00035.

12 Optimisation of Parameters for Enhanced Oil Recovery and the Role of Nanotechnology

Uma Sankar Behera

12.1 INTRODUCTION

The world's energy demand is rising exponentially and is going to increase by 28% by the end of 2040, according to the Energy Information Administration (EIA) (Mead, 2017). EIA also predicted that developing countries of Asia, like China and India, will consume 60% of the total increase in energy demand due to the rapid rise in population and fast-growing economy, digitalisation, automation of industries, etc. In the present scenario, non-conventional energy sources, such as solar, wind, tidal, biomass, etc., are not self-sufficient to fulfil the huge global energy demand due to a lack of technological advancement in the respective fields. The energy-demand challenges are mostly met with the hydrocarbon sources (coal, crude oil, natural gas), which are limited on the earth. Among the hydrocarbons, crude oil plays an essential role in meeting the challenges of energy needs. The oil- and gas-producing companies are under pressure to increase the production rate. Hence, oil- and gas-producing companies are trying to explore new reservoirs. However, exploration costs and uncertainty in finding crude oil reservoirs are very high. Therefore, priority is given to the recovery of oil from existing reservoirs over the development of new reservoirs, employing advanced techniques. As the crude oil reservoirs are heterogeneous, no single method can be applied to all the reservoirs for oil recovery.

Crude oil is recovered from the reservoirs utilising three methods (e.g., primary, secondary, tertiary or Enhanced Oil Recovery [EOR]). Figure 12.1 shows the various oil recovery methods adopted for oil recovery at different stages of the crude oil reservoir. The reservoir pressure is usually high at the early stage of crude oil recovery, which helps the oil flow towards the wellbore due to the difference in reservoir and bottom-hole pressure. The technique is referred to as the primary recovery method. Crude oil of 10%–15%, the original oil in place (OOIP), is reported to be recovered using the primary recovery method. Various drive mechanisms prevail during the immediate recovery process, such as the expansion of reservoir fluid, gravity segregation, gas cap expansion, and water influx (Satter & Iqbal, 2016). Subsequently, the crude oil production rate decreases due to a decrease in natural reservoir pressure,

DOI: 10.1201/9781032650685-12

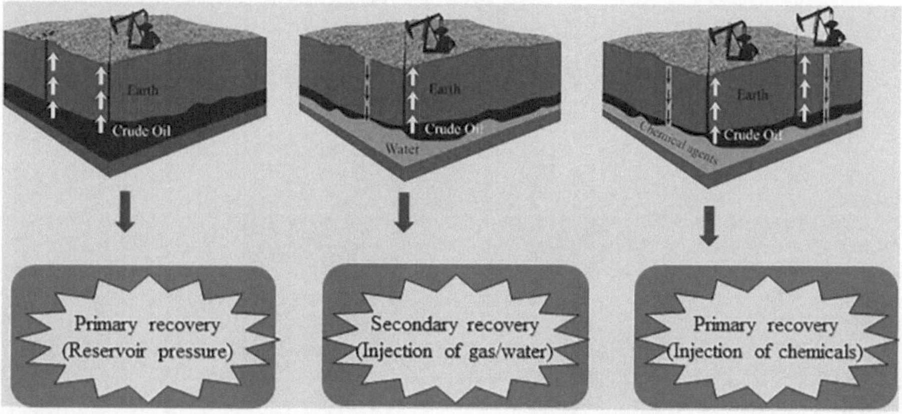

FIGURE 12.1 Various methods adopted for oil recovery at different stages of crude oil reservoir.

and eventually, the primary oil recovery method becomes infeasible. Hence, various fluids (e.g., natural gas, CO_2, water) are injected into the reservoir through the injection well to maintain the reservoir pressure. The increased reservoir pressure led to an increase in the rate of oil recovery. Secondary recovery refers to the oil recovery process that involves maintaining reservoir pressure through the use of external sources. During the secondary oil recovery method, an additional 15%–20% of crude oil can be recovered. Water injection is widely used in secondary oil recovery techniques started in the 1930s.

The water injection process consists of a producing well surrounded by injection wells. The well patterns commonly used are the nine-spot, seven-spot, five-spot, four-spot, three-spot, and line-drive patterns. The five-spot pattern is the most common well pattern and consists of four injection wells and one production well with a well spacing of nearly 330 ft. Note that an increase in the well spacing decreases the production rate (15%–20% lower at a spacing of 660 ft.) (Crawford, 1960). In the five-spot pattern, the four injection wells are located at the four corners of the square, and the production well is at the centroid. The crude of the targeted area is displaced by the water injected through the injection well towards the producing well. The water used for injection is generally water from the same reservoir or various reservoirs or seawater. Immiscible gas injection is also generally practised during the secondary recovery method. The injected gas is collected from the nearby gas field or gas cap of the same reservoir. At the end stage of the secondary recovery, i.e., when it reaches its limit, the maximum amount of the injection fluid is produced from the production well, and after that, the oil production becomes no longer economical. The crude oil left over in the reservoir after secondary recovery is called residual oil saturation.

After the secondary recovery process, the residual oil remains in the pores of the reservoir rock due to capillary force (Muggeridge et al., 2014; Nwidee et al., 2016). Hence, to recover trapped oil from the reservoir rock, the rock and fluid properties of the reservoir need to be altered. The EOR method involves altering the

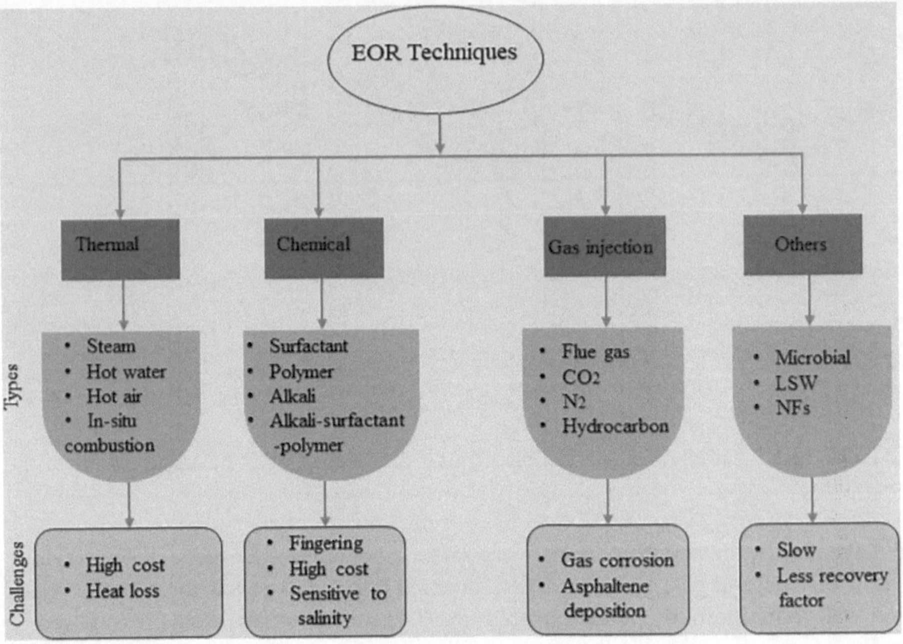

FIGURE 12.2 Various EOR techniques for matured crude oil reservoirs with challenges.

reservoir rock and fluid properties to recover residual oil left after secondary recovery techniques. The various EOR techniques and their challenges are represented in Figure 12.2. Some EOR techniques commercially available are thermal, chemical, and gas injection. The EOR techniques can yield an additional 15%–20% of oil after the secondary recovery process. Thermal methods are applied to the heavy crude oil reservoir. Approximately 70% of crude oil located globally are heavy oil reservoirs, whereas 30% is light oil (Prats, 1986). In a thermal method, heat is introduced into the reservoir to lower the viscosity of the crude oil to increase mobility. The heat energy is presented through various means, such as steam, hot water, hot air, and in-situ combustion. The heat energy increases the reservoir temperature significantly, decreasing oil viscosity and increasing mobility. However, the cost of oil production is very high, and heat loss also takes place.

Chemical EOR involves injecting surfactants, alkalis, and polymers into reservoirs to enhance oil recovery by improving sweep efficiency through reduced interfacial tension (IFT). Challenges include viscous fingering and surfactant sensitivity to salinity. Gas injection methods utilise natural gas, nitrogen, or CO_2 to reduce oil viscosity and increase flow rates, applicable to intermediate-to-light crude oil reservoirs. Biological EOR with microorganisms faces challenges in culturing at reservoir conditions, limiting its popularity. Figure 12.3 shows the global EOR market size related to various technology between 2014 and 2025. The global EOR market is expected to increase by 70% by 2025, dominated by thermal, gas injection, and chemical methods, with rising interest in low-salinity water flooding and nanofluid (NF) flooding.

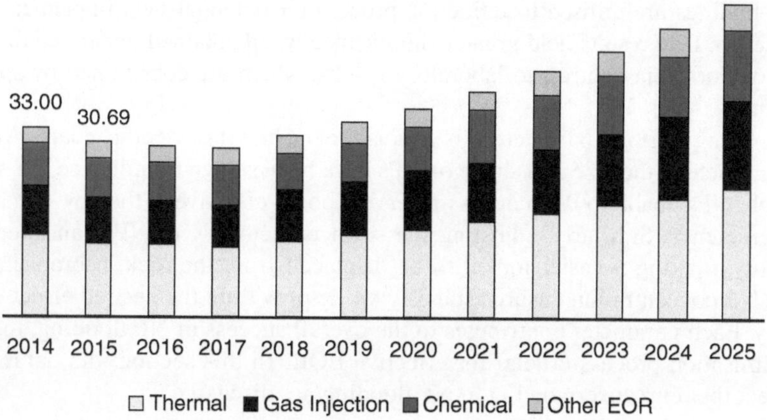

FIGURE 12.3 The EOR market size with respect to technology from 2014 to 2025 (USD Billion). Data were extracted from Hexa Research in July 2017 and reproduced (Statista Research Department, 2023). "Other EOR" is the topmost layer of each graph bar. Under that is "Chemical," then "Gas Injection," and finally "Thermal" on the bottommost layer.

In recent years, the application of nanotechnology has significantly increased in all sectors, including the oil field, owing to the unique properties of the nanoparticles (NPs). Loss of chemicals due to adsorption on the rock surface during chemical flooding is one of the significant problems of chemical EOR, which can be minimised by combined flooding of surfactant and NPs (Yousefvand & Jafari, 2018). It has been reported that NPs have the potential to reduce the IFT of an oil-water system if injection fluid containing NPs is flooded into the porous medium (Idogun et al., 2016; Zargartalebi et al., 2015). Modification of the rock surface's wettability stands as a principal factor influencing EOR, and this can potentially be realised through the integration of NPs with the injection fluid. Researchers have reported that the distinctive self-structuring characteristics of NPs contribute to the alteration of rock wettability (Behera et al., 2022; Hendraningrat et al., 2013; Panda & Gangawane, 2023; Sun et al., 2017). The researcher has reported additional oil recovery due to the injection fluid carrying NPs (Sedaghat et al., 2020). Formulating stable NFs is a challenge when using NPs in the oil field because they tend to aggregate and settle, reducing their effectiveness. Adding stabilising agents like polymers or surfactants prevents this clumping. Adjusting the pH of the NFs and modifying the NP surfaces also help make them stable. Combining surfactants, polymers, and NPs improves stability, making it easier for the NFs to interact with rocks, oil, and water for EOR. The chapter thoroughly examines the optimisation of different parameters, including the performance of NPs in altering three-phase interfacial properties, the stability of NFs, and economic considerations. These aspects are covered in detail to provide a comprehensive understanding.

12.1.1 Experimental Studies for Parameter Optimisation

Experimental studies play a pivotal role in the optimisation of parameters for the practical application of a technique. Therefore, initial experimental efforts are undertaken, and the findings are utilised to refine the process for potential field implementation. Nevertheless, field results hold greater significance when obtained under genuine reservoir conditions, in contrast to laboratory results, which are constrained by specific limitations.

Optimising various parameters is crucial for achieving specific goals. Among these parameters, the size and shape of NPs have been shown to influence NF stability. Stable NFs enable NPs to access reservoir pores effectively, thereby enhancing sweep efficiency. Similarly, adjusting the surface chemistry of NPs can alter rock wettability, making it easier for oil to be displaced from the rock. Moreover, controlling NP concentration ensures stability while providing the desired effect on oil recovery. Each parameter contributes to the overall success of NF flooding, making the optimisation process crucial for effective EOR. In this section, the parameters that affect the oil recovery owing to NF flooding are discussed.

12.1.2 Shape and Size of NPs

The shape and size of NPs play a significant role in influencing several key physical parameters of NFs, including stability, thermal conductivity, and viscosity (Kumar & Gangawane, 2022; Panda & Gangawane, 2023). These factors, in turn, hold considerable significance in the context of utilising NFs for EOR applications. Although a limited number of studies have been conducted concerning the impact of NP shape and size on the properties above, a concise discussion follows below.

The majority of EOR techniques are centred on enhancing sweep efficiency through control of the mobility ratio between the injected fluid and the reservoir fluids. The viscosity of the injected fluid significantly influences the mobility ratios of the fluids involved in displacement. To investigate the impact of NPs' size and shape on NFs' viscosity, Tengku Mohd et al. (2016) examined spherical solid and spherical porous NPs at a size of 15 nm for NF formulation. Additionally, the authors explored the effect of NP size by considering NPs of 15 nm and 60 nm. The study revealed that decreasing the NP size led to increased viscosity, with solid spherical NPs exhibiting higher viscosity compared to porous ones. The rise in viscosity due to solid spherical particles might be attributed to increased fluid density, impeding fluid flow.

Furthermore, when the size of particles is reduced by half, their surface area increases twofold. This is because when particles get smaller, their outer layer makes up a more significant portion compared to their inner volume. On the other hand, when the diameter of NPs grows from 1 to 100 nm, the ratio of surface area to volume decreases significantly. This reduction happens because the volume of the particle grows faster than its surface area. This decrease in the surface-area-to-volume ratio is not ideal, as it can lead to NPs sticking together, a phenomenon known as aggregation. Aggregation can limit the effectiveness of NPs, especially in applications where their dispersed and separate nature is crucial, such as the EOR field (Li et al., 2018).

Scholars have documented that the formulation of stable NFs using NPs of smaller dimensions (generally around 100 times less than the size of pore throats) tends not to obstruct these pore throats physically. However, due to physiochemical interactions with the pore walls, notable retention occurs, leading to alterations in wettability and a decrease in permeability (Binshan et al., 2002; Zhang et al., 2015). In some instances, larger metallic NPs have been reported to contribute to the actual blockage of pore channels (Zamani et al., 2012). Therefore, in the formulation of NFs, the shape and size of NPs play an essential role in determining their suitability for applications in the oil and gas field.

12.1.3 Nature of Dispersing Medium

The viscosity of the injection fluid is contingent upon the properties of the dispersing medium. A study examining four different dispersing fluids – namely distilled water, ethylene glycol, ethanol, and a brine solution (30,000 ppm) containing 0.1 wt% of surfactant – was conducted to investigate viscosity. In Figure 12.4, the viscosity of various NFs is displayed based on the nature of the dispersing medium. At room temperature, the recorded fluid viscosities were 20 cP (distilled water), 36.3 cP (ethylene glycol), 19.52 cP (ethanol), and 20.43 cP (brine) when 5 wt% of NPs were dispersed. The study concludes that the viscosity of resulting NFs correlates with the inherent viscosity of the dispersing medium (Mohd et al., 2016).

The viscosity of NFs is influenced not just by the dispersing medium but also by the fluid's pH. Particles with similar charges repel each other, thus reducing particle-particle interactions that impact fluid viscosity (Jia-Fei et al., 2009). It was found that when particle size in the base fluid is under 20 nm, the viscosity becomes more reliant on the suspension's pH, showing notable fluctuations primarily between pH 5 and pH 7. In the study conducted by Timofeeva et al. (2009), it was observed that by adjusting the pH of the suspension, the viscosity of the NFs increased by 31%, while there was no corresponding alteration in thermal conductivity.

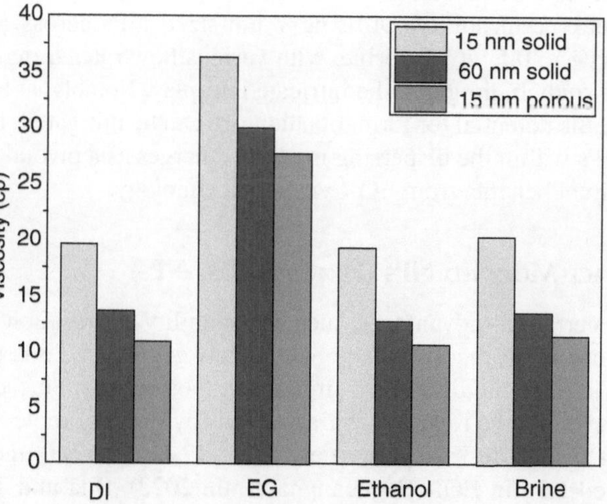

FIGURE 12.4 The viscosity of the various nanofluids based on the dispersing medium. Data are extracted from Tengku Mohd et al. (2016). The leftmost bar in each graph is "15 nm solid." The middle one of each is "60 nm solid," and the rightmost is "15 nm porous."

Li et al. (2018) reported that the chemicals utilised in EOR processes remain unrecoverable, and the metallic NPs integrated into the formulation of NFs exhibit the potential for recovery from both oil and water phases, allowing for subsequent reuse. This characteristic holds regardless of the base fluid selected for the preparation of NFs. This outcome suggests that the choice of base fluid has a negligible influence in recovering the magnetic NPs, which shows an advantage in comparison to chemical EOR methods.

12.1.4 Concentration of NPs

The concentration of NPs within the dispersing fluid affects both the resulting fluid's viscosity and stability. The subsequent section explores how NP concentration impacts NF stability. Research indicates that as NP concentration in the base fluid increases, viscosity also increases (Das et al., 2007; Putra et al., 2003). Investigating metal oxide NPs (Al_2O_3 and CuO), Nguyen et al. (2007) found that, for particles sized 36 nm and 47 nm, NF viscosity remained mostly identical up to 4 wt% NP concentration. However, with higher concentrations, viscosity increased significantly, particularly for larger particles. Comparing Al_2O_3 and CuO NFs, CuO NFs exhibited significantly higher viscosity; this could be due to a smaller particle size and higher density (29 nm) than that of alumina NFs. According to Mohd et al. (2016) an increase in concentration of the NPs results in higher fluid viscosity due to increased resistance to flow. At elevated NP concentrations within the NFs, particles tend to collide, generating resistance in the water flow. Conversely, at low concentrations, fewer particles are present within the water, allowing the water's flow characteristics to dominate the resistance arising from particle collisions.

The investigation led by Hendraningrat et al. (2013) unveiled a noteworthy insight: elevating the NF concentration for crude oil recovery via flooding could lead to the detriment of rock permeability. Utilising 7 nm-sized NPs across a concentration range of 0.01 wt% to 0.5 wt% in brine, with various flow rates (ranging from 0.5 to 4 cm^3/min), the study highlighted the intricate interplay. Notably, at lower flow rates (<0.5 cm^3/min), the potential for formation damage can be mitigated. Hence, the concentration of NPs within the dispersing medium emerges as a pivotal determinant in harnessing optimal benefits from EOR via nanotechnology.

12.1.5 Surface-Modified NPs (Functionalise NPs)

Bare NPs have certain disadvantages, such as instability aggregation, which reduces their performance in oil recovery as pore channels get blocked due to NP aggregation. However, surface-modified NPs circumvent those issues. Surface modification is done to achieve specific required characteristics by grafting molecules, polymers, or ligands that confer desired properties. Table 12.1 represents the role of some surface-modified NPs in EOR (Ngouangna et al., 2022). Bila and Torsæter (2021) reported additional oil recovery of approximately 6% after secondary water flooding by employing polymer-coated NPs. In the research by Choi et al. (2017), it was observed that the injection of polymer-coated silica NPs into water-wet core samples resulted in the extraction of 74.1% of the oil, in contrast to 68.9% of the initial oil in

TABLE 12.1
Represents the contribution of functionalised NPs in modifying surface/fluid properties relevant to EOR application

NP types	Surface modifying agent	Impact on EOR mechanism	Analysis technique	References
SiO_2	Surfactant SDS	IFT reduction	Spinning drop	Zargartalebi et al. (2014)
ZrO_2	Surfactant SDS & CTAB	IFT reduction	Pendant drop	Esmaeilzadeh et al. (2014)
SiO_2	Polyethylene glycol	Wettability alteration	Glass beads	Daneshmand et al. (2021)
Al_2O_3	Polymer	IFT reduction, contact angle reduction, viscosity increase	Glass beads	Hu et al. (2021)
SiO_2	Surfactant TX100	IFT reduction	Spinning drop	Adil et al. (2020)
Fe_3O_4	Chitosan	Wettability alteration	Sand-pack rector	Rezvani et al. (2018)

Abbreviations: Al_2O_3, aluminium oxide; CTAB, cetyltrimethylammonium bromide; EOR, enhanced oil recovery; Fe_3O_4, iron(II, III) oxide; IFT, interfacial tension; NPs, nanoparticles; SDS, sodium dodecyl sulphate; SiO_2, silicon dioxide; ZrO_2, zirconium dioxide.

place (OOIP) achieved through water flooding alone and 72.7% of OOIP through the use of bare silicon dioxide (SiO_2) NPs. Dai et al. (2015) studied how silica NPs with modified surfaces could help in EOR. They found that when these modified SiO_2 NPs were added, rocks changed from hydrophilic to hydrophobic, making it easier for water to flow through them. It could be helpful for oil field development as it reduces the pressure needed to push water in, which is really important. Ahmed et al. (2020) conducted a study focusing on reducing IFT using pure and surfactant-coated SiO_2 NFs. Their findings elucidated that surface-modified SiO_2 NFs exhibited the capacity to enhance ultimate oil recovery from sandstone cores, showcasing improvements of 9.1% and 21.7%, respectively.

12.1.6 Stability of NFs

A stable NF refers to a mixture of NPs and a base fluid (e.g., water or oil) that maintains a uniform dispersion of NPs throughout the fluid without significant settling or aggregation over time. Unstable NFs can lead to aggregation or settling of NPs, resulting in uneven distribution within the reservoir. It can lead to clogging of pore spaces, reducing permeability and hindering fluid flow. Stable NFs ensure consistent performance over time and maintain the desired properties for interaction with reservoir rock and fluids. Moreover, stable NFs can flow smoothly through the reservoir and have the ability to access hard-to-reach oil packets. It enhances sweep efficiency and ultimately improves oil recovery.

Various methods and strategies are adopted to circumvent the aggregation or settling of NPs. Figure 12.5 illustrates the commonly employed techniques to obtain

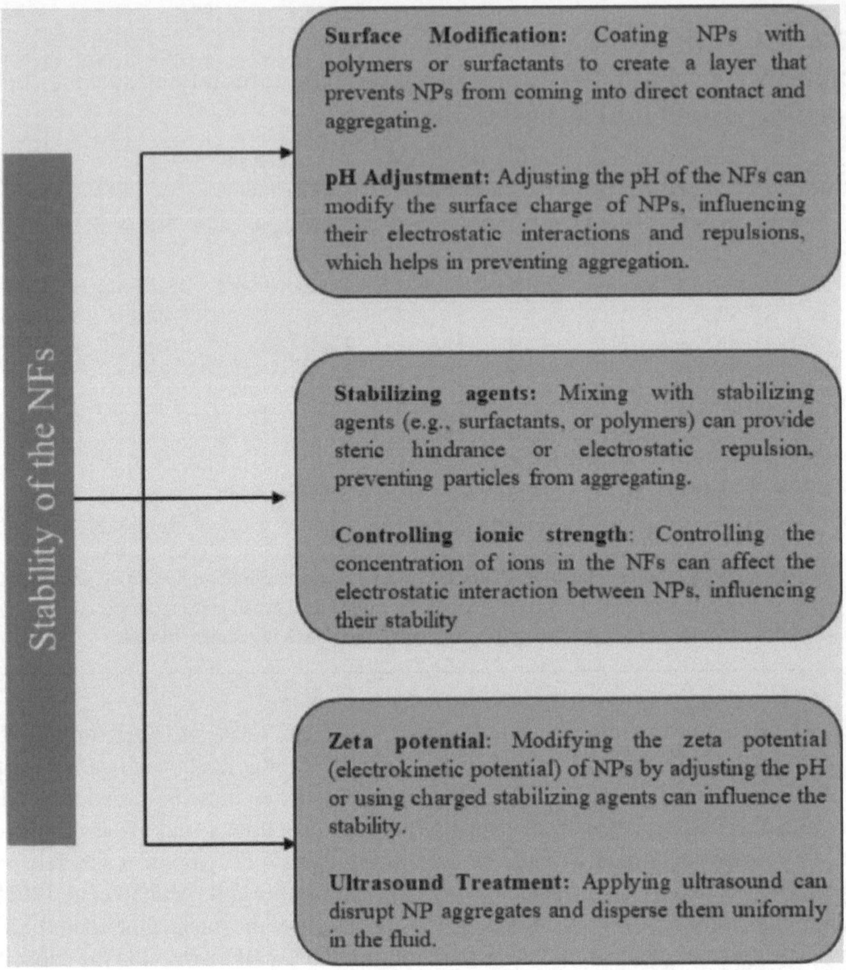

FIGURE 12.5 Illustrates the commonly employed techniques to obtain stabilised NFs

stabilised NFs. Apart from the methods mentioned above, proper solvent selection (i.e., base fluid) can minimise the attractive interaction and improve stability; the appropriate size and surface properties of NPs also help to minimise attractive forces to enhance stability.

They have highlighted that the stability of NPs is significantly influenced by the concentration of ions and surfactants in the base fluid. The researchers noted that when the salinity of the base fluid is low (0 wt% sodium chloride [NaCl]), the best stability (with a zeta potential of −43 mV) was attained at a concentration of 2,450 mg/L of the anionic surfactant sodium dodecyl sulphate (SDS). However, higher concentrations of the surfactant resulted in reduced stability of the solution. With an elevation in salinity to 1 wt%, the zeta potential reached approximately 32 mV at a surfactant concentration of 980 mg/L. However, further escalation in salinity led to a decrease in stability (Al-Anssari et al., 2017). Likewise, Choudhary et al. (2017) affirmed that the stability of alumina NFs exhibited a direct correlation with both the concentration of surfactant and the duration of sonication. Improved stabilisation of water-based graphene NFs was achieved by applying a silica coating to the graphene surface (Li et al., 2014). The researchers observed that the NFs exhibited sustained stability, with no observable stratification even after one week. In their research, Behera and Sangwai reported that the stability of NPs is influenced by a combination of factors, including the presence of stabilising agents (such as surfactants and polymers) and the techniques employed for dispersing the NPs within the base fluid (Behera & Sangwai, 2021b; Behera et al., 2022).

It's important to note that the choice of method will depend on the specific NPs, the base fluid, and the intended application. Experimentation and optimisation are often necessary to achieve the desired level of stability.

12.1.7 Effect on Interfacial Properties (IFT/Wettability)

Research indicates that altering the IFT and wettability within rock-oil-water systems is pivotal for effectively recovering residual crude oil post-secondary recovery. The EOR techniques focus on modifying the interfacial properties of these systems. The NPs of the NFs occupy the liquid-liquid interface during interactions between two immiscible fluids, thereby leading to a reduction in IFT. They have conducted dissipative particle dynamics simulations, illustrating a decrease in IFT within the water-decane system due to NP presence at the interface. The authors also noted a strong alignment between experimental and simulated results.

Nevertheless, specific instances of experimental data disparity at higher concentrations might stem from NP-NP interactions, as reported by Fan and Striolo (2012). Roustaei and Bagherzadeh (2015) reported an increase in the IFT of the oil-water system with increasing NF concentration, and the reason for this phenomenon was unknown to the authors. However, improved oil recovery was noticed due to an improvement in the capillary pressure owing to the NF flooding and a further decrease in the IFT of the n-decane-water system when zinc oxide (ZnO) NPs were added to the aqueous solution containing anionic surfactant. They observed a drop in the IFT of crude oil (API 16) and aqueous solution system by increasing the concentration of the silica NPs from 0.01 wt% to 0.5 wt% (i.e., from 25.1 to 9.3 mN/m) (Keykhosravi et al., 2021); however, beyond 0.5 wt% no significant effect was noticed. This study reports the impact of NP concentration of IFT in a liquid-liquid system. Shi et al. (2022) conducted research involving silica NPs (DP9711) and anionic surfactants (calimulse AOS) to investigate their ability to reduce the IFT in a crude oil-water system. The authors found that the presence of silica NPs and surfactant led to IFT values of 27.8 mN/m and 0.87 mN/m, respectively, in the crude oil-water system. The IFT of the crude oil-water system with NPs matched that of a brine solution, i.e., 28 mN/m. This study indicates that the effectiveness of silica NPs in reducing IFT within the crude oil-water system is relatively minor compared to the impact of the anionic surfactant employed in their research.

They have experimented with investigating how the concentration of a surfactant, in the presence and absence of spherical ZnO NPs with a diameter of 50 nm, affects the contact angle of n-decane droplets on a quartz surface (Moghadam & Azizian, 2014). The authors observed a reduction in the contact angle as the surfactant concentration increased (ranging from 0.002 to 0.01 M), both when NPs were absent and present in the aqueous solution. Behera and Sangwai (2021a) investigated the use of NPs (kaolinite and silica) along with different types of model oils (acidic and basic) and dead crude oil (light crude oil). The findings showed that NFs containing surfactants led to a reduction in the contact angle between oil droplets on quartz surfaces. Notably, the impact was substantial with the silica NFs relative to kaolinite for all acidic, basic, and crude systems. However, in all instances, a shift in wettability was observed with these systems from intermediate-wet to water-wet conditions. In their research, Keykhosravi et al. (2021) conducted contact angle experiments involving crude oil droplets on dolomite surfaces. They examined how the concentration of silica NPs in injection fluids influenced these angles. The researchers also observed that the salinity of the injection fluid had an effect on contact angles – lower salinity led to smaller contact angles compared to higher salinity conditions.

Furthermore, as NP concentrations increased up to 0.3 wt%, a significant dynamic change in contact angle over time (spanning seven days) was noted. After 168 hours, the equilibrium contact angle decreased from 147° to 24°. The equilibrium contact angle against the NP concentration of 0.1 wt% was more significant than the 0.3 wt% solution. This study illustrates the impact of time on contact angles and concentrations of NPs as well. In their research, Shi et al. (2022) asserted that silica NPs (i.e., DP9711) exhibited superior proficiency in maintaining wettability subsequent to exposure to the test solution, as opposed to inducing alterations. The authors have conducted the wettability test on the carbonate plate.

The above discussion reveals that the interfacial properties of rock-oil-water systems are influenced by NP concentrations, which aid in enhancing the efficacy of the chemical flooding, leading to higher oil recovery. The efficiency of surfactants in reducing the IFT of oil-water systems was found to be greater than that of NPs; however, further investigations with various surface-modified NPs were required to strengthen the statement.

12.2 COMPATIBILITY WITH OTHER EOR

NP-assisted EOR is a cutting-edge approach that uses NPs to improve oil recovery from reservoirs. The compatibility of NP-assisted EOR with other EOR methods like chemical flooding, polymer flooding, thermal EOR (TEOR), and low-salinity EOR (LSEOR) can vary based on specific conditions and objectives. In this section, the compatibility of NP-assisted EOR with other techniques is discussed briefly. Figure 12.6 shows the effectiveness of the nano-assisted chemical EOR technique in modifying fluid-fluid and fluid-rock properties.

12.2.1 Chemical Flooding

Chemical flooding methods, like surfactant and polymer flooding, involve injecting chemicals into the reservoir to alter fluid properties and mobilise trapped oil. NP-assisted EOR can be compatible with these methods if the NPs do not interfere

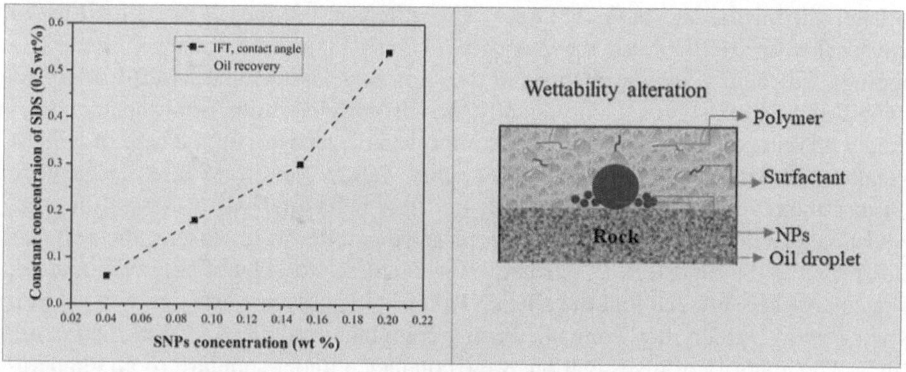

FIGURE 12.6 Effect of the NPs on interfacial properties in combination with other chemicals for EOR. Data taken from Rezk and Allam (2019), copyright ACS, 2019.

with the performance of injected chemicals or compromise their effectiveness. It is essential to ensure that NPs and chemicals do not interact negatively, potentially leading to reduced recovery efficiency (Bera & Belhaj, 2016).

When surfactants are mixed with NPs and injected into the underground rock formations, they can alter rock-oil-water properties. This change can be beneficial for getting more oil out. It happens because the mixture can make the rocks less resistant to letting oil flow, reduce the IFT between oil and water, and weaken the forces that hold oil trapped. Researchers have used computer simulations to show that this combination of effects is like a perfect recipe for recovering additional oil from the residual oil saturation (Zhao & Wen, 2017). Silica NPs were extensively examined alongside SDS surfactant for EOR. A glass micromodel was employed to comprehend the synergistic effect of NPs and surfactants. The addition of 2.2 wt% silica NPs to SDS resulted in a 13% increase in heavy-oil recovery compared to using SDS alone. The wettability of the system was reported to be altered significantly (shift from 73° to 11°). The increase in oil recovery was attributed to the increased viscosity of the injection fluid (Cheraghian et al., 2017). The impact of 0.05 wt% alumina and ZnO NPs with SDS surfactant was investigated to understand the oil recovery efficacy using a glass beads model and tensiometer. Quantitative analysis revealed that ZnO NPs exhibited lower IFT compared to alumina NPs. However, ZnO NPs demonstrated more significant IFT effects, leading to a higher cumulative oil recovery when compared to alumina NPs (Zaid et al., 2014).

12.2.2 Polymer Flooding

Polymer flooding aims to increase the viscosity of injected water, displacing oil more effectively. NP-assisted EOR can be compatible with polymer flooding as long as the NPs do not interact unfavourably with the polymers, affecting their ability to increase viscosity. The selection of NPs and polymers should consider their potential interactions to ensure both methods work optimally together. Using nanomaterials with polymers helps make the polymers stronger and more stable when exposed to heat, chemicals, and physical stress. The influence of silica NPs on hydrolysed polyacrylamide (HPAM) in the presence of 3 wt% salinity was examined. The findings indicated viscosity changes of 8 cP for pure HPAM and 35 cP for HPAM combined with silica NPs. Consequently, the viscosity elevation prominently correlated with reduced fingering effects and improved oil recovery in the case of HPAM blended with silica NPs (Yousefvand & Jafari, 2015). Moreover, researchers were curious about silica NPs' effect on polymer flow behaviour when flowing through sandstone. Their finding was that when these silica NPs were present, the polymer didn't stick to the sandstone as much, even when there was a lot of polymers around. The authors also noticed that the silica NPs made the polymer less likely to get thin when flowing, which they confirmed using computer models (Khalilinezhad et al., 2017). The majority of investigations were conducted under standard room temperature conditions. However, for a comprehensive comprehension of the intricate interactions among polymers, surfactants, and NPs, further studies are required under elevated and challenging reservoir conditions.

12.2.3 Thermal Enhanced Oil Recovery

TEOR methods, such as steam injection or in-situ combustion, rely on heat to lower oil viscosity and improve mobility. NP-assisted EOR might have limited compatibility with TEOR because high temperatures involved in thermal methods can potentially alter NP properties or lead to aggregation. Compatibility studies and careful selection of NPs are crucial to prevent any negative impact on either technique.

Nanomaterials have extensive applications in TEOR, primarily as nanocatalysts. Nanocatalysts based on nickel, iron, and cobalt are employed in nano-assisted TEOR to decrease the viscosity of heavy oil. Nickel NPs were utilised in a method called cyclic steam simulation with sandstone core samples (Yi et al., 2018). They showed that nickel NPs could assist in breaking certain bonds in the oil using water at high temperatures (aquathermolysis). However, this effect becomes weaker over time. The only drawback is that it increases the amount of water produced along with the oil.

Nevertheless, they suggested that placing the nickel NPs deeper in the reservoir, away from where they're injected, might lead to getting more oil out in the end. A study provided fresh insights into lowering water content in the steam-assisted gravity drainage (SAGD) process using ZnO NPs in sand packs. Through two sets of experiments, ZnO NPs were found to boost oil recovery by 35.5% compared to regular SAGD. Additionally, they noticed that the viscosity drop happened at low NP concentrations, around 0.2 to 0.5 wt%. The authors also reported considerable amount of water-cut reduction due to ZnO NPs. In the same way, the outcomes were linked to the NPs causing a chemical reaction that breaks apart certain bonds between carbon and sulphur (Tajmiri & Ehsani, 2016).

Notable reductions in viscosity and water content have been reported by integrating nanomaterials into TEOR. Consequently, this innovative approach holds the potential for practical field applications, which, however, necessitate further implementation and evaluation.

12.2.4 Low-Salinity Enhanced Oil Recovery

LSEOR involves injecting low-salinity water to alter reservoir rock wettability and enhance oil recovery. Low salinity water is essentially seawater or formation water that has been diluted with freshwater. This diluted water can be achieved by adding freshwater to seawater. Freshwater is river/ground water or can be produced by treating water containing various impurities employing advanced technologies, which are detailed in other studies (Kumar & Gangawane, 2023a, 2023b), and beyond the scope of the present study. The compatibility of NP-assisted EOR with LSEOR largely depends on the NPs' behaviour in common salinity conditions. It's essential to ensure that NPs do not aggregate or react in a way that undermines the effectiveness of LSEOR-induced wettability changes.

LSWF induces the migration of fine particles, subsequently diminishing oil recovery due to pore blockage in the rock. However, a strategy involving NF flooding prior to LSWF has been proposed as an effective means to mitigate fine migration, thereby enabling sustained long-term oil recovery (Yuan & Moghanloo, 2018). Chemical EOR presents a potential avenue for oil recovery. However, economic viability has been hampered by surfactant depletion due to adsorption onto the reservoir rock. Additionally, surfactants deteriorate under harsh reservoir conditions and elevated temperatures. By introducing stable NFs into previously inaccessible reservoir regions, the interfacial characteristics of solid-liquid-liquid systems can be altered, establishing a more favourable environment for EOR. The incorporation of a displacing fluid composed of low-salinity water, NPs, and chemicals could potentially diminish IFT and modify rock-oil-water wettability, thus leading to heightened oil recovery. Nevertheless, comprehensive understanding necessitates ongoing research efforts, considering the novelty of this approach. Given its nascent nature, further investigation is imperative to grasp the full potential of this innovative technique (Behera & Sangwai, 2022; Maghzi et al., 2011; Murshed et al., 2008; Pal & Sen, 2023a, 2023b).

In all cases, compatibility assessments and laboratory testing are essential to ensure that the combination of NP-assisted EOR with other EOR techniques delivers the desired results without compromising the effectiveness of either method. The specific reservoir characteristics, fluid properties, and NP behaviour will influence the degree of compatibility with other EOR methods.

12.3 CASE STUDIES AND FIELD TRIAL

Field tests are conducted after laboratory experiments to validate the performance of EOR technologies under real-world conditions, including the complex and dynamic nature of reservoirs. These tests assess scalability, economic viability, and compliance with environmental and safety standards, providing critical data for practical implementation in the oil industry. Field tests also uncover practical challenges that may not be evident in laboratory settings. The practical implementation of NFs in EOR plays a vital role in confirming their effectiveness, evaluating their economic feasibility, and improving our understanding of this technology. However, relatively few field tests have been conducted using this technology. Table 12.2 illustrates a compilation of these field tests, accompanied by their corresponding results.

TABLE 12.2
The field application of nanotechnology and the associate outcomes

Country	Field	Purpose	Result	References
Saudi Arabia	Ghawar field	To test the stability of the NPs under harsh reservoir condition (100°C temperature and 120,000 ppm salinity)	NPs (A-dot carbon based fluorescent) were found stable under challenging reservoir condition	Kanj et al. (2011)
China	Offshore field	To reduce the production lost due to fine migration	Surface modified (by polymer) cross-linked NPs were used	Bera and Belhaj (2016)
Colombia	Kupiagua field	To mitigate the precipitation of asphaltenes.	Alumina NPs were used, and an increase in oil recovery was observed	Zabala et al. (2014)
Russia	Volga-Urals region	To enhance production	Nanofluids treated with surfactants resulted decrease in water cut and an increase up to 7.3% oil recovery	Bodrova et al. (2020)

The above field's results suggest that NPs show promise as a potential agent for EOR. However, further research is needed to assess their economic viability and address environmental concerns comprehensively. It's crucial to emphasise that drawing definitive conclusions solely based on a limited study is not advisable. Hence, more studies are required to explore the application of NPs in the oil and gas industries.

12.4 TRADITIONAL AND NANO-ASSISTED CHEMICAL EOR

Analysing the performance of traditional chemical EOR alongside NF-assisted chemical EOR is essential for evaluating the merits and limitations of each approach. This comparative analysis yields insights into their effectiveness, efficiency, and economic feasibility. Such comprehensive studies provide valuable insights that can guide the future development, optimisation, and selection of the most suitable EOR strategy for specific reservoir conditions.

Maghzi et al. (2014) performed an experimental investigation involving the incorporation of 0.1 wt% silica NPs into a polymer solution to augment its pseudoplastic characteristics. Subsequently, this modified solution was employed for oil recovery within a micromodel. The researchers noted a supplementary 10% enhancement in oil recovery in comparison to employing chemical flooding exclusively. This improvement was attributed to the enhanced rheological attributes of the injection fluid. Similarly, Yousefvand and Jafari (2015) experimented with silica NPs on HPAM, revealing significant viscosity enhancement from 8 cP for pure HPAM to 35 cP for HPAM combined with silica NPs. This increased viscosity was closely associated with reduced fingering effects and EOR when HPAM was combined with silica NPs. While similar discussions have been covered in preceding sections,

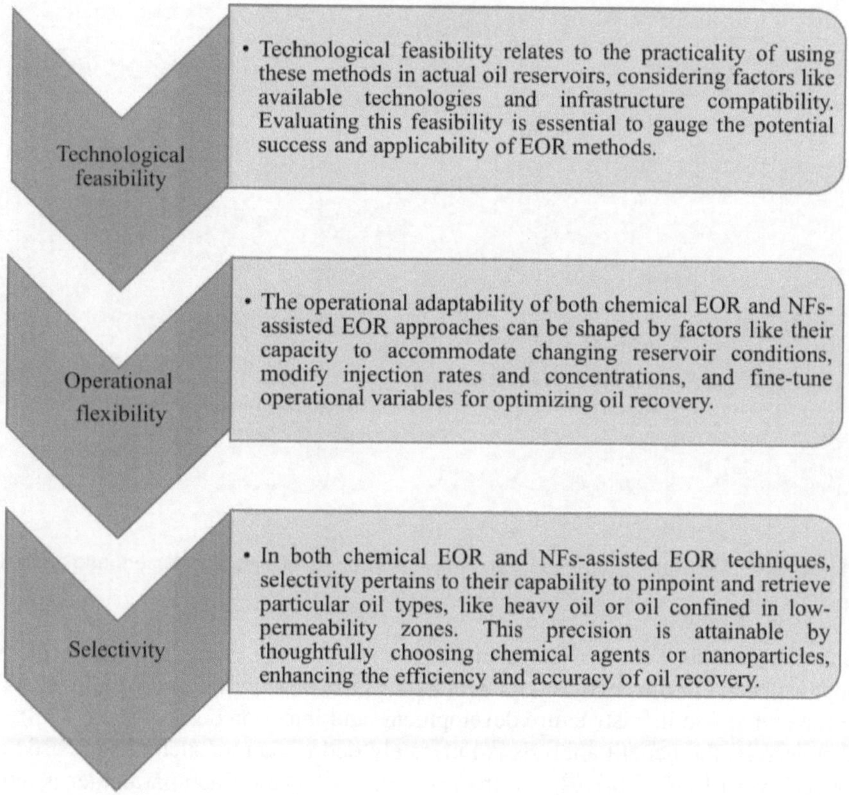

FIGURE 12.7 Major factors required to consider for feasibility and scalability of the technique.

the following additional technical insights are presented in Figure 12.7 to explore other aspects pertaining to feasibility and scalability.

Furthermore, operational risks and anticipated challenges must also be taken into account when assessing the feasibility of this technology. The prolonged use of NPs may lead to formation damage and other unforeseen phenomena, necessitating further exploration before adaptation of the technology.

12.5 OPTIMISATION THROUGH TECHNICAL AND ECONOMIC ANALYSIS

Optimising things involves looking at lots of factors such as the size and shape of particles, concentration of the NPs, stability of the NFs, ability to alter reservoir fluids and rock interfacial properties, oil recovery efficiency, economic viability, etc. Those factors can be optimised through various analysis techniques provided in Figure 12.8. The impact of these factors on oil recovery aimed at optimising parameters has been thoroughly discussed in the previous section (Aliabadian et al., 2020).

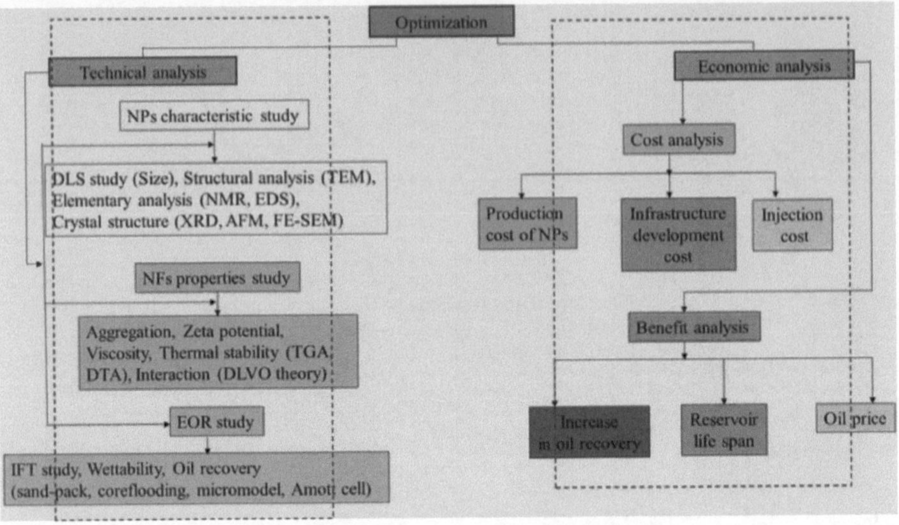

FIGURE 12.8 Optimisation steps for parameters through technical and economic analysis.

Economic analysis entails a comprehensive examination involving both cost and benefit aspects. On the cost side, it encompasses the production cost of NPs or NFs, expenses related to infrastructure development, and injection costs (Figure 12.8).

Conversely, the benefit analysis is intricately tied to factors such as increased oil recovery, prevailing oil prices, and the reservoir's lifespan. Each parameter is interrelated and must be carefully balanced to achieve the desired improvements in oil recovery while ensuring economic viability and minimising environmental impact. The detailed process is outlined below.

- *Particle size and shape*: The size and shape of NPs are fundamental considerations. Smaller NPs often have a larger surface area relative to their volume, which can impact their interactions with fluids and surfaces. Different shapes of NPs, such as spherical, cylindrical, or rod-like, can have varying effects on fluid behaviour and interactions with reservoir rocks.
- *NP concentration*: Determining the optimal concentration of NPs in NFs is crucial. A concentration that is too low may not yield the desired effects, while excessively high concentrations can lead to stability issues and potential formation damage.
- *NF stability*: Ensuring the stability of NFs under reservoir conditions is essential. Unstable NFs can result in particle aggregation and pore blockage, which can hinder fluid flow and reduce the effectiveness of EOR techniques.
- *Alteration of reservoir fluids*: One of the objectives of nano-assisted EOR is to modify the properties of reservoir fluids. This may involve enhancing the viscosity of injection fluids, reducing IFT between fluids, or promoting wettability alteration. The choice of NPs and their concentrations directly impact these alterations.

- *Rock interfacial properties*: The interaction between NPs and rock surfaces is pivotal. NPs can affect the wettability of reservoir rocks, making them more oil-wet or water-wet. Understanding these interactions is crucial for successful EOR applications.
- *Oil recovery efficiency*: Ultimately, the primary goal is to improve oil recovery. This necessitates evaluating how the chosen NPs and their concentrations influence the displacement of oil within the reservoir. This may involve core flooding experiments to simulate reservoir conditions and assess recovery rates.
- *Economic viability*: Assessing the economic feasibility of nano-assisted EOR involves considering the cost of producing NPs, developing the necessary infrastructure for field application, injection costs, and potential gains in oil production. Economic analyses are vital for decision-making and project justification.

12.6 ENVIRONMENTAL IMPACT

In today's world, optimising any technology is not complete without considering its environmental impact. It's vital to be mindful of government regulations because even if we formulate an efficient EOR agent, it will not see the light of day if it does not meet regulatory criteria (Pal & Sen, 2023a, 2023b). The global trend is clear – countries are taking environmental protection seriously, making compliance a top priority. Table 12.3 displays the impact of frequently studied NPs in the oil and gas industry on living organisms and the environment.

TABLE 12.3
Effect of NPs on the living organism and environment

Nanoparticle	Impact	References
Carbon NPs (CNT, MWCNTs)	Potential hazards to humans and biological systems, including inducing inflammatory and apoptosis responses in human T-cells	Ding et al. (2005)
TiO_2	These NPs can cross various barriers, like the blood-brain barrier and placental barrier, posing potential risks to the central nervous system, neurons, glial cells, and liver cells	Ding et al. (2005)
SiO_2	Adverse effects on cardiovascular system based on their size, structure, and surface modifications	Guo et al. (2021)
ZnO	They can trigger testicular cell apoptosis by causing DNA damage through reactive oxygen species, potentially impacting spermatogenesis and male fertility	Han et al. (2016)
Al_2O_3	Severe effects on fish health can lead to the accumulation of metal granules in human livers	Murali et al. (2017)
Fe_2O_3	The most significant impact on bacterial growth, biofilm formation, and impairment of antimicrobial peptide function	Borcherding et al. (2014)

Abbreviations: CNT, carbon nanotube; Fe_2O_3, ferric oxide; MWCNTs, multiwall carbon nanotubes; TiO_2, titanium dioxide; ZnO, zinc oxide.

12.6.1 Strategies to Mitigate Environmental Impact

In EOR, choosing eco-friendly NPs like silica and clay can make a big difference in reducing harm to the environment. Silica NPs are considered safe because they are common, are not very toxic, and found in abundance. Similarly, clay NPs, specifically a type called montmorillonite, are eco-friendly because they are naturally abundant, are not very toxic, and work well in oil and gas production. These environmentally friendly NPs help in two ways.

As they are common and not very harmful, they are safe to use during production and application.

They are good at keeping the EOR fluids stable, which means they are less likely to leak into the environment.

This stability not only bolsters the safety aspect of EOR practices but also ensures the containment of NPs within the reservoir matrix. Furthermore, by adhering to recommended regulations and employing advanced techniques such as polymer-coated NPs or core-shell structures, these NPs exhibit enhanced compatibility with reservoir conditions. Conducting a comprehensive review of rules and guidelines pertaining to NP utilisation in oil recovery processes can facilitate the informed selection of appropriate NP candidates. This not only promotes sustainability but also enhances the efficiency of oil recovery methodologies.

12.7 CONCLUSION

Multiple parameters affecting nano-assisted EOR have been thoroughly examined, including the fundamental mechanisms of crude oil recovery and the pivotal factors influencing it. The optimisation of these parameters is contingent upon comprehensive technical and economic analyses. To enhance the effectiveness of the process, the following key considerations should be taken into account.

- Ensuring the stability of NFs under harsh reservoir conditions, characterised by elevated temperatures and salinity levels, is imperative. The instability of NFs can result in pore blockages and formation damage. Therefore, it is essential to explore the formulation of long-lasting stable NFs, potentially in conjunction with appropriate stabilising agents.
- Silica-based NPs are extensively studied in EOR due to their favourable, cost-effective, and environmentally friendly characteristics, including non-toxicity and the ability to switch between hydrophobic and hydrophilic properties.
- Conducting studies focused on optimising various parameters that influence nano-assisted EOR is crucial, especially under reservoir conditions. These studies aim to enhance both oil recovery and cost efficiency. However, only a limited number of studies have been conducted under certain limitations to optimise a few parameters for nano-assisted EOR.
- As the cost of NPs in present situation is expensive, their efficient recovery and subsequent reuse represent cost-saving opportunities in nano-assisted EOR. Therefore, it is crucial to gain a comprehensive understanding of NPs'

adsorption behaviour in porous media, necessitating a thorough investigation involving both experimental and simulation studies.

- Research into the impact of NP shapes has been relatively limited so far. Exploring various shapes of amphiphilic Janus NPs, featuring both hydrophilic and hydrophobic surfaces, could hold substantial promise for achieving significant improvements in EOR performance.
- The lack of field trials for nano-assisted EOR may be attributed to the high cost of NPs. Therefore, developing cost-effective NP production methods is crucial to incentivise field trials, which can uncover valuable insights into the factors affecting oil recovery.
- Injection fluid containing a combination of two or more different types of NPs could be beneficial. The interaction between hydrophilic and partially hydrophobic NPs may result in prosperous outcomes.
- Comprehensive research on the safety and environmental impact of NPs is imperative to safeguard both human health and our ecosystems.

REFERENCES

Adil, M., Mohd Zaid, H., Raza, F., & Agam, M. A. (2020). Experimental evaluation of oil recovery mechanism using a variety of surface-modified silica nanoparticles: Role of in-situ surface-modification in oil-wet system. *PLOS ONE*, *15*(7), e0236837. https://doi.org/10.1371/journal.pone.0236837.

Ahmed, A., Saaid, I. M., Ahmed, A. A., Pilus, R. M., & Baig, M. K. (2020). Evaluating the potential of surface-modified silica nanoparticles using internal olefin sulfonate for enhanced oil recovery. *Petroleum Science*, *17*(3), 722–733. https://doi.org/10.1007/s12182-019-00404-1.

Al-Anssari, S., Arif, M., Wang, S., Barifcani, A., & Iglauer, S. (2017). Stabilising nanofluids in saline environments. *Journal of Colloid and Interface Science*, *508*, 222–229. https://doi.org/10.1016/J.JCIS.2017.08.043.

Aliabadian, E., Sadeghi, S., Rezvani Moghaddam, A., Maini, B., Chen, Z., & Sundararaj, U. (2020). Application of graphene oxide nanosheets and HPAM aqueous dispersion for improving heavy oil recovery: Effect of localised functionalisation. *Fuel*, *265*(December 2019), 116918. https://doi.org/10.1016/j.fuel.2019.116918.

Behera, U. S., Kumar, G., & Sangwai, J. S. (2022). Pore-scale investigation and performance evaluation of SMART LowSal flooding for enhanced oil recovery from matured reservoirs using a lab-on-a-chip. *Energy & Fuels*, *36*(15), 8115–8127. https://doi.org/10.1021/acs.energyfuels.2c01009.

Behera, U. S., & Sangwai, J. S. (2021a). Nanofluids of kaolinite and silica in low saline seawater (LowSal) with and without surfactant: Interfacial tension and wettability alteration of oil–water–rock system for low salinity-enhanced oil recovery. *Industrial & Engineering Chemistry Research*, *60*(1), 291–313. https://doi.org/10.1021/acs.iecr.0c04853.

Behera, U. S., & Sangwai, J. S. (2021b). Nanofluids of silica nanoparticles in low salinity water with surfactant and polymer (SMART LowSal) for enhanced oil recovery. *Journal of Molecular Liquids*, *342*, 117388. https://doi.org/10.1016/j.molliq.2021.117388.

Behera, U. S., & Sangwai, J. S. (2022). Silica nanofluid in low salinity seawater containing surfactant and polymer: Oil recovery efficiency, wettability alteration and adsorption studies. *Journal of Petroleum Science and Engineering*, *211*, 110148. https://doi.org/10.1016/j.petrol.2022.110148.

Bera, A., & Belhaj, H. (2016). Application of nanotechnology by means of nanoparticles and nanodispersions in oil recovery - A comprehensive review. *Journal of Natural Gas Science and Engineering*, *34*, 1284–1309. https://doi.org/10.1016/j.jngse.2016.08.023.

Bila, A., & Torsæter, O. (2021). Article experimental investigation of polymer-coated silica nanoparticles for EOR under harsh reservoir conditions of high temperature and salinity. *Nanomaterials*, *11*(3), 1–17. https://doi.org/10.3390/nano11030765.

Binshan, J., Shugao, D., Zhian, L., Tiangao, Z., Xiantao, S., & Xiaofeng, Q. (2002). A study of wettability and permeability change caused by adsorption of nanometer structured polysilicon on the surface of porous media. In *SPE - Asia Pacific Oil and Gas Conference* (pp. 915–926). https://doi.org/10.2118/77938-ms.

Bodrova, E. V., Kalinov, V. V., & Krasivskaya, V. N. (2020). The beginning of the development of oil fields in ural-volga region in 1930s—1940s. *Nauchnyi Dialog*, *1*(10), 323–336. https://doi.org/10.24224/2227-1295-2020-10-323-336.

Borcherding, J., Baltrusaitis, J., Chen, H., Stebounova, L., Wu, C.-M., Rubasinghege, G., Mudunkotuwa, I. A., Caraballo, J. C., Zabner, J., Grassian, V. H., & Comellas, A. P. (2014). Iron oxide nanoparticles induce Pseudomonas aeruginosa growth, induce biofilm formation, and inhibit antimicrobial peptide function. *Environmental Science: Nano*, *1*(2), 123. https://doi.org/10.1039/c3en00029j.

Cheraghian, G., Kiani, S., Nassar, N. N., Alexander, S., & Barron, A. R. (2017). Silica nanoparticle enhancement in the efficiency of surfactant flooding of heavy oil in a glass micromodel. *Industrial & Engineering Chemistry Research*, *56*(30), 8528–8534. https://doi.org/10.1021/acs.iecr.7b01675.

Choi, S. K., Son, H. A., Kim, H. T., & Kim, J. W. (2017). Nanofluid enhanced oil recovery using hydrophobically associative zwitterionic polymer-coated silica nanoparticles. *Energy & Fuels*, *31*(8), 7777–7782. https://doi.org/10.1021/acs.energyfuels.7b00455.

Choudhary, R., Khurana, D., Kumar, A., & Subudhi, S. (2017). Stability analysis of Al2O3/water nanofluids. *Journal of Experimental Nanoscience*, *12*(1), 140–151. https://doi.org/10.1080/17458080.2017.1285445.

Crawford, P. B. (1960). Factors affecting waterflood pattern performance and selection. *Journal of Petroleum Technology*, *12*(12), 11–15. https://doi.org/10.2118/1503-G.

Dai, C., Wang, S., Li, Y., Gao, M., Liu, Y., Sun, Y., & Zhao, M. (2015). The first study of surface modified silica nanoparticles in pressure-decreasing application. *RSC Advances*, *5*(76), 61838–61845. https://doi.org/10.1039/c5ra09883a.

Daneshmand, H., Rezaeinasab, M., Asgary, M., & Karimi, M. (2021). Wettability alteration and retention of mixed polymer-grafted silica nanoparticles onto oil-wet porous medium. *Petroleum Science*, *18*(3), 962–982. https://doi.org/10.1007/s12182-021-00552-3.

Das, S. K., Choi, S. U. S., Yu, W., & Pradeep, T. (2007). Nanofluids: Science and technology. *Nanofluids: Science and Technology*, 1–397. https://doi.org/10.1002/9780470180693.

Ding, L., Stilwell, J., Zhang, T., Elboudwarej, O., Jiang, H., Selegue, J. P., Cooke, P. A., Gray, J. W., & Frank Chen, F. (2005). Molecular characterisation of the cytotoxic mechanism of multiwall carbon nanotubes and nano-onions on human skin fibroblast. *Nano Letters*, *5*(12), 2448–2464. https://doi.org/10.1021/nl051748o.

Esmaeilzadeh, P., Hosseinpour, N., Bahramian, A., Fakhroueian, Z., & Arya, S. (2014). Effect of ZrO2 nanoparticles on the interfacial behavior of surfactant solutions at air-water and n-heptane-water interfaces. *Fluid Phase Equilibria*, *361*, 289–295. https://doi.org/10.1016/j.fluid.2013.11.014.

Fan, H., & Striolo, A. (2012). Nanoparticle effects on the water-oil interfacial tension. *Physical Review E - Statistical, Nonlinear, and Soft Matter Physics*, *86*(5), 1–11. https://doi.org/10.1103/PhysRevE.86.051610.

Guo, C., Liu, Y., & Li, Y. (2021). Adverse effects of amorphous silica nanoparticles: Focus on human cardiovascular health. *Journal of Hazardous Materials*, *406*(August 2020), 124626. https://doi.org/10.1016/j.jhazmat.2020.124626.

Han, Z., Yan, Q., Ge, W., Liu, Z.-G., Gurunathan, S., De Felici, M., Shen, W., & Zhang, X.-F. (2016). Cytotoxic effects of ZnO nanoparticles on mouse testicular cells. *International Journal of Nanomedicine*, *11*, 5187–5203. https://doi.org/10.2147/IJN.S111447.

Hendraningrat, L., Li, S., & Torsæter, O. (2013). A coreflood investigation of nanofluid enhanced oil recovery. *Journal of Petroleum Science and Engineering*, *111*, 128–138. https://doi.org/10.1016/j.petrol.2013.07.003.

Hu, Y., Zhao, Z., Dong, H., Mikhailova, M. V., & Davarpanah, A. (2021). Hybrid application of nanoparticles and polymer in enhanced oil recovery processes. *Polymers*, *13*(9), 1–13. https://doi.org/10.3390/polym13091414.

Idogun, A. K., Iyagba, E. T., Ukwotije-Ikwut, R. P., & Aseminaso, A. (2016). A review study of oil displacement mechanisms and challenges of nanoparticle enhanced oil recovery. In *Society of Petroleum Engineers - SPE Nigeria Annual International Conference and Exhibition*. https://doi.org/10.2118/184352-ms.

Jia-Fei, Z., Zhong-Yang, L., Ming-Jiang, N., & Ke-Fa, C. (2009). Dependence of nanofluid viscosity on particle size and pH value. *Chinese Physics Letters*, *26*(6), 066202. https://doi.org/10.1088/0256-307X/26/6/066202.

Kanj, M. Y., Rashid, M. H., & Giannelis, E. P. (2011). Industry first field trial of reservoir nanoagents. *All Days*, *3*, 1883–1892. https://doi.org/10.2118/142592-MS.

Keykhosravi, A., Vanani, M. B., Daryasafar, A., & Aghayari, C. (2021). Comparative study of different enhanced oil recovery scenarios by silica nanoparticles: An approach to time-dependent wettability alteration in carbonates. *Journal of Molecular Liquids*, *324*, 115093. https://doi.org/10.1016/j.molliq.2020.115093.

Khalilinezhad, S. S., Cheraghian, G., Roayaei, E., Tabatabaee, H., & Karambeigi, M. S. (2017). Statement of removal. *Energy Sources, Part A: Recovery, Utilisation, and Environmental Effects*, *45*(1), 1779–1779. https://doi.org/10.1080/15567036.2017.1302521.

Kumar, A., & Gangawane, K. M. (2022). Synthesis and effect on the surface morphology & magnetic properties of ferrimagnetic nanoparticles by different wet chemical synthesis methods. *Powder Technology*, *410*(June), 117867. https://doi.org/10.1016/j.powtec.2022.117867.

Kumar, A., & Gangawane, K. M. (2023a). Advanced computational approaches for water treatment. In *Advanced Computational Approaches for Water Treatment* (pp. 1–12). CRC Press. https://doi.org/10.1201/9781003325147-1.

Kumar, A., & Gangawane, K. M. (2023b). Molecular dynamics for the membrane process. In *Advanced Computational Approaches for Water Treatment* (pp. 13–38). CRC Press. https://doi.org/10.1201/9781003325147-2. *(IFCoN)*.

Li, K., Wang, D., & Jiang, S. (2018). Review on enhanced oil recovery by nanofluids. *Oil & Gas Science and Technology – Revue d'IFP Energies Nouvelles*, *73*, 37. https://doi.org/10.2516/ogst/2018025.

Li, X., Chen, Y., Mo, S., Jia, L., & Shao, X. (2014). Effect of surface modification on the stability and thermal conductivity of water-based SiO2-coated graphene nanofluid. *Thermochimica Acta*, *595*, 6–10. https://doi.org/10.1016/j.tca.2014.09.006.

Maghzi, A., Kharrat, R., Mohebbi, A., & Ghazanfari, M. H. (2014). The impact of silica nanoparticles on the performance of polymer solution in presence of salts in polymer flooding for heavy oil recovery. *Fuel*, *123*, 123–132. https://doi.org/10.1016/j.fuel.2014.01.017.

Maghzi, A., Mohebbi, A., Kharrat, R., & Ghazanfari, M. H. (2011). Pore-scale monitoring of wettability alteration by silica nanoparticles during polymer flooding to heavy oil in a five-spot glass micromodel. *Transport in Porous Media*, *87*(3), 653–664. https://doi.org/10.1007/s11242-010-9696-3.

Mead, I. (2017). *International Energy Outlook 2017 Key Takeaways : IEO2017 Reference Case*. U.S. Energy Information Administration.

Moghadam, T. F., & Azizian, S. (2014). Effect of ZnO nanoparticles on the interfacial behavior of anionic surfactant at liquid/liquid interfaces. *Colloids and Surfaces A: Physicochemical and Engineering Aspects*, *457*(1), 333–339. https://doi.org/10.1016/j.colsurfa.2014.06.009.

Mohd, S. M., Wan, R. W. S., Abdul, R. I., Mohd, Z. J., & Issham, I. (2016). Wettability alteration of dolomite rock using nanofluids for enhanced oil recovery. *Materials Science Forum*, *864*, 194–198. https://doi.org/10.4028/www.scientific.net/MSF.864.194.

Muggeridge, A., Cockin, A., Webb, K., Frampton, H., Collins, I., Moulds, T., & Salino, P. (2014). Recovery rates, enhanced oil recovery and technological limits. *Philosophical Transactions of the Royal Society A: Mathematical, Physical and Engineering Sciences*, *372*(2006), 20120320. https://doi.org/10.1098/rsta.2012.0320.

Murali, M., Suganthi, P., Athif, P., Sadiq Bukhari, A., Syed Mohamed, H. E., Basu, H., & Singhal, R. K. (2017). Histological alterations in the hepatic tissues of Al2O3 nanoparticles exposed freshwater fish Oreochromis mossambicus. *Journal of Trace Elements in Medicine and Biology*, *44*(July), 125–131. https://doi.org/10.1016/j.jtemb.2017.07.001.

Murshed, S. M. S., Tan, S. H., & Nguyen, N. T. (2008). Temperature dependence of interfacial properties and viscosity of nanofluids for droplet-based microfluidics. *Journal of Physics D: Applied Physics*, *41*(8). https://doi.org/10.1088/0022-3727/41/8/085502.

Ngouangna, E. N., Zaidi Jaafar, M., Norddin, M. N. A. M., Agi, A., Oseh, J. O., & Mamah, S. (2022). Surface modification of nanoparticles to improve oil recovery mechanisms: A critical review of the methods, influencing parameters, advances and prospects. *Journal of Molecular Liquids*, *360*, 119502. https://doi.org/10.1016/j.molliq.2022.119502.

Nguyen, C. T., Desgranges, F., Roy, G., Galanis, N., Maré, T., Boucher, S., & Angue Mintsa, H. (2007). Temperature and particle-size dependent viscosity data for water-based nanofluids – Hysteresis phenomenon. *International Journal of Heat and Fluid Flow*, *28*(6), 1492–1506. https://doi.org/10.1016/j.ijheatfluidflow.2007.02.004.

Nwidee, L. N., Theophilus, S., Barifcani, A., Sarmadivaleh, M., & Iglauer, S. (2016). EOR processes, opportunities and technological advancements. In *Chemical Enhanced Oil Recovery (cEOR) - A Practical Overview*. InTech. https://doi.org/10.5772/64828.

Pal, D., & Sen, S. (2023a). Emerging petroleum pollutants and their adverse effects on the environment. In I. D. Behera, & A. P. Das (Eds.), *Impact of Petroleum Waste on Environmental Pollution and Its Sustainable Management through Circular Economy. Environmental Science and Engineering* (pp. 103–137). Springer Nature Switzerland. https://doi.org/10.1007/978-3-031-48220-5_5.

Pal, D., & Sen, S. (2023b). In-depth coverage of petroleum waste sources, characteristics, environmental impact, and sustainable remediation process. In I. D. Behera, & A. P. Das (Eds.), *Impact of Petroleum Waste on Environmental Pollution and Its Sustainable Management through Circular Economy* (pp. 1–38). Springer Nature Switzerland. https://doi.org/10.1007/978-3-031-48220-5_1.

Panda, D., & Gangawane, K. M. (2023). Hybrid NiFe2O4-NiCO4O4 nanoparticles-based eutectic phase change materials for enhancement of thermal efficiency of pin-fin heat sink arrangement. *Journal of Energy Storage*, *60*(January), 102417. https://doi.org/10.1016/j.est.2023.106644.

Prats, M. (1986). *Thermal Recovery*. SPE.

Putra, N., Roetzel, W., & Das, S. K. (2003). Natural convection of nano-fluids. *Heat and Mass Transfer/Waerme- Und Stoffuebertragung*, *39*(8–9), 775–784. https://doi.org/10.1007/s00231-002-0382-z

Rezk, M. Y., & Allam, N. K. (2019). Impact of nanotechnology on enhanced oil recovery: A mini-review. *Industrial and Engineering Chemistry Research, 58*(36), 16287–16295. https://doi.org/10.1021/acs.iecr.9b03693.

Rezvani, H., Riazi, M., Tabaei, M., Kazemzadeh, Y., & Sharifi, M. (2018). Experimental investigation of interfacial properties in the EOR mechanisms by the novel synthesised Fe3O4@ chitosan nanocomposites. *Colloids and Surfaces A: Physicochemical and Engineering Aspects, 544*(February), 15–27. https://doi.org/10.1016/j.colsurfa.2018.02.012.

Roustaei, A., & Bagherzadeh, H. (2015). Experimental investigation of SiO2 nanoparticles on enhanced oil recovery of carbonate reservoirs. *Journal of Petroleum Exploration and Production Technology, 5*(1), 27–33. https://doi.org/10.1007/s13202-014-0120-3.

Satter, A., & Iqbal, G. M. (2016). Primary recovery mechanisms and recovery efficiencies. In *Reservoir Engineering* (pp. 185–193). Elsevier. https://doi.org/10.1016/B978-0-12-800219-3.00011-5.

Sedaghat, M., Esfandiarian, A., & Azdarpour, A. (2020). Experimental investigation of synergism between low saline water and surfactant-silica nanoparticle in enhanced oil recovery processes. In *82nd EAGE Annual Conference & Exhibition* (Vol. 2020, no. 1, pp. 1–5). European Association of Geoscientists & Engine.

Shi, Y., Wang, X., & Mohanty, K. (2022). Effect of a nanoparticle on wettability alteration and wettability retainment of carbonate reservoirs. *Journal of Petroleum Science and Engineering, 215*(PB), 110684. https://doi.org/10.1016/j.petrol.2022.110684.

Statista Research Department. (2023). *Market Value of Enhanced Oil Recovery in the United States from 2014 to 2025, by Technology (in Billion U. S. Dollars)* (pp. 3–4).

Sun, X., Zhang, Y., Chen, G., & Gai, Z. (2017). Application of nanoparticles in enhanced oil recovery: A critical review of recent progress. *Energies, 10*(3). https://doi.org/10.3390/en10030345.

Tajmiri, M., & Ehsani, M. R. (2016). The potential of ZnO nanoparticles to reduce water consuming in Iranian heavy oil reservoir. *Journal of Water and Environmental Nanotechnology, 1*(12), 84–9084. https://doi.org/10.7508/jwent.2016.02.002.

Tengku Mohd, T. A., Baco, J., Bakar, N. F. A., & Jaafar, M. Z. (2016). Effects of particle shape and size on nanofluid properties for potential Enhanced Oil Recovery (EOR). *MATEC Web of Conferences, 69*, 03006. https://doi.org/10.1051/matecconf/20166903006.

Timofeeva, E. V., Routbort, J. L., & Singh, D. (2009). Particle shape effects on thermophysical properties of alumina nanofluids. *Journal of Applied Physics, 106*(1). https://doi.org/10.1063/1.3155999.

Yi, S., Babadagli, T., & Li, H. A. (2018). Use of nickel nanoparticles for promoting aquathermolysis reaction during cyclic steam stimulation. *SPE Journal, 23*(1), 145–156. https://doi.org/10.2118/186102-pa.

Yousefvand, H., & Jafari, A. (2015). Enhanced oil recovery using polymer/nanosilica. *Procedia Materials Science, 11*(2010), 565–570. https://doi.org/10.1016/j.mspro.2015.11.068.

Yousefvand, H. A., & Jafari, A. (2018). Stability and flooding analysis of nanosilica/NaCl/HPAM/SDS solution for enhanced heavy oil recovery. *Journal of Petroleum Science and Engineering, 162*(March 2017), 283–291. https://doi.org/10.1016/j.petrol.2017.09.078.

Yuan, B., & Moghanloo, R. G. (2018). Nanofluid precoating: An effective method to reduce fines migration in radial systems saturated with two mobile immiscible fluids. *SPE Journal, 23*(3), 998–1018. https://doi.org/10.2118/189464-pa.

Zabala, R., Mora, E., Botero, O. F., Cespedes, C., Guarin, L., Franco, C. A., Cortes, F. B., Patino, J. E., & Ospina, N. (2014). Nano-technology for asphaltenes inhibition in Cupiagua South Wells. *IPTC 2014: International Petroleum Technology Conference, 2*(7), 1260–1274. https://doi.org/10.3997/2214-4609-pdb.395.IPTC-17344-MS.

Zaid, H. M., Ahmad Latiff, N. R., & Yahya, N. (2014). The effect of zinc oxide and aluminum oxide nanoparticles on interfacial tension and viscosity of nanofluids for enhanced oil recovery. *Advanced Materials Research*, *1024*, 56–59. https://doi.org/10.4028/www.scientific.net/AMR.1024.56.

Zamani, A., Maini, B., & Pereira-Almao, P. (2012). Flow of nanodispersed catalyst particles through porous media: Effect of permeability and temperature. *The Canadian Journal of Chemical Engineering*, *90*(2), 304–314. https://doi.org/10.1002/cjce.20629.

Zargartalebi, M., Barati, N., & Kharrat, R. (2014). Influences of hydrophilic and hydrophobic silica nanoparticles on anionic surfactant properties: Interfacial and adsorption behaviors. *Journal of Petroleum Science and Engineering*, *119*, 36–43. https://doi.org/10.1016/j.petrol.2014.04.010.

Zargartalebi, M., Kharrat, R., & Barati, N. (2015). Enhancement of surfactant flooding performance by the use of silica nanoparticles. *Fuel*, *143*, 21–27. https://doi.org/10.1016/j.fuel.2014.11.040.

Zhang, K., Li, Y., Hong, A., Wu, K., Jing, G., Torsæter, O., Chen, S.., & Chen, Z. (2015). Nanofluid alternating gas for tight oil exploitation. *All Days*, 10–13. https://doi.org/10.2118/176241-MS.

Zhao, J., & Wen, D. (2017). Pore-scale simulation of wettability and interfacial tension effects on flooding process for enhanced oil recovery. *RSC Advances*, *7*(66), 41391–41398. https://doi.org/10.1039/C7RA07325A.

13 Nanotechnology in Enhanced Oil Recovery

Challenges and Future Directions

Krishna Raghav Chaturvedi, Abhishek Kumar and Anand Raghav Chaubey

13.1 INTRODUCTION

In the previous chapters, it was established that the use of nanoparticles (and nanomaterials) is promising for subsurface applications as they influence oil recovery through various mechanisms. In the previously reported studies, the nanomaterials have been found to alter the interfacial properties of oil, water, and rock, enhancing displacement and recovery efficiency. The use of nanomaterials in combination with polymers and surfactants can create tailored solutions for specific reservoir conditions. When combined with chemicals like polymers and surfactants, synergistic effects can occur, improving oil recovery by modifying fluid behaviour, reducing interfacial tension, and increasing sweep efficiency in reservoirs. This integrated approach addresses challenges such as wettability alteration and mobility control, contributing to Enhanced Oil Recovery (EOR) strategies. Furthermore, in terms of foam formation and stability, nanomaterials can impact these processes positively. Nanoparticles can stabilize foam by providing a reinforcing effect on the liquid lamellae, thereby improving foam longevity (Kumar et al., 2022, 2023a, 2023b; Kumar & Gangawane, 2022a, 2022b).

Additionally, nanomaterials can influence the rheological properties of foaming solutions, affecting bubble size distribution and foam stability during oil recovery operations (Ahumada et al., 2021; Chaturvedi & Sharma, 2021). Overall, the integration of nanoparticles and nanomaterials, along with other chemicals, offers a promising avenue for optimizing oil recovery processes, improving efficiency, and addressing challenges associated with foam stability in the reservoir (Panda et al., 2023; Panda & Gangawane, 2023b). However, the broader field applicability of these promising materials in subsurface applications is challenged due to a plethora of issues that will be discussed in this chapter of the book.

DOI: 10.1201/9781032650685-13

13.2 CHALLENGES BEFORE NANOMATERIAL APPLICATIONS IN EOR

While promising, the broader applicability of nanomaterials in subsurface applications is constrained by the following challenges inherent to the material properties and the field conditions observed in the oilfield (Figure 13.1). These challenges include:

- *Agglomeration*: Nanomaterials tend to agglomerate, reducing their effectiveness. This occurs due to attractive forces such as van der Waals interactions and electrostatic forces between nanoparticles (Chaturvedi et al., 2021b; Kumar et al., 2021). The accumulation of the nanomaterials minimizes the available subsurface area, lowering their propensity to reduce the interfacial tension and altering the wettability of the underground reservoir, directly reducing their oil recovery potential.
- *Temperature and Salinity Impact*: Temperature and salinity variations in reservoir conditions can exacerbate agglomeration (Chaturvedi et al., 2021b; Kumar et al., 2021). Changes in these factors, especially high temperature and salinity, can affect the stability of colloidal suspensions, leading to increased agglomeration of nanomaterials. Temperature influences the kinetic energy of particles, affecting their Brownian motion and collision frequency. Salinity alters the electrostatic interactions, impacting repulsion or attraction forces between nanoparticles. Both factors can lead to

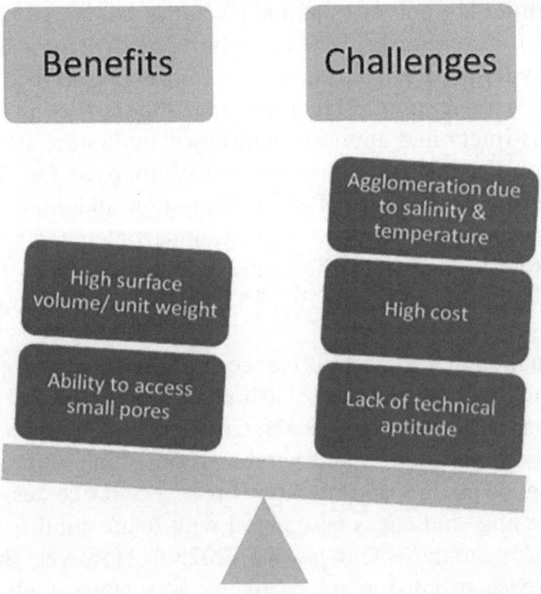

FIGURE 13.1 Benefits and challenges before nanomaterial utilization in subsurface applications.

increased agglomeration under certain conditions (Singh et al., 2023). These factors will be explored in detail in the subsequent sections of this chapter.

- *Reservoir Compatibility*: Ensuring the compatibility of nanomaterials with diverse reservoir conditions poses a challenge. Factors like pH, pressure, and rock mineralogy need consideration for successful EOR application.
- *Agglomeration Mechanisms*: Nanomaterials agglomerate due to attractive forces between particles stemming from factors like surface charge, size distribution, and chemical composition. Controlling these factors is crucial for preventing agglomeration. These factors are influenced by the same agents that affect the reservoir compatibility of the nanomaterials (Kumar et al., 2021; Singh et al., 2023).
- *Field Applicability*: Transitioning from lab-scale success to field-scale implementation is complex. Challenges arise in maintaining the stability and performance of nanomaterials in the dynamic and harsh conditions of oil reservoirs.
- *Cost Considerations*: The cost of nanomaterials is a significant barrier. High production costs may hinder widespread adoption in the oil industry, impacting the economic feasibility of EOR projects. The high cost of nanomaterials poses economic challenges for oil recovery applications. Striking a balance between performance benefits and economic viability is essential to encourage the widespread use of nanomaterials in EOR.

Addressing these challenges requires interdisciplinary research, focusing on material engineering, reservoir chemistry, and economic feasibility to unlock the full potential of nanomaterials in enhancing oil recovery processes.

13.3 IMPACT OF HIGH SALT CONCENTRATION ON NANOMATERIAL STABILITY

Oilfields tend to have an elevated level of salts dissolved in the reservoir fluids (brines). When the nanomaterials encounter the salts, they tend to exhibit agglomeration, causing an increase in particle size and lowering the efficiency of the nanomaterials when used in subsurface applications (Singh et al., 2023). Salt (ions) in a solution influences the stability of nanoparticles through electrostatic interactions. The stability of colloidal systems often relies on repulsion between charged particles, preventing agglomeration or precipitation (Panda & Gangawane, 2023a). The various mechanisms identified behind the role of salt in the agglomeration of nanomaterials have been listed as follows:

- *Debye Screening Length*: Salt increases the ionic strength of a solution, reducing the Debye screening length. This reduces the effective range of electrostatic forces, diminishing repulsion between nanoparticles and facilitating agglomeration.
- *Charge Neutralization*: Ions in the salt solution can neutralize the surface charges of nanoparticles. When the surface charge is reduced or

eliminated, the electrostatic repulsion between particles diminishes, leading to agglomeration.

- *Surface Complexation*: Salt ions can adsorb onto the nanoparticle surface, forming a layer that modifies the surface charge. This alteration in surface properties affects the interparticle forces and can contribute to agglomeration.
- *Ion-Mediated Bridging*: In some instances, ions in the salt solution may act as bridges between nanoparticles, promoting particle agglomeration. This ion-mediated bridging increases the likelihood of nanoparticle contact and subsequent agglomeration.

Understanding these mechanisms is crucial for tailoring nanoparticle formulations to specific conditions, such as those encountered in EOR or other applications. Adjusting the nanoparticle surface charge or employing stabilizing agents can help mitigate the impact of salt on stability and optimize the performance of nanomaterials in various environments (Pal & Sen, 2023a, 2023b; Panda et al., 2024; Panda & Gangawane, 2023c).

13.4 IMPACT OF INCREASING TEMPERATURE ON NANOMATERIAL STABILITY

Furthermore, the high-temperature conditions in the reservoir also influence the nanomaterials (nanoparticles) to agglomerate (or, in most cases, accelerate agglomeration; Figure 13.2).

The mechanisms behind this have been identified as under:

- *Thermal Agitation*: As temperature rises, the thermal energy of the system increases. This heightened thermal agitation can disrupt the stability of nanomaterial dispersions by promoting particle collisions and overcoming repulsive forces.
- *Brownian Motion*: Elevated temperatures enhance Brownian motion, causing nanoparticles to move more rapidly. This increased motion can lead to enhanced collisions between particles, increasing the likelihood of agglomeration.
- *Ostwald Ripening*: At higher temperatures, Ostwald ripening becomes more pronounced. This phenomenon involves the preferential growth of larger nanoparticles at the expense of smaller ones. This process can result in the coalescence and agglomeration of nanoparticles.
- *Surface Chemistry Changes*: Temperature can alter the surface chemistry of nanoparticles. Changes in surface charge or functional groups can impact the electrostatic and steric stabilization mechanisms, influencing the overall stability of nanomaterials.
- *Solvent Evaporation*: Elevated temperatures can accelerate solvent evaporation, affecting the stability of nanomaterial dispersions. As the solvent content decreases, the colloidal stability may be compromised, leading to agglomeration.

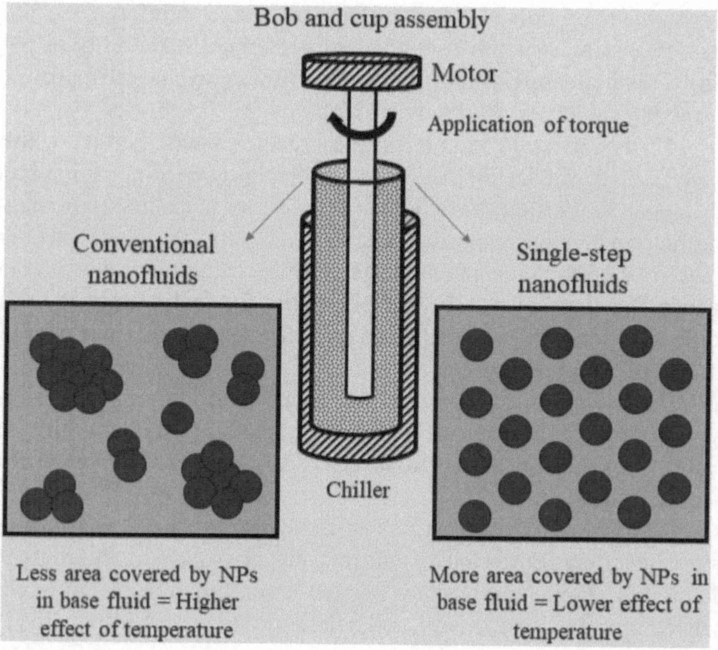

FIGURE 13.2 Comparative effect of nanomaterials with two different nanomaterial solutions. Adapted with permission from Chaturvedi et al. (2021a).

Understanding these mechanisms is essential for designing nanomaterials that can maintain stability under high-temperature conditions, as encountered in various industrial processes or applications such as EOR. Strategies to address these challenges may include optimizing surface coatings, adjusting dispersant formulations, or incorporating stabilizing agents to counteract the destabilizing effects of increased temperature.

13.5 FUTURE APPLICATION: IMPROVEMENT OF THERMAL STABILITY IN POLYMER MATERIALS WITH NANOMATERIALS

One potential area of nanomaterial use is their application as insulating agents to improve the thermal stability of polymers in EOR. The inclusion of nanomaterials in polymer matrices contributes to improved thermal stability through mechanisms such as enhanced thermal conductivity, reinforcement, barrier effects, catalytic activity, reduced flammability, and synergistic interactions (Chaturvedi & Sharma, 2020; Chaturvedi et al., 2021c). It makes nanocomposite materials promising for applications where thermal stability is a critical factor. The following mechanisms can understand this:

- *Enhanced Thermal Conductivity*: Nanomaterials, such as nanoparticles or nanotubes, can improve thermal stability by enhancing the thermal conductivity of the polymer matrix. This helps dissipate heat more efficiently, reducing the risk of thermal degradation.

- *Reinforcement of Polymer Matrix*: Nanomaterials reinforce the polymer matrix, providing structural support and reducing the mobility of polymer chains. This increased structural integrity can contribute to improved thermal stability under elevated temperatures.
- *Barrier Effect*: Certain nanomaterials create a barrier effect within the polymer matrix, hindering the diffusion of volatile decomposition products that can lead to thermal degradation. This barrier helps maintain the stability of the polymer at higher temperatures.
- *Catalytic Activity*: Some nanomaterials exhibit catalytic properties that can influence the thermal stability of polymers. By facilitating or inhibiting specific reactions, these nanomaterials can alter the thermal degradation pathways of the polymer.
- *Reduced Flammability*: Incorporating flame-retardant nanomaterials into polymers can enhance thermal stability by reducing flammability. This is especially relevant in applications where exposure to high temperatures may pose a fire risk.
- *Synergistic Effects*: Combinations of different nanomaterials can exhibit synergistic effects, offering a multifaceted approach to improving thermal stability. This synergy may involve a combination of reinforcement, enhanced thermal conductivity, and barrier properties. Additionally, in several previous studies, the inclusion of nanomaterials has been found to have a positive impact on the shear rheology of polymer solutions (Figure 13.3), influencing oil recovery (Chaturvedi et al., 2021b).

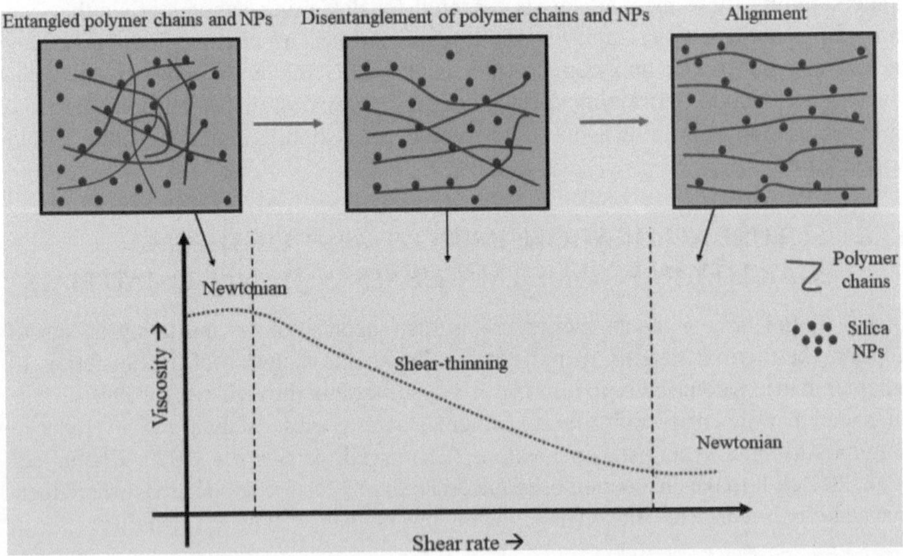

FIGURE 13.3 Rheological behaviour of nanomaterials and polymer solution. Adapted with permission from Chaturvedi et al. (2021a).

13.6 FUTURE APPLICATION: USE OF SURFACTANTS TO IMPROVE NANOMATERIAL STABILITY IN SALINE ENVIRONMENTS

Surfactants can be effectively employed to enhance the stability of nanomaterials in saline environments. Utilizing surfactants tailored to specific nanomaterials and saline conditions is crucial for optimizing stability. The choice of surfactant type, concentration, and surface chemistry considerations plays a significant role in achieving effective stabilization in saline environments. The mechanisms behind this stabilization involve several factors:

- *Surface Modification*: Surfactants adsorb onto the surface of nanoparticles, modifying their surface chemistry. This alteration can create a protective layer, reducing the likelihood of agglomeration or precipitation in the presence of salts (Chaturvedi et al., 2021c).
- *Electrostatic Stabilization*: Surfactants can provide electrostatic stabilization by introducing charges on the nanoparticle surface. This helps to counteract the attractive forces between particles, preventing agglomeration in saline solutions.
- *Steric Hindrance*: Surfactant molecules can create a steric hindrance effect, forming a physical barrier around the nanoparticles. This hinders close approach and contact between particles, thereby preventing agglomeration induced by salt ions (Singh et al., 2023).
- *Improved Dispersion*: Surfactants aid in dispersing nanomaterials uniformly in the solution. This enhanced dispersion minimizes the chances of particle agglomeration and promotes stability in the presence of salts (Chaturvedi et al., 2021b).
- *Salting-Out Prevention*: Surfactants can prevent the salting-out effect, where high salt concentrations lead to reduced solubility of nanoparticles. By maintaining proper dispersion and preventing phase separation, surfactants contribute to stability in saline environments.
- *Hydration Shell Formation*: Some surfactants form a hydration shell around nanoparticles, shielding them from the influence of salt ions. This hydration layer helps maintain colloidal stability by preventing direct interactions between nanomaterials and salts.
- *Controlled Wettability*: Surfactants can influence the wettability of nanoparticles, impacting their interaction with saline solutions (Singh et al., 2023). Controlled wettability helps maintain the stability of nanomaterials by regulating the interfacial properties.

13.7 FUTURE APPLICATION: NANOPARTICLES AND FOAM STABILITY

Nanoparticles can enhance the stability of foam in various applications, including EOR (Chaturvedi & Sharma, 2021). The addition of nanoparticles to foaming agents or solutions can influence the foam's structure and longevity. Mechanisms for improved foam stability with nanoparticles are as follows:

- *Particle Bridging*: Nanoparticles can act as bridges between adjacent bubbles, reinforcing the liquid lamellae and preventing coalescence. This bridging effect contributes to enhanced foam stability.
- *Surface Modification*: Nanoparticles can be designed with surface properties that interact favourably with the foam components, strengthening the interfaces and hindering drainage and coarsening processes.
- *Reduced Ostwald Ripening*: Nanoparticles can inhibit the Ostwald ripening process, where smaller bubbles tend to shrink, and their mass transfers to more giant bubbles. By controlling this phenomenon, nanoparticles contribute to more stable foams.
- *Electrostatic Stabilization*: Surface-charged nanoparticles can provide electrostatic repulsion between adjacent bubbles, counteracting coalescence forces and promoting foam stability.

Stable foam increases the viscosity of the displacing fluid, improving its ability to sweep through the reservoir and displace oil (Chaturvedi et al., 2021b). This enhanced sweep efficiency contributes to higher oil-recovery rates. Furthermore, foams containing nanoparticles can alter the wettability of reservoir rocks, facilitating the detachment of oil from rock surfaces and improving oil mobilization (Chaturvedi et al., 2021c). Stable foam impedes the mobility of injected gas, allowing it to stay in contact with the reservoir for a more extended period. This prolonged contact enhances gas-liquid interactions, promoting oil displacement (Chaturvedi et al., 2020).

Moreover, the stable foam resists gravity drainage, preventing the rapid rise of gas to the upper parts of the reservoir (Chaturvedi & Sharma, 2021). This resistance helps maintain efficient gas injection at lower levels, contributing to improved oil displacement. In summary, stable foams with the incorporation of nanoparticles play a crucial role in enhancing oil recovery by improving sweep efficiency, altering wettability, reducing gas mobility, and resisting gravity drainage. These effects contribute to a more effective displacement of oil in reservoirs during EOR processes.

13.8 FUTURE APPLICATION: INFLUENCE OF NANOPARTICLES ON OIL RECOVERY IN CARBONATED WATER INJECTION

Carbonated water injection, also known as carbon dioxide (CO_2) injection or carbonated water flooding, is a tertiary oil-recovery method employed in the oil and gas industry to enhance oil production from reservoirs. This technique involves injecting carbonated water or CO_2 into oil reservoirs to alter the physical and chemical properties of the reservoir fluids, thereby improving oil recovery and lowering the carbon footprint of the oil (Azzolina et al., 2016; Budinis et al., 2018; Metz et al., 2005). Initially, carbonated water is injected into the reservoir through injection wells. The CO_2 then dissolves in the oil, reducing the oil viscosity and interfacial tension. Then, this dissolved CO_2 causes the oil to swell, increasing reservoir sweep efficiency (Ekwurzel et al., 2017; Ferguson et al., 2009). The altered properties facilitate the movement of oil towards production wells, enhancing overall oil recovery. Carbonated water injection can also be used for Gas Cap Management, preventing the premature breakthrough of gas to production wells. However, there are some challenges associated with carbonated water injection, as identified under:

- *Reservoir Compatibility*: The success of carbonated water injection depends on the reservoir characteristics; not all reservoirs are suitable for this method.
- *Corrosion Risk*: CO_2 injection may increase the risk of corrosion in pipelines and well equipment.
- *Monitoring and Control*: Precise monitoring and control of injection rates and reservoir conditions are crucial for effective implementation.
- *Infrastructure Modifications*: Existing infrastructure may require modifications to accommodate carbonated water injection.
- *Economic Viability*: The cost of implementing carbonated water injection, including the capture and transportation of CO_2, can be significant (Raza et al., 2018; Shakiba et al., 2019).

As observed, while carbonated water injection holds the potential for enhancing oil recovery by modifying reservoir fluid properties, its successful application requires careful consideration of reservoir characteristics, infrastructure, and economic viability. Addressing these challenges is essential to unlock the full potential of carbonated water injection in the oil and gas industry.

- *Wettability Alteration*: Nanoparticles can modify the wettability of reservoir rocks, promoting a wetter water condition. This alteration facilitates improved contact between water and oil, enhancing oil recovery in carbonated water injection processes.
- *Enhanced Displacement Efficiency*: Nanoparticles can improve the displacement efficiency of carbonated water by altering the fluid-rock interactions. This leads to better mobilization of oil, especially in tight reservoir spaces where conventional displacement might be less effective.
- *Stabilization of CO_2 Bubbles*: Nanoparticles can be used to stabilize CO_2 bubbles in the injected water. This stabilization helps maintain the desired phase behaviour of the injected fluid, ensuring a more effective displacement of oil in the reservoir.
- *Improved Sweep Efficiency*: The addition of nanoparticles to carbonated water can enhance sweep efficiency by preventing channelling and improving the distribution of injected fluids within the reservoir. This leads to a more thorough displacement of oil from the reservoir rock.
- *Reduced Interfacial Tension*: Nanoparticles can reduce the interfacial tension between oil and water, facilitating the detachment of oil droplets from the rock surfaces. This reduction in interfacial tension contributes to improved oil mobilization and recovery.
- *Particle Transport*: Nanoparticles can be designed to transport through the reservoir with the injected fluids. Their presence can influence the

reservoir's permeability and alter fluid flow patterns, impacting oil recovery dynamics (Kumar et al., 2021).

- *Foam Stabilization*: Nanoparticles can enhance the stability of foam generated during carbonated water injection. Stable foam improves displacement efficiency by providing better contact between gas and oil phases, leading to increased oil recovery (Chaturvedi & Sharma, 2021).

Incorporating nanoparticles in carbonated water injection processes offers a multifaceted approach to improving oil recovery. The interactions between nanoparticles, fluids, and reservoir rocks can be tailored to optimize the displacement efficiency and maximize the overall recovery of oil from reservoirs (Kumar & Gangawane, 2023a, 2023b).

Involving nanomaterials to improve the performance of carbon sequestration and oil recovery methods is essential for a low-carbon/no-carbon future (Ferguson et al., 2009; Fernandez et al., 2013; Gunderson et al., 2020; House et al., 2006). This remains a significant area of research and future commercial deployment.

13.9 CONCLUSION

In conclusion, the exploration of nanotechnology in EOR has unveiled a spectrum of challenges, ranging from agglomeration induced by salt and time to high costs and low technical proficiency. However, amidst these obstacles lies a promising path forward. The envisioned future of this field involves a strategic alliance between nanomaterials and conventional oilfield chemicals like polymers and surfactants. By harnessing the synergistic potential of these components, we pave the way for more efficient, cost-effective, and technically proficient solutions. Furthermore, the use of nanotechnology enhances the efficacy of other oil recovery methods like foam flooding and carbonated water injection. As we navigate the uncharted territory of nanotechnology in EOR, these future directions offer a beacon of innovation, propelling the industry towards sustainable and advanced practices.

REFERENCES

Ahumada ZB, Soltero-Martínez JFA, Castillo R. Aqueous foams and emulsions stabilized by mixtures of silica nanoparticles and surfactants: A state-of-the-art review. *Chem Eng J Adv* 2021;7:100116. doi:10.1016/j.ceja.2021.100116.

Azzolina NA, Peck WD, Hamling JA, Gorecki CD, Ayash SC, Doll TE, et al. How green is my oil? A detailed look at greenhouse gas accounting for CO2-enhanced oil recovery (CO2-EOR) sites. *Int J Greenh Gas Control* 2016;51:369–79. doi:10.1016/j.ijggc.2016.06.008.

Budinis S, Krevor S, Mac DN, Brandon N, Hawkes A. An assessment of CCS costs, barriers and potential. *Energy Strateg Rev* 2018;22:61–81. doi:10.1016/j.esr.2018.08.003.

Chaturvedi KR, Fogat M, Sharma T. Low temperature rheological characterization of single-step silica nanofluids: An additive in refrigeration and gas hydrate drilling applications. *J Pet Sci Eng* 2021a:108742. doi:10.1016/j.petrol.2021.108742.

Chaturvedi KR, Narukulla R, Amani M, Sharma T. Experimental investigations to evaluate surfactant role on absorption capacity of nanofluid for CO2 utilization in sustainable crude mobilization. *Energy* 2021b;225:120321. doi:10.1016/j.energy.2021.120321.

Chaturvedi KR, Narukulla R, Sharma T. CO2 capturing evaluation of single-step silica nanofluid through rheological investigation for nanofluid use in carbon utilization applications. *J Mol Liq* 2020;304:112765. doi:10.1016/j.molliq.2020.112765.

Chaturvedi KR, Narukulla R, Trivedi J, Sharma T. Effect of single-step silica nanoparticle on rheological characterization of surfactant based CO2 foam for effective carbon utilization in subsurface applications. *J Mol Liq* 2021c:116905. doi:10.1016/j.molliq.2021.116905.

Chaturvedi KR, Sharma T. Carbonated polymeric nanofluids for enhanced oil recovery from sandstone reservoir. *J Pet Sci Eng* 2020;194:107499. doi:10.1016/j.petrol.2020.107499.

Chaturvedi KR, Sharma T. In-situ formulation of pickering CO2 foam for enhanced oil recovery and improved carbon storage in sandstone formation. *Chem Eng Sci* 2021;235. doi:10.1016/j.ces.2021.116484.

Ekwurzel B, Boneham J, Dalton MW, Heede R, Mera RJ, Allen MR, et al. The rise in global atmospheric CO2, surface temperature, and sea level from emissions traced to major carbon producers. *Clim Change* 2017;144:579–90. doi:10.1007/s10584-017-1978-0.

Ferguson RC, Nichols C, Leeuwen TV, Kuuskraa VA. Storing CO2 with enhanced oil recovery. *Energy Procedia* 2009;1:1989–96. doi:10.1016/j.egypro.2009.01.259.

Fernandez M, Leask C, Arnota C. Liability for sequestered CO2: The path forward for Alberta. *Energy Procedia* 2013;37:7709–22. doi:10.1016/j.egypro.2013.06.717.

Gunderson R, Stuart D, Petersen B. The fossil fuel industry's framing of carbon capture and storage: Faith in innovation, value instrumentalization, and status quo maintenance. *J Clean Prod* 2020;252:119767. doi:10.1016/j.jclepro.2019.119767.

House KZ, Schrag DP, Harvey CF, Lackner KS. Permanent carbon dioxide storage in deep-sea sediments. *Proc Natl Acad Sci U S A* 2006;103:12291–5. doi:10.1073/pnas.0605318103.

Kumar A, Gangawane KM. Effect of precipitating agents on the magnetic and structural properties of the synthesized ferrimagnetic nanoparticles by co-precipitation method. *Powder Technol* 2022a;401:117298. doi:10.1016/j.powtec.2022.117298.

Kumar A., Gangawane KM. Synthesis and effect on the surface morphology & magnetic properties of ferrimagnetic nanoparticles by different wet chemical synthesis methods. *Powder Technol* 2022b;410;117867. doi:10.1016/j.powtec.2022.117867.

Kumar A., Gangawane KM. Advanced computational approaches for water treatment. In *Advanced Computational Approaches for Water Treatment* (pp. 1–12). CRC Press, 2023a. doi:10.1201/9781003325147-1.

Kumar A, Gangawane KM. Molecular dynamics for the membrane process. In *Advanced Computational Approaches for Water Treatment* (pp. 13–38). CRC Press, 2023b. doi:10.1201/9781003325147-2.

Kumar A, Gangawane KM, Ramanjaneyulu B. Ferrofluids for waste-water treatment. In *International Conference on Chemical, Bio and Environmental Engineering* (pp. 723–44), 2022. https://link.springer.com/chapter/10.1007/978-3-030-96554-9_48.

Kumar, A., Jena, H. M., & Gangawane, K. M. (2023a). Effect of ferrimagnetic nanoparticles as an oxygen absorber on the shelf life of tomatoes. Poster presented at the *9th International Food Convention (IFCoN), CSIR-CFTRI*, Mysuru, India, 9 December 2023.

Kumar, A., Jena, H. M., & Gangawane, K. M. (2023b). Variations in magnetic behavior of ferrimagnetic nanoparticles with temperature. Poster presented at the *75th Indian Chemical Engineering Congress (CHEMCON 2023)*, Dr. H L Roy Building, Jadavpur University Campus, Kolkata, India, 28 December 2023.

Kumar RS, Al-Arbi Ganat T, Sharma T. Performance evaluation of silica nanofluid for sand column transport with simultaneous wettability alteration: An approach to environmental issue. *J Clean Prod* 2021;303:127047. doi:10.1016/j.jclepro.2021.127047.

Metz B, Davidson O, de Coninck H, Loos M, Meyer L. *Carbon Dioxide Capture and Storage — IPCC* (pp. 431 and 481). Cambridge University Press, 2005.

Pal D, Sen S. Emerging petroleum pollutants and their adverse effects on the environment. In ID Behera, AP Das (Eds.), *Impact of Petroleum Waste on Environmental Pollution and Its Sustainable Management through Circular Economy. Environmental Science and Engineering* (pp. 103–37). Springer Nature Switzerland, 2023a. doi:10.1007/978-3-031-48220-5_5.

Pal D, Sen S. In-depth coverage of petroleum waste sources, characteristics, environmental impact, and sustainable remediation process. In ID Behera, AP Das (Eds.), *Impact of Petroleum Waste on Environmental Pollution and Its Sustainable Management through Circular Economy* (pp. 1–38). Springer Nature Switzerland, 2023b. doi:10.1007/978-3-031-48220-5_1.

Panda D, Gangawane KM. Development of superhydrophobic hybrid silica-cellulose aerogel as promising thermal insulation and sound absorption. *Int J Mater Res* 2023a;114(4–5):405–17. doi:10.1515/ijmr-2022-0066.

Panda D, Gangawane KM. Hybrid NiFe2O4-NiCO4O4 nanoparticles-based eutectic phase change materials for enhancement of thermal efficiency of pin-fin heat sink arrangement. *J Energy Storage* 2023b;60:102417. doi:10.1016/j.est.2023.106644.

Panda D, Gangawane KM. Recycled cellulose–silica hybrid aerogel for effective oil adsorption: optimization and kinetics study. *Sadhana - Acad Proc Eng Sci* 2023c;48(3). doi:10.1007/s12046-023-02161-9.

Panda D, Sahu AK, Gangawane KM. Effect of molar concentration and drying methodologies on monodispersed silica sol for synthesis of silica aerogels with temperature-resistant characteristics. *Mater Sci Eng B* 2024;299:117062. doi:10.1016/j.mseb.2023.117062.

Panda D, Sharma S, Gangawane KM. Synthesis of novel biocatalyst from organic waste protonated with acid treatment for hydrogen production. *Int J Hydrogen Energy* 2023;48(63):24242–54. doi:10.1016/j.ijhydene.2023.03.117.

Raza A, Gholami R, Rezaee R, Bing CH, Nagarajan R, Hamid MA. CO2 storage in depleted gas reservoirs: A study on the effect of residual gas saturation. *Petroleum* 2018;4:95–107. doi:10.1016/j.petlm.2017.05.005.

Shakiba M, Riazi M, Ayatollahi S, Takband M. The impact of connate water saturation and salinity on oil recovery and CO2 storage capacity during carbonated water injection in carbonate rock. *Chinese J Chem Eng* 2019;27:1699–707. doi:10.1016/j.cjche.2018.09.008.

Singh A, Chaturvedi KR, Sharma T. An experimental evaluation of green surfactants to stabilize silica nanofluids in saline conditions and its application in CO2 absorption. *Int J Chem Reactor Eng* 2023. doi:10.1515/ijcre-2022-0241.

Index

Note: **Bold** page numbers refer to tables and *italic* page numbers refer to figures.

acoustic 7
adsorbents 21
adsorption 14, 20, 30, *31*, 40, 41, 45, 46, 50–53, 58, 62, 64, 66, 77, 78, 81, 82, 85, 86, 93, 94, 96, 105, 114, 127, 128, 167, 193, 203, 209
adsorption layer 62, 105
advanced computational fluid dynamics 167
ANSYS 165
Ansys fluent 165

biological treatment 6, 129, *130*, 145–146
boundary condition 164, 168, 169, 179, 180

carbon nanotube 17–18, 45, 61, **69**, 78, 94, **124**
centrifugation 42, 141–145
CFD 162–171
CFD modelling 162, 163, 165–167
chemical reaction kinetics 95, 129, 202
computational 162–170
computational fluid dynamics 162–171
concentration difference 11, 106, 185
crystallisation 136

Darcy's law 2, 169, 177
diffusion coefficient 180

eco-friendly 18, 21, 23, 65, 66, 113, 114, 128, 129, 139, 145, 208
eco-system 99, 123, 209
electric double layer 107, 114
electrode ionization 132
electromagnetic field 17

Fe_2O_3 15, 45, **69**, 105, 106, 133, 144, **207**
ferromagnetic 132
filtration 141–143, 145, 164, 165
finite difference 183
finite volume 183
flocculation processes 62

governing equation 48, 162, 164, 177–179
groundwater 21, 163, 165

heavy species transport 164, 169

industrial applications 9
industrial processes 34, 130, 162, 219

machine learning 185
macroscopic 5, 6, 39, 60, 176, 185
magnetic field 17, 128
magnetic nanoparticles 16–17, 22
magnetic separation 9
magnetization effect 8, 128, 143, 145
mathematical model 21, 22, 162, 165, 169, 175, 177–179, 183–186
membrane 141
membrane filtering 141, 143
membrane filtration 141, 165
meshing 141, 163, 166, 168, 169
micromixer 143
microorganisms 6, 129, 142, 143, 192
mobility 3–8, 10, 14, 20, 27, 30, 31, 41, 60–62, 64, 81, 96, 97, 105–107, 113, 125, 128, 174, 176, 178, 192, 194, 202, 215, 220, 222
modelling process 167

nanometers 32, 61
nanoparticles 8, 9, 11, 12, 14–23, 27–35, 40, *44*, *50*, 58, 74–86, 92–100, 104–115, 122–148, 162–171, 174–188, 193, **207**, 215–219, 221–224
Navier-Stokes 162, 164–166, 168, 169
Newtonian 64
NF 61–66, 192, 194, 196–199, 203, 206
non-Newtonian 64
numerical errors 184

osmotic pressure model 12

physicochemical effects 93
plasma 7, 132, 133
pollutant 21
polymerization 20, 138
porosity 1–2, 52, 59, 60, 67, 164–166, 168, 169, 178, 181, 184, 186
precipitation 13–15, 22, 50, 59, 60, 62, 105, 106, 123, 134, 217, 221

rearrangement of particle 63, *64*

sedimentation 11, 35
separation 32, 34, 145, 162, 165, 221
simulation 162–171, 174–188, 199, 201, 202, 209

simultaneous 15, 31, 52, 74, 167, 181, 183
supersaturation 133

thermal 2, 3, 8, 9, 18, 22, 40, 46, 58, 62, 66, 74, 105, 113, 122, 123, 127, 138, 192, 194, 200, 202, 218–220

wastewater plant 21
wastewater treatment 21
water-contact surfaces 12, 46

zeta potential 64, 65, 78, 80, 83, 85, 198